Current Topics in Behavioral Neurosciences

Volume 14

For further volumes:
http://www.springer.com/series/7854

About this Series

Current Topics in Behavioral Neurosciences provides critical and comprehensive discussions of the most significant areas of behavioral neuroscience research, written by leading international authorities. Each volume offers an informative and contemporary account of its subject, making it an unrivalled reference source. Titles in this series are available in both print and electronic formats.

With the development of new methodologies for brain imaging, genetic and genomic analyses, molecular engineering of mutant animals, novel routes for drug delivery, and sophisticated cross-species behavioral assessments, it is now possible to study behavior relevant to psychiatric and neurological diseases and disorders on the physiological level. The *Behavioral Neurosciences* series focuses on "translational medicine" and cutting-edge technologies. Preclinical and clinical trials for the development of new diagostics and therapeutics as well as prevention efforts are covered whenever possible.

Philip J. Cowen · Trevor Sharp
Jennifer Y. F. Lau
Editors

Behavioral Neurobiology of Depression and Its Treatment

Springer

Editors
Philip J. Cowen
Department of Psychiatry
Warneford Hospital
University of Oxford
Oxford
UK

Jennifer Y. F. Lau
Department of Experimental Psychology
University of Oxford
Oxford
UK

Trevor Sharp
Department of Pharmacology
University of Oxford
Oxford
UK

ISSN 1866-3370 ISSN 1866-3389 (electronic)
ISBN 978-3-642-35424-3 ISBN 978-3-642-35425-0 (eBook)
DOI 10.1007/978-3-642-35425-0
Springer Heidelberg New York Dordrecht London

Library of Congress Control Number: 2013930537

Printed on acid-free paper

Springer is part of Springer Science+Business Media (www.springer.com)

Preface

Major depressive disorder (MDD) is remarkably common. At any one time, worldwide, over 120 million people may be suffering from depression. Lifetime risks vary from 10 to 25 % for women and 5 to 12 % for men for Major Depression, with some indication that these estimates have risen over the past two decades. MDD is defined by episodes of low mood and/or anhedonia with changes in physical and psychological function. These episodes are highly recurrent. At least 60 % of individuals with a single episode develop a second episode, and of these, 70 % may have a third. Interestingly, many adult cases of depression emerge in adolescence, suggesting not only a stable and chronic lifelong course, but also one that has its roots early in development. MDD is disabling for individuals and families and is costly for society, too. Developing effective treatments and identifying early targets for preventative interventions are long-term goals for MDD researchers. These goals can be reached with greater understanding of the etiology of MDD, which involves identifying the risk (and resilience) factors involved and the mechanisms by which these effects are expressed. The question of why some individuals develop debilitating conditions, such as depression, while others remain relatively unscathed by affective symptoms has attracted many answers. These include susceptibility genes, neurochemical imbalances, brain circuitry dysfunction, faulty information processing, negative cognitions, and social or environmental stressors. It has become increasingly apparent over the last few decades of depression research that any holistic understanding of this complex phenotype relies on the recognition of multiple levels of analysis and causation. As a result, current research efforts are aimed toward understanding the synergy between these different factors. This knowledge can then be used to better develop intervention tools that can target the disorder from different causal levels.

The goal of this book, the Behavioral Neuroscience of Depression, which is part of the Current Topics in Behavioral Neurosciences, is to bring together scholarly reviews from experts working across these disciplines. Our book is divided into four sections.

The first part discusses aspects of the depressive phenotypes, and begins with a chapter by Philip Cowen on the classification of mood disorders. Operational criteria have made the diagnosis of depression more reliable but the question of where unhappiness ends and clinical disorder begins is unresolved. In the second chapter of the book, Jennifer Lau highlights an important developmental angle in explaining the emergence of depressive symptoms, and how intervention during the developmental process might reduce the risk of depressive outcomes. The argument here is that, should adolescence reflect a critical period for the emergence of depressive symptoms—owing to the various psychosocial challenges and maturational changes experienced in this age range—then perhaps one needs to also ask whether it is a period of opportunity for targeted interventions. While different depressive subtype diagnoses have some limited implications for treatment, the heterogeneity apparent in both treatment and pathophysiology studies of "major depression" indicates that syndromal diagnosis probably covers a number of different entities. Although the hallmark features of depression—loss of pleasure and enjoyment, hopelessness—is very much associated with the human condition, this has not stopped the use of elegant animal models to mimic other features of depression to enhance understanding of its etiology. Thus, in the final contributions to this section, Jaanus Harro and Christopher Barkus separately discuss animal models of depression. Much focus here is on the validity of measuring core depressive endophenotypes in rodent models; the models being created through the application of factors that underpin depression vulnerability in humans, or by pharmacological or genetic manipulation.

Part II of the book focuses on cellular and molecular mechanisms of depression. This begins with a chapter by Karen Hodgson and Peter McGuffin on the genetic basis of depression, drawing on the proliferation of human studies dedicated to investigating genetic influences on depression—an expansion that has been driven by cheap methods to collect, sequence, and analyze DNA variants. Molecular studies of relevant brain mechanisms in depressed patients are more challenging but PET imaging provides an exquisite delineation of relevant targets as described by Jeff Meyer. Of particular interest are the correlations now possible between in vivo neurochemistry and psychological measures of crucial relevance to depression such as dysfunctional attitudes. In addition, the substantial increase in monoamine oxidase binding apparent in PET studies of depressed patients enables a new way of understanding the long-established monoamine abnormalities in depressed patients. Another important development in molecular approaches to depression comes from increasing knowledge of the role of inflammation in pathophysiology. Patricia Zunszain and colleagues provide a fascinating account of how inflammatory mediators can lead to depression and the inflammatory profile apparent in a significant subgroup of depressed patients. These discoveries have important implications for novel methods of treatment, particularly in patients refractory to current approaches. Our understanding of the biochemical and cellular basis of depression has been transformed by the realisation that neural plasticity offers novel insights to understanding the depression pathophysiology. In their chapter, Margherita Lee and colleagues discuss the topical but increasingly controversial theory that a deficit in the production of newly formed neurons contributes to depression. This article reviews neurogenesis data from both human postmortem specimens and animal models. In a final part of this section, Lara Foland-Ross and colleagues investigate the extent to which some of these diverse neurobiological anomalies found in relation to depression may reflect inherited (or environmentally-acquired) familial risks. Describing findings from some elegant offspring studies, which have followed-up the never depressed children of parents with recurrent depression, these authors identify widespread disruptions in key emotional and stress-related systems as possible biomarkers of depression.

In Part III, we visit cognitive accounts of depression. Catherine Harmer makes a compelling case that emotional relearning is an important part of the therapeutic mechanism of monoamine potentiating drug in depression. Her theory represents an exciting synthesis of neuropsychology and neuropharmacology and an excellent example of the benefits of integrating the different levels of enquiry that characterise research in depression. Continuing in the line of integrative approaches, Katherine Collins and Daniela Schiller offer novel insights into the associated information-processing difficulties that emerge through MDD disrupted neurophysiology. Their focus is on basic emotion-processing examining the notion that cognitive processing mechanisms of reward and fear learning may form the cognitive markers of underlying neurobiological disruptions. This chapter provides a neat foundation for the subsequent contribution by Michael Browning and colleagues. This discusses the emergence of novel cognitive interventions, which attempt to improve symptoms by targeting the basic emotional processing difficulties from which they arise. As well as providing an excellent example of mechanism-driven interventions, where cognitive science meets clinical psychology, the Chapter ends by also raising some interesting questions on new cognitive perspectives about the role of mental imagery on depressive disorders.

The final part of the book is about treatment initiatives. Ian Anderson provides a masterly account of the pharmacological treatment of major depression in an article which is an exemplar of the "evidence based approach" to treatment, now so important in clinical management. Although the major emphasis of this book concerns major depressive disorder, we have included a chapter on bipolar depression because of the recent clinical advances in management. Kate Saunders and Guy Goodwin demonstrate that the optimum pharmacological management of depression in bipolar disorder has important differences from the treatment of unipolar (major) depression. In fact the development of ideas about the 'bipolar spectrum' means that the clinical cut-off between unipolar and bipolar disorder is no longer clear-cut which makes matching individual depressed patients to the correct treatment ever more important and challenging. Novel pharmacological treatments have been slow in coming in depression; however, it is now easier to use the treatments we have as well a possible. Trevor Sharp discusses molecular and cellular mechanisms of antidepressant drugs. Theories developed in the 1960s that the drugs act simply by increasing monoamine availability, have advanced remarkably to concepts which posit that this change is just the starting point for a complex sequence of neuroadaptive changes that interact with neuropsychological processes ultimately to elevate mood. Neurosurgical treatment for psychiatric disorder, initiated by Egas Moniz in 1936, is at once an old treatment for depression and a new one particularly in the form of deep brain stimulation. The latter is currently taking advantage of the explosion of knowledge about the neural circuitry underpinning the experience of depression as well as technical developments in electrode design and implantation. In the final chapter Yasin Temel and Lee Wei Lim explain the pre-clinical science behind modern neurosurgical approaches and the present state of clinical progress.

Oxford, UK

Philip J. Cowen
Trevor Sharp
Jennifer Y. F. Lau

Contents

Part I
Phenotypic Features

Classification of Depressive Disorders

Phillip J. Cowen

Abstract The classification of depression is well established in major diagnostic systems and operational definitions now make the diagnosis of depression reasonably reliable. However, classification and diagnosis continue to be based on clinical presentation and course and are not currently informed by aetiological or pathophysiological considerations. It is still unclear, for example, whether or not categories such as melancholic depression represent distinct subforms of illness or whether a dimensional classification based on severity can capture clinical presentation adequately and more economically. Despite these caveats, there is some evidence that distinctions between various subforms of depression can predict treatment response and this can be useful for clinical practice as well as guiding research strategies.

Keywords Depression · Classification · Diagnosis · ICD · DSM · Melancholia · Atypical depression · Psychotic depression · Treatment

Contents

P. J. Cowen (✉)
University Department of Psychiatry, Warneford Hospital, Oxford, OX3 7JX, UK
e-mail: cowen@psych.ox.ac.uk

Curr Topics Behav Neurosci (2013) 14: 3–13
DOI: 10.1007/7854_2012_218

Published Online: 26 August 2012

Mood disorders are common psychiatric conditions where the main feature is a persistent disturbance in affect in either a positive or negative direction. In the past, states of anxiety were included in discussions of clinical mood disorders, but in current psychiatric classifications, anxiety disorders are considered separately. Nevertheless, most depressed patients have symptoms of anxiety, and anxiety and depressive and anxiety disorders are frequently co-morbid (Joyce 2009).

1 "Clinical" Depression

It is part of normal experience to feel unhappy at times of adversity. What makes a depressive condition clinically significant is still a somewhat arbitrary distinction, because depressive symptoms in the population exist over a continuum and statistically there is no obvious point at which a satisfactory clinical cut-off can be made. Current approaches, therefore, make a diagnosis of clinical depression on pragmatic grounds using criteria such as severity, duration and functional impairment (Maj 2011).

It is also important to note that, the symptom of depressed mood is found in many different psychiatric syndromes and is also commonly found in certain physical diseases; for example, in infections such as hepatitis and some neurological disorders. Depression can also occur as a result of drug treatment, for example with steroids. Such depressive conditions are often referred to as "secondary" depression and are not usually classified with primary depressive disorders (Parker 2009).

2 Clinical Symptoms of Depression

The central symptomatology of a depressive syndrome consists of depressed mood, pessimistic thinking, lack of enjoyment and lowered energy. Other common symptoms include poor concentration and memory, reduced confidence and self-esteem and feelings of guilt and self-blame. Disturbed sleep and appetite are also frequent. Characteristically in moderate to severe depression, sleep is reduced and appetite for food is impaired, though in certain forms of depression, sometimes called "atypical depression" (see below), appetite and sleep may be increased.

Table 1 Symptoms needed to meet criteria for "depressive episode" in ICD–10

A	Depressed mood
	Loss of interest and enjoyment
	Reduced energy and decreased activity
B	Reduced concentration
	Reduced self-esteem and confidence
	Ideas of guilt and unworthiness
	Pessimistic thoughts
	Ideas of self-harm
	Disturbed sleep
	Diminished appetite

Mild depressive episode: at least two of A and at least two of B
Moderate depressive episode: at least two of A and at least three of B
Severe depressive episode: all three of A and at least four of B
Symptoms should be present every day for at least 2 weeks and should be associated with at least some impairment in social and occupational spheres

Table 2 Criteria for major depressive episode—DSM-IV

A Five (or more) of the following symptoms have been present during the same 2-week period and represent a change from previous functioning; at least one of the symptoms is either (1) depressed mood or (2) loss of interest or pleasure

(1) Depressed mood most of the day, nearly every day, as indicated by either subjective report (e.g. feels sad or empty) or observation made by others (e.g. appears tearful)
(2) Markedly diminished interest or pleasure in all, or almost all, activities most of the day, nearly every day (as indicated by either subjective account or observation made by others)
(3) Significant weight loss when not dieting or weight gain (e.g. a change of more than 5 % of body weight in a month), or decrease or Increase in appetite nearly every day
(4) Insomnia or hypersomnia nearly every day
(5) Psychomotor agitation or retardation nearly every day (observable by others, nor merely subjective feelings of restlessness or being slowed down)
(6) Fatigue or loss of energy nearly every day
(7) Feelings of worthlessness or excessive or inappropriate guilt (which may be delusional) nearly every day (not merely self-reproach or guilt about being sick)
(8) Diminished ability to think or concentrate, or indecisiveness, nearly every day (either by subjective account or as observed by others)
(9) Recurrent thoughts of death (not just fear of dying), recurrent suicidal ideation without a Specific plan or a suicide attempt or a specific plan for committing suicide

Particularly important are thoughts of life being futile and not being worth living because such thoughts can be linked to overt suicidal ideation and behaviour.

In more severe depression, psychomotor changes are prominent, usually seen as various kinds of psychomotor slowing. Interviewing may become difficult because of lengthy gaps between questions and answers. Psychomotor agitation can also occur and is experienced by the patient as inability to relax and by the observer as restless, purposeless activity (see Cowen et al. 2012).

Table 3 Clinical features of melancholic and somatic depression

Melancholic features (DSM-IV)
Loss of interest or pleasure in usual activities[a]
Lack of reactivity to pleasurable stimuli[a]
Plus at least three of the following:
Distinct quality of mood (unlike normal sadness)
Morning worsening of mood[a]
Early morning waking[a]
Psychomotor agitation or retardation[a]
Significant anorexia or weight loss[a]
Excessive guilt
Marked loss of libido[a] (ICD–10 only)

[a] Somatic symptoms of depression in ICD–10 (at least four required for diagnosis)

The various symptoms of depression needed to make a diagnosis of depressive episode in the International Classification of Diseases (ICD-10) (World Health Organisation 1992) are shown in Table 1. Similar symptoms are required to make a diagnosis of major depression in the other major classificatory system of psychiatric disorders, the Diagnostic and Statistical Manual of Mental Disorders (DSM-IV) (American Psychiatric Association 1994) (Table 2).

3 Some Distinctions in Acute Clinical Presentation

3.1 Melancholic Depression

As noted above, some patients with depression can experience what have been called "biological" symptoms. These symptoms include loss of appetite, psychomotor changes, weight loss, constipation, reduced libido, amenorrhoea and early morning waking. Diurnal mood variation, in which patients feel at their worst in the morning improving somewhat by the evening, can also be a feature (Parker 2009; Cowen et al. 2012).

These symptoms have been sometime called melancholic and they have been used to delineate a specific subgroup of depressive disorders, namely major depression with melancholia in DSM-IV, or depressive episode with somatic symptoms in ICD-10 (Table 3). The difficulty with this classification is that most patients have melancholic symptoms of some kind, though it may require a careful search to reveal them. Therefore, the number of symptoms needed to fulfil the criterion for depression with melancholia is somewhat arbitrary. Despite this caveat, it is generally agreed that clear-cut melancholic depression is associated with a number of clinical correlates (see Coryell 2007):

- More severe symptomatology
- Family history of depression

- Poor response to placebo medication
- More evidence of neurobiological abnormalities (decreased latency to rapid eye movement sleep, cortisol hypersecretion).

It is still not clear whether melancholic depression is a distinct depressive subtype or rather represents a point on a continuum of severity, toward the more severe end. Kendler (1997) attempted to answer this question using a population sample of twins. He found evidence that melancholic depression did represent a valid subtype in that it identified a group of individuals with a particularly high familial risk of depression. However, the diagnosis of melancholia indicated the presence of a quantitatively more severe form of depression rather than a distinct aetiological subtype.

3.2 Psychotic Depression

As depressive disorders become increasingly severe, there is complete loss of function in social and occupational spheres. Inattention to basic hygiene and nutrition may give rise to fears for the patient's safety. Psychomotor retardation may make interview impossible. In addition, there may be delusions and hallucinations; the disorder is then called psychotic depression.

The delusions of severe depressive disorders are concerned with the same negative themes as the non-delusional thinking of moderate depressive disorders. Therefore they are termed "mood congruent". The patient with a delusion of guilt may believe that some act of minor dishonesty will be publicly revealed and he will be punished and humiliated. He is likely to believe that such punishment is deserved. A patient with hypochondriacal delusions may be convinced that he has cancer or venereal disease. Hallucinations, when they occur, are usually heard in the second person and also have themes of guilt and denigration (see Cowen et al. 2012).

3.3 Atypical Depression

The term *atypical depression* is generally applied to disorders of moderate clinical severity. The meaning of the term has varied over the years, but currently it is applied to disorders characterised by:

- variably depressed mood with mood reactivity to positive events;
- overeating and oversleeping;
- extreme fatigue and heaviness in the limbs (leaden paralysis);
- pronounced anxiety.

Many patients with these clinical symptoms have a lifelong tendency to react in an exaggerated way to perceived or real rejection (rejection sensitivity), though this character trait can be exacerbated by the presence of a depressive disorder. Also patients with atypical depression have an earlier onset of illness and a more chronic course (Stewart et al. 2009). The importance of recognizing atypical depression is that, because of their interpersonal sensitivity, patients with this disorder can be hard to manage and may be regarded as having "difficult" personalities rather than depressive disorder. A good history from an informant about the personality of the patient when well will help with diagnosis.

4 Distinctions in Depressive Course

4.1 Recurrent Versus Single Episodes of Depression

The majority of depressive disorders (about 80 %) are recurrent but it is customary to divide depressive conditions on the basis of their longitudinal course as single episodes (so far at least) and recurrent. This often requires a careful search for earlier episodes which may not be readily recalled by the patient (Angst 2009).

4.2 Bipolar Versus Unipolar Illness

Modern classifications of mood disorders make an important distinction according to whether patients with depression have ever previously experienced mania or hypomania. If they have, then the disorder is a so-called bipolar illness and the diagnosis is bipolar depression. In contrast, patients who have never experienced mania or hypomania have a unipolar depressive disorder (Parker 2009).

Whether or not unipolar and bipolar depression show important differences in their clinical presentation is controversial. Epidemiological studies suggest that depression with a bipolar course is associated more with psychotic features, diurnal mood variation and hypersomnia (Forty et al. 2008). However, the differences are not great. There must be some overlap between the two groups, because a patient classified as having unipolar depression at one time may have a manic disorder later. In other words, the unipolar group inevitably contains some bipolar cases that have not yet declared themselves. Despite this limitation, the division into unipolar and bipolar cases is a useful classification because it has implications for treatment (see below).

4.3 Seasonal Affective Disorder

Some patients repeatedly develop a depressive disorder at the same time of year, usually the *autumn or winter*. In some cases, the timing reflects extra demands placed on the person at a particular season of the year, either in work or in other aspects of their life. In other cases, there is no such cause and it has been suggested that seasonal affective disorder is related in some way to the changes in the seasons, for example, to the length of daylight. Although these seasonal affective disorders are characterised mainly by the time at which they occur, some symptoms are said to occur more often than in other mood disorders. These symptoms are:

- increased sleep
- increased appetite with craving for carbohydrate
- afternoon slump in energy.

The most common pattern is onset of depression in autumn or winter, and recovery in spring or summer. Some patients show evidence of hypomania or mania in the summer, suggesting that they have a seasonal bipolar illness. This pattern has led to the suggestion that shortening of daylight is important in the pathophysiology of winter depression and treatment methods include exposure to bright artificial light during hours of darkness (Eagles 2003).

4.4 Brief Recurrent Depression

Some individuals experience recurrent depressive episodes of short duration, typically 2–7 days that are not of sufficient duration to meet criteria for major depression or depressive episode. These episodes recur with some frequency, about once a month on average. There is no apparent link with the menstrual cycle in female sufferers. Whilst the depressive episodes are short, they are as severe as the more enduring depressive disorders and can be associated with suicidal behaviour. Hence, recurrent brief depression is associated with much personal distress and social and occupational impairment. There appears to be little co-morbidity with bipolar illness (Pezawas et al. 2003).

5 Milder Depressive Disorders

Both DSM-IV and ICD-10 have categories of mild depression where criteria for a depressive episode are met, but the depressive symptoms are fewer and less severe (see Table 4). However, in the community and in primary care, milder depressive disorders, not meeting formal diagnostic criteria for the depressive conditions outlined above are common and often clinically significant. For example, in The Adult Psychiatric Morbidity Survey in England, McManus et al. (2007) found that mixed anxiety depression was the most common clinically significant mental disorder in the community with a 1-week prevalence of about 9 %.

It might be expected that mild depressive disorders would present with symptoms similar to those of the depressive disorders described already, but with

Table 4 Classification of depressive disorders

ICD–10	DSM-IV
Depressive episode	*Major depressive episode*
Mild	Mild
Moderate	Moderate
Severe	Severe
Severe with psychosis	Severe with psychosis
Other depressive episodes	
Atypical depression	
Recurrent depressive	*Major depressive disorder*
Disorders	*Recurrent*
Currently mild	
Currently moderate	
Currently severe	
Currently severe with psychosis	
In remission	
Persistent mood disorders Dysthymic disorder	
Cyclothymia	
Dysthymia	
Other mood disorders	*Depressive disorders not*
	Otherwise specified
Recurrent brief depression	Recurrent brief depression

less intensity. To some extent this is so, but in mild depressive disorder there are frequently additional symptoms that are less prominent in severe disorders. These symptoms have been characterised in the past as "neurotic", and they include anxiety, phobias, obsessional symptoms and, less often, dissociative symptoms (Cowen et al. 2012).

Apart from the "neurotic" symptoms found in some cases, mild depressive disorders are characterised by the expected symptoms of low mood, lack of energy and interest and irritability. There is sleep disturbance, but not the early morning waking that is characteristic of more severe depressive disorders. Instead, there is more often difficulty in falling asleep and periods of waking during the night, usually followed by a period of sleep at the end of the night. "Biological" features (poor appetite, weight loss and low libido) are not usually found. Although mood may vary during the day, it is usually worse in the evening than in the morning. The patient may not be obviously dejected in his appearance or slowed in his movement (Cowen et al. 2012).

Many of these mild depressive disorders are brief, starting at a time of personal misfortune and subsiding when fortunes have changed or a new adjustment has been achieved. In these cases, a diagnosis of adjustment disorder may be appropriate. However, some cases persist for months or years, causing considerable suffering, even though the symptoms do not increase. These chronic depressive states are called dysthymia. The term cyclothymic disorder refers to a persistent instability of mood in which there are numerous periods of mild elation or mild depression. It is seen as a milder variant of bipolar disorder. It is not unusual for episodes of more severe mood disorder to supervene in patients who experience dysthymia or cyclothymia and the converse can also occur (Akiskal 2009). This suggests that in some people milder symptoms are a limited expression of an underlying major mood disturbance.

6 Classification in ICD-10 and DSM-IV

The main categories in the sections on depressive disorders in DSM-IV and the ICD-10 are shown in Table 3. Broad similarities are evident, together with some differences. The first similarity is that both systems contain categories for single episodes of depression as well as categories for recurrent episodes. The second is that both recognise mild but persistent mood disturbances in which there is either a repeated alternation of high and low mood (cyclothymia) or a sustained depression of mood (dysthymia). In neither case are the mood disturbances sufficiently severe to meet criteria for a hypomanic or depressive episode. In DSM-IV, mood disorders judged to be secondary to a medical condition are included as a subcategory of mood disorders, whereas in ICD-10 these conditions are classified as mood disorders under "Organic mental disorders".

Both ICD-10 and DSM-IV classify depressive episodes on the basis of severity and whether or not psychotic features are present. It is also possible to specify whether the depressive episode has melancholic (DSM-IV) or somatic (ICD-10) features. In DSM-IV, an episode of major depression with appropriate clinical symptomatology (see above) can be specified as atypical depression. In ICD-10, atypical depression is classified separately under 'Other depressive episodes'. Both ICD-10 and DSM-IV allow the diagnosis of recurrent brief depression, but under slightly different headings.

7 Implications and Limitations of Classification of Depression

It will be noted that, the current classification of depression is essentially clinical and aetiological and pathophysiological factors do not play a significant role. For example, many depressive episodes are triggered by adverse life events but this does not seem to have a strong influence on clinical presentation or response to treatment. Similarly, depression with melancholic features is more likely to be associated with cortisol hypersecretion (Coryell 2007) but, as noted above, this may be a matter of severity rather than the delineation of a separate pathophysiological entity. The development of operational criteria means that the diagnosis of various clinical forms of depression is now reasonably reliable, but doubts about validity can be resolved only by a better understanding of pathophysiology.

There is also criticism of how current diagnostic systems identify clinical depression with the implication that common human unhappiness is being medicalised and drug

Table 5 Classification of depression and implications for treatment

Depressive disorder	*Best drug treatment*
Melancholic	TCA or SNRI
Atypical	SSRI or MAOI
Psychotic	TCA or SNRI and antipsychotic drug
Bipolar depression	Quetiapine or lamotrigine

TCA Tricyclic antidepressant
SNRI Serotonin and noradrenaline re-uptake inhibitor
SSRI Selective serotonin re-uptake inhibitor
MAOI Monoamine oxidase inhibitor

treatment offered inappropriately at the prompting of profit seeking drug companies (Horwitz and Wakefield 2007). As noted above, at present a clinical cut-off for depressive illness has to be decided on pragmatic grounds and can be informed by the knowledge that even mild depressive symptomatology is liable to cause functional impairment (Broadhead et al. 1990). Of course, the presence of depressive symptomatology does not mean that drug treatment is necessarily indicated; indeed, the evidence is that a certain threshold of symptomatic severity needs to be exceeded before drug treatment is worthwhile, though where that threshold is has been disputed (Horder et al. 2011). It is also worth noting that, drug treatment has been shown to be effective in dysthymia which is characterised by relatively mild, albeit persistent and depressive symptomatology (Lima and Moncrieff 2000). This suggests that duration of symptoms as well as severity is a useful criterion in deciding treatment.

The current classification of depression does have some implications for treatment response. These notions are sometimes based on clinical experience rather than reliable evidence from randomised trials, however, they do appear to have clinical utility. For example, it is established from controlled trials that patients with depressive psychosis respond poorly to antidepressant medication alone but can often be helped by a combination of antipsychotic and antidepressant treatment (Spiker et al. 1985; Wijkstra et al. 2010).

There is also a strong clinical impression that patients with melancholic features do better with tricyclic antidepressants (TCAs) and serotonin-noradrenaline re-uptake inhibitors (SNRIs) than with selective serotonin re-uptake inhibitors (SSRIs) (Shelton and Tomarken 2001), suggesting that for the alleviation of more severe depressive states, recruitment of both catecholamine and serotonergic mechanisms may be necessary. Conversely in atypical depression, SSRIs and monoamine oxidase inhibitors (MAOIs) appear more effective than TCAs (Stewart et al. 2009). Finally, the treatment of bipolar depression differs in a number of important respects from that of unipolar depression. For example, standard antidepressant treatments do not appear very helpful and may increase the risk of mania or mood cycling. However, useful effects in bipolar depression may be obtained from atypical antipsychotic drugs such as quetiapine or the mood stabilizer, lamotrigine (Goodwin 2009) (for a summary see Table 5).

References

Akiskal HS (2009) Dysthymia, cyclothmia and hyperthymia. In: Gelder MG, Andreasen NC, Geddes JR (eds) New oxford textbook of psychiatry. Oxford University Press, Oxford, pp 680–691

American Psychiatric Association (1994) Diagnostic and statistical manual of mental disorders, 4th edn. American Psychiatric Association, Washington

Angst J (2009) Course and prognosis of mood disorders. In: Gelder MG, Andreasen NC, Geddes JR (eds) New oxford textbook of psychiatry. Oxford University Press, Oxford, pp 665–669

Broadhead WE, Blazer DG, George LK et al (1990) Depression, disability days, and days lost from work in a prospective epidemiologic survey. JAMA 264:2524–2528

Coryell W (2007) The facets of melancholia. Acta Psychiatr Scand 115:31–36

Cowen PJ, Harrison PJ, Burns T (2012) Shorter Oxford Textbook of Psychiatry, Ch 11, Mood disorders. Oxford University Press, Oxford

Eagles JM (2003) Seasonal affective disorder. Br J Psychiatry 182:174–176

Forty L, Smith D, Jones I et al (2008) Clinical differences between bipolar and unipolar depression. Br J Psychiatry 192:388–389

Goodwin GM (2009) Evidence-based guidelines for treating bipolar disorder: recommendations from the British association for psychopharmacology. J Psychopharmacol 23:346–388

Horder J, Matthews P, Waldmann R (2011) Placebo, prozac and PloS: significant lessons for psychopharmacology. J Psychopharmacol 25:1277–1288

Horwitz AV, Wakefield JC (2007) The loss of sadness: how psychiatry transformed normal sorrow into depressive disorder. Oxford University Press, Oxford

Joyce PR (2009) Epidemiology of mood disorders. In: Gelder MG, Andreasen NC, Lopez-Ibor JJ, Geddes JR (eds) New Oxford Textbook of Psychiatry. Oxford University Press, Oxford, pp 645–650

Kendler KS (1997) The diagnostic validity of melancholic major depression in a population-based sample of female twins. Arch Gen Psychiatry 54:299–304

Lima MS, Moncrieff J (2000) Drugs verses placebo for dysthymia. Cochrane Database Syst Rev 4:CD001130

Maj M (2011) When does depression become a mental disorder? Br J Psychiatry 199:85–86

McManus S, Meltzer H, Brugha TS et al (2007) Adult psychiatric morbidity in England: results of a household survey. The NHS Information Centre for Health and Social Care, Leeds

Parker G (2009) Diagnosis, classification and differential diagnosis of mood disorders. In: Gelder MG, Andreason NC, Lopez-Ibor JJ, Geddes JR (eds) New oxford textbook of psychiatry. Oxford University Press, Oxford, pp 637–645

Pezawas L, Wittchen HU, Fister H et al (2003) Recurrent brief depressive disorder reinvestigated: a community sample of adolescents and young adults. Arch Gen Psychiatry 33:407–418

Shelton RC, Tomarken AJ (2001) Can recovery from depression be achieved? Psychiatr Serv 52:1469–1478

Spiker DG, Cofsky Weiss J, Dealey RS et al (1985) The pharmacological treatment of delusional depression. Am J Psychiatry 142:430–436

Stewart JW, McGraith PJ, Quitkin FM et al (2009) DSM-1V depression with atypical features: is it valid. Neuropsychopharmacology 34:2625–2632

Wijikstra J, Burger H, Van den Broek WW et al (2010) Treatment of unipolar psychotic depression: a randomized double-blind study comparing imipramine, venlafaxine and venlafaxine plus quetiapine. Acta Psychiatr Scand 121:190–200

World Health Organization (1992) The ICD-10 classification of mental and behavioural disorders-clinical descriptions and diagnostic guidelines. World Health Organisation, Geneva

Developmental Aspects of Mood Disorders

Jennifer Y. F. Lau

Abstract Rates of depression double in the transition to adolescence. Symptoms of depression in adolescence also predict adult mood disorders. From a preventative perspective, research into the risk factors for adolescent depression is critical. Simultaneously, there is also growing interest in understanding why adolescence reflects a particularly critical period for the onset of early mood symptoms. Here, we review (a) the role of social, genetic, neural and cognitive factors in explaining individual differences in the propensity to develop symptoms in adolescence, and (b) whether changes in these factors may explain why adolescence is a particularly sensitive period for the onset of depressive symptoms. Finally, we speculate on whether novel social stressors interact with 'genetic innovation' to produce changes in brain circuitry structure and function and associated changes in emotion regulation abilities and social understanding—to increase emergence of symptoms in adolescence.

Keywords Adolescence · Depression · Genetic · Social environment · Brain maturation · Cognitive biases

Contents

J. Y. F. Lau (✉)
Department of Experimental Psychology, University of Oxford,
South Parks Road,
Oxford, OX1 3UD, UK
e-mail: jennifer.lau@psy.ox.ac.uk

Curr Topics Behav Neurosci (2013) 14: 15–27
DOI: 10.1007/7854_2012_214

Published Online: 26 August 2012

Depression is common, disabling and costly, emphasising the need for a better understanding of the root causes of depression and the development of preventive interventions to target these causes. Late childhood and adolescence may reflect an important focal point for studying the onset of depression. Data show increases in the rates of depression in the transition to and across adolescence, particularly among girls (Bartels et al. 2011; Hankin et al. 1998). In turn, adolescent depressive symptoms predict adult mood disorders (Pine et al. 1999) suggesting that those with a propensity to develop persistent, perhaps lifelong mood problems are more likely to develop these early. There has therefore been a surge of interest in not only identifying risk factors for *individual differences* in the propensity to develop depression in adolescence but also in identifying *developmental differences* in the propensity to develop depressive symptoms across life.

The main goal of this chapter is to review biological, cognitive and social changes that occur in adolescence, which may explain the 'developmental sensitivity' for the onset of mood problems, and indeed other emotion-related conditions such as anxiety. To do this, factors that have typically been linked to *individual* differences in depressive problems are considered with a particular focus on whether changes in these factors account for *developmental* differences in risk for these internalising conditions. A multifactorial account of developmental sensitivity to mood and anxiety problems in adolescence is explored. More particularly, the extent to which novel social stressors interact with 'genetic innovation' to produce changes in brain circuitry structure and function, and associated changes in emotion regulation abilities and social understanding are speculated on. Some work has focussed on gender differences in why adolescence is a developmentally significant juncture in the emergence of mood symptoms—from differences in the role of pubertal hormones to gender-specific increases in interpersonal stressors. However, as these studies have not systematically compared multifactorial aetiological models in adolescent males versus females, the gender differences in differential rises in depression rates remain elusive. Thus, the present chapter focuses on a gender-general model for explaining adolescent susceptibility to mood and other emotion-related problems in adolescence.

1 Changes in the Social World

Are environmental factors implicated in adolescent depression? One of the most consistent findings to emerge regarding the aetiology of depression is the importance of environmental influences (Brown et al. 1996; Kendler et al. 2003), a finding that also characterises adolescents with depression (Grant et al. 2004). These include proximal life events, which often precipitate the onset of symptoms, but also aversive early life experiences, such as maltreatment and separation from parents which may have lasting effects on stress responding (Brown et al. 1993; Levitan et al. 2003). Daily hassles and chronic stressors have also been implicated in early depressive symptoms (Eley and Stevenson 2000). These include financial strains (Gilman et al. 2003) and interpersonal strains in relationships with parents and peers (see Arseneault et al. 2010; Hawker and Boulton 2000 for reviews). While less well studied, romantic dating has in some young people been associated with higher depression symptoms (Davila et al. 2004). More generally, adolescent depression has been associated with an absence of social support, which may serve to buffer against negative experiences (Barrera and Garrisonjones 1992; Quamma and Greenberg 1994; Cauce et al. 1992).

Are there age-related changes in these environmental factors? Changes in the social environment of adolescents have been found. The focus of most of this research has been on young people growing up in Western societies. As children and adolescents mature, they may experience *more* stressful life events (Hoffman et al. 1992) and daily hassles (Compas et al. 1985), although not all of these reports have been replicated (Hoffmann and Cerbone 1999). Adolescents may experience new chronic stressors; the transition to secondary school brings alterations in educational demands, peer groups and daily routines, and is often associated with a period of anticipatory anxiety and subsequent readjustment (Puskar and Rohay 1999; Fenzel 2000). During this period, adolescents compared to pre-transition children experience greater stress responses towards performance-related stressors (Stroud et al. 2009).

Another source of upheaval is that adolescents begin to spend more time with their peers than with their families (Larson et al. 1996), with some suggestion that parent–child conflicts also increase (Sallinen et al. 2007). Romantic relationships may also develop for the first time, impacting on existing social networks (Zimmer-Gembeck 2002). These changes may serve to enhance the emotional salience of peer feedback. Indeed, adolescents exert a great deal of energy forming peer networks and soliciting peer approval (Steinberg and Morris 2001). While having positive experiences with peers is generally beneficial, the need to gain peer approval can also result in negative outcomes. For example, some data point to more risky decisions being taken in the presence of peers (compared to playing alone) in adolescents, than in children and adults (Gardner and Steinberg 2005). But some other data also show that resistance to peers increases across adolescence (Steinberg and Monahan 2007).

As well as finding peers more rewarding, adolescents may also experience greater distress to negative peer experiences than adults. This has been demonstrated with the Cyberball paradigm. During this task, participants play a computer-based ball-tossing game with two other "players" that are in actuality computer generated. Under conditions of exclusion, the participant is rarely tossed the ball by either of the two co-players who toss to one another; under conditions of inclusion, the participant receives the ball roughly a third of the time. Recent studies exposing adolescent and adult participants to these conditions found that adolescents tended to report lower mood and more distress following ostracism relative to inclusion and baseline, than adults, who showed no difference across conditions (Sebastian et al. 2011b; Sebastian et al. 2010). Other studies have also demonstrated that adolescents undergoing the transition to secondary school may be more responsive to peer rejection than children prior to transition (Stroud et al. 2009). Changes in the social world of young people could also bring increasingly complex social exchanges, which require adolescents to acquire a more sophisticated understanding of the minds of others. Experimental studies in our lab show that adolescents deploy distinct strategies for establishing and maintaining reciprocity in social exchanges than adults, particularly during social rupture (Belli et al. in press).

In summary, certain chronic stressors known to predict mood problems in young people may increase in the transition from late childhood to adolescence. These may include the experience of peer rejection and/or the experience of reduced peer acceptance, and perhaps the encountering of more ambiguous, perplexing social situations. These may enhance risks for developing mood and anxiety problems.

2 Genetic Innovation Across Development

Are genetic factors implicated in adolescent depression? Twin studies provide a powerful tool for estimating the heritability of behavioural phenotypes such as depression. These have reported moderate genetic effects on early depression symptoms (Lau and Eley 2006; Lau et al. 2007). However, attempts to identify specific genes have been largely unsuccessful despite advances in technology. While the short form of the serotonin transporter gene polymorphism has been implicated in adult depression, findings in children and adolescents are more mixed. Some findings have been replicated, but not others. The pattern of replications and non-replications of results is not just limited to developmental samples, but across the field (Risch et al. 2009) and it is becoming increasingly apparent that complex disorders such as depression are influenced by multiple, interacting genetic and environmental influences of small effect size (see Plomin et al. 2008). Thus, conclusions of specific genes in general and the few drawn from younger samples in particular await further verification in larger and more powerful studies, such as that offered by genome-wide association studies (Boomsma et al. 2008; Flint et al. 2012).

Researchers have begun to turn to the question of *how* genetic influences identified in twin studies take effect. Twin studies have shown how genetic risk factors may influence exposure towards particular pathological environments, and/or influence emotional responses towards these (Lau and Eley 2008b; Lau et al. 2007). We have also begun to explore how genetic factors may contribute to negative cognitive styles linked to adolescent depressive symptoms (Lau and Eley 2008a; Lau et al. 2006). These data show that not only are cognitive styles heritable, these genetic influences overlap with those contributing to depressive symptoms, suggesting shared genes. Finally, we have also found associations between the serotonin transporter gene variant and increased neural responses to the appraisal of fear in depressed (and anxious) adolescent patients (Lau et al. 2009) but as with all candidate gene studies, these will be verified in larger samples.

Are there age-related changes in genetic factors? While most readers will be open to the idea that environmental influences on difficulties such as depression change throughout development, it is not obvious how genes could become more or less important influences on difficulties over time given that the sequence of our DNA is stable across the life course. These suggestions appear less provocative when considered in the context of long-standing 'maturationist' theories of development, such as that of Arnold Gesell (1932). These theories point to the presence of a genetically determined biological programme, in which genes contributing to depression could become 'switched on' at particular developmental milestones.

A plethora of literature shows exactly that. Genetic influences on depressive symptoms have generally been shown to be greater in adolescents than in children (Rice et al. 2002; Lamb et al. 2010; Scourfield et al. 2003; Thapar and McGuffin 1994). But the picture is more complicated, with findings of substantial genetic influences in early childhood compared to middle childhood (Boomsma et al. 2005; van der Valk et al. 2003), and decreasing genetic effects on depressive symptoms over time during adolescence (O'Connor et al. 1998a, b). Thus, changes in heritability may not be linear.

Studies exploring change and continuity in genetic factors have reported novel genetic influences on symptoms of depression in adolescents (Kendler et al. 2008; Lau and Eley 2006). These data do not, however, tell us much about the precise 'developmental' genes that are likely to be involved or how these may take effect. One speculation is that these new genetic effects reflect maturational changes in the brain circuits involved in emotional and social processing, and in emerging cognitive styles that regulate emotions. The growing field of epigenetics (the study of chemical processes that modify gene expression, sometimes through exposure to environmental events) promises to help address some of these questions (Champagne et al. 2003).

3 Continuing Maturation of the Emotional Brain

Are neural factors implicated in child and adolescent depression? The most robust group differences to have emerged between depressed and anxious youth and their psychiatrically healthy counterparts have been found in the so-called 'emotion' circuits of the brain. Symptomatic children and adolescents show greater response in the amygdala during viewing or appraisal of emotional (Beesdo et al. 2009) or social stimuli (Lau et al. in press-b; Guyer et al. 2008). There may also be reduced activity in reward-related regions, such as the striatum during reward anticipation and outcome in symptomatic and even at-risk individuals (Forbes et al. 2006, 2009; Gotlib et al. 2010). Regions of the prefrontal cortex (PFC) including ventromedial, dorsolateral and orbitofrontal regions—all of which are thought to regulate basic affective responses through higher order cognitive processes, such as fear extinction and cognitive appraisals (Hartley and Phelps 2010)—have also attracted growing attention. However, while some studies in adults show reduced responses in PFC regions during emotion processing (e.g. Halari et al. 2009), not all adolescent studies of patients replicate these (Monk et al. 2006). Similarly, findings on the degree of co-activation between the amygdala, and PFC regions have also been mixed. Adult studies typically find reduced connectivity between these regions in patients, supporting the hypothesis that larger amygdala responses towards emotional and social stimuli result from poorer top-down regulation. But the direction of this relationship is complex in adolescents, varying across tasks (McClure et al. 2007).

Are there age-related changes in neural factors? Neuroimaging research over the last decade has revealed that the human brain does not stop developing beyond infancy and childhood but in fact continues to mature throughout adolescence and young adulthood (Giedd and Rapoport 2010). Striking age-related changes in cerebral volume, including linear increases in white matter and regionally specific inverted u-shaped trajectories for grey matter brain structures have been reported (Giedd and Rapoport 2010). Such changes may result in an overall increase in the efficiency of the brain, particularly in communication across structures and circuits, by increasing signal to noise in the system and allowing for the strengthening of relevant connections (Durston et al. 2006).

A proliferation of studies has begun to assess age-related changes in brain activation patterns during the appraisal of negative and positive stimuli; cognitive regulation of emotional responses; and the appraisal and understanding of salient social interactions using functional neuroimaging techniques. These studies suggest a general trend of enhanced subcortical responding, both in the amygdala and striatum during the respective processing of threat- and reward-related emotional stimuli in adolescents relative to children and adults (see Somerville et al. 2010 for a review). These same studies also suggest age-linked differences in prefrontal regional responses during the control and inhibition of basic emotional responses. Sometimes these differences manifest as a reduction in activation, but other times as an increase in responding, depending on task demands. In one of our studies, we compared adolescent and adult patterns of neural responding during the acquisition of fear to threat cues and the acquisition of a reduced fear response to safety cues using a fear-learning task (Lau et al. 2011a). Compared to adults, adolescents engaged their amygdala to a greater degree to threat cues. However interestingly, only adults showed a positive correlation between a reduction of fear to safety cues and activation of the dorsolateral PFC. These findings may suggest that the degree to which prefrontal regions, such as the dorsolateral PFC inhibited emotional responding of the amygdala, particularly under non-threatening, i.e. safe contexts varies with age.

Several studies have also investigated age differences in areas of the 'social' brain Burnett et al. (2011). During appraisal of peer exclusion, these have found differences between adolescents and adults, emerging in emotion-response and regulation regions such as the striatum and lateral PFC (Sebastian et al. 2011b; Guyer et al. 2009) and in areas implicated in social understanding, such as the anterior cingulate cortex (Gunther Moor et al. 2012; Masten et al. 2009) and medial PFC (Sebastian et al. 2011b).

In summary, there may be changes in brain structure and function in regions implicated in the processing of emotional and social information. While it is unclear how the direction of these differences can be interpreted, many of these studies support the notion that differences in functional activity may explain differences in young people's responses towards emotional situations such as peer rejection and/or understanding the new social landscape. While these data are intriguing, most have recruited broadly-defined age groups, which may mask more subtle, continuous developmental changes. Second, age reflects a rather more crude measure of development compared to the stage of puberty, which is often not measured.

4 Developmental Aspects of Information-Processing Biases

Are information-processing biases and cognitive vulnerability factors implicated in adolescent depression? Processing biases and distorted cognitions characterise depressed adolescents (Gladstone et al. 1997; Haeffel et al. 2008; Jacobs et al. 2008; Gotlib and Joormann 2010). Drawing on many different experimental tasks, these studies are persuasive in suggesting that young people with depressive symptoms vary in how they attend to and recall negative and positive information, and interpret ambiguous information (Gotlib and Joormann 2010). In turn, these difficulties may explain poorer emotion regulation capacities in adolescents with depression. While an earlier criticism of this work was the inability to disentangle whether these biases were a cause or an effect of symptoms, more recent work has used cognitive training techniques to manipulate and induce particular biases and then assess their effects on symptoms. These data, while few and conducted in healthy volunteers, show reasonably convincing support for a temporal relationship between induced biases and subsequent effects on negative affect (Lothmann et al. 2011; Lau et al. 2011b; Lester et al. 2010; Muris et al. 2008, 2009) and emotional responsiveness to a psychological challenge (Lau et al. in press-a).

Depressive symptoms in children and adolescents have also been linked to other cognitive distortions, such as a negative attributional style, the tendency to endorse internal, stable and global reasons for negative events, dysfunctional attitudes and negative interpersonal cognitions, such as distorted perceptions and expectations of the world and other people (Jacobs et al. 2008). These cognitive biases are thought to predict symptoms in the presence of stress, with generally robust support for this 'diathesis-stress' hypothesis.

Are there age-related changes in information-processing and cognitive vulnerability? As children mature, gradual changes in cognitive vulnerability factors, such as attributional style have been noted. Thus, attributional style may play a more consistent role in the development of depressive symptoms in adolescents compared to children. Indeed, a recent review of existing child and adolescent studies suggested that the interaction of life events and attributional style was a more potent predictor of depression as children aged (Jacobs et al. 2008). There are also data to suggest subtle changes in the nature of the relationship between negative attributions, stress and depressive symptoms. Thus, while adolescent negative attributional style acts as a trait-like diathesis, elevating depressive responses to stress, negative attributions in children have been found to mediate the effects of early stress on symptoms (Cole and Turner 1993; Turner and Cole 1994). However, these developmental dissociations have also been contradicted by reports of an interaction between negative attributions and stress in earlier childhood when using more developmentally sensitive measures of attributional style (Conley et al. 2001).

Cognitive vulnerability factors are also thought to become increasingly stable and trait-like in the transition to adolescence. An analysis of test–retest correlations on multi-wave measures of depressogenic cognitive style, dysfunctional attitudes, rumination, neuroticism and depressive mood in adolescents has been used in one study to support a trait-like model rather than a contextual model in adolescents. In this model, an enduring vulnerability factor explains stability over time (Hankin et al. 2005). In contrast, in the younger half of their sample, both the trait-like and contextual model fit the data well. Finally, there is some suggestion that with development, cognitive vulnerability factors also may become more interrelated. Indeed, factor analyses do point to increasingly common, overlapping factors in adolescents, rather than more distinct cognitive vulnerability factors in childhood (Hankin et al. 2007). These ideas align well with proposed age trends in the consolidation of cognitive schema (Nolen-Hoeksema et al. 1992), in which repeated learning experiences and increasing cognitive capacity serve to reinforce and possibly integrate the various facets of cognitive style.

In summary, there is some suggestion that cognitive distortions such as negative attributional style and dysfunctional attitudes become increasingly stable, cohesive and trait-like in the transition to adolescence. The developmental emergence of these enduring and more habitual vulnerability factors are thought to reflect increasing experience but may also point to (neuro)cognitive maturation.

5 Concluding Remarks

In this chapter, we have discussed some of the age-related changes that can occur in the transition to adolescence. Interestingly, as these changes co-occur with the time when the rates of mood symptoms double, we suggest that these developmental changes may explain why adolescence is a critical period for the emergence of persistent mood problems in vulnerable individuals. In particular, we suggest that changes in the social environment provide young people with new challenges that they may not previously have encountered such as negotiating more complex peer networks and/or changes to the school routine. Difficulties in regulating their emotions towards some of these challenges, or even understanding these new challenges may arise in part because of limited experience, but also from a protracted development of neural circuits involved in emotion and social processing. In contrast to subcortical regions, which mature early, prefrontal regions may be slower to mature. In fact, the maturing of these circuits could reflect the sources of new genetic variance identified by quantitative genetic studies. How might these patterns of brain activity then affect information processing? We suggest that strong affective responses could serve to bias information processing, resulting in greater difficulties to ignore negative information and/or to misinterpret ambiguous cues. With repeated experiences and consolidation of cognitive schema, these information processing styles may become habitual, manifesting as cognitive vulnerability factors such as negative attributional style and dysfunctional attitudes. In turn, these negative cognitions act as trait-like diatheses, elevating depressive responses in the presence of the new changes and stresses of adolescence.

References

Arseneault L, Bowes L, Shakoor S (2010) Bullying victimization in youths and mental health problems: 'much ado about nothing'? Psychol Med 40:717–729

Barrera M, Garrisonjones C (1992) Family and peer social support as specific correlates of adolescent depressive symptoms. J Abnorm Child Psychol 20:1–16

Bartels M, Van de Aa N, Van Beijsterveldt CE, Middeldorp CM, Boomsma DI (2011) Adolescent self-report of emotional and behavioral problems: interactions of genetic factors with sex and age. J Can Acad Child Adolesc Psychiatry 20:35–52

Beesdo K, Lau JY, Mcclure-Tone EB, Guyer AE, Monk CS, Nelson EE, Fromm SJ, Goldwin MA, Wittchen HU, Leibenluft E, Ernst M, Pine DS (2009) Common and specific amygdala engagement perturbations in depressed versus anxious adolescents. Arch Gen Psychiatry 66:275–285

Belli SR, Rogers RD, Lau JY (in press) Adult and adolescent social reciprocity: experimental data from the trust game. J Adolesc [Epub ahead of print: 10 June, 2012]

Boomsma DI, Van Beijsterveldt CE, Hudziak JJ (2005) Genetic and environmental influences on anxious/depression during childhood: a study from the Netherlands twin register. Genes Brain Behav 4:466–481

Boomsma DI, Willemsen G, Sullivan PF, Heutink P, Meijer P, Sondervan D, Kluft C, Smit G, Nolen WA, Zitman FG, Smit JH, Hoogendijk WJ, Van Dyck R, de Geus EJ, Penninx BW (2008) Genome-wide association of major depression: description of samples for the GAIN major depressive disorder study: NTR and NESDA biobank projects. Eur J Hum Genet 16:335–342

Brown GW, Harris TO, Eales MJ (1993) Etiology of anxiety and depressive-disorders in an inner-city population. 2. comorbidity and adversity. Psychol Med 23:155–165

Brown GW, Harris TO, Eales MJ (1996) Social factors and comorbidity of depressive and anxiety disorders. Br J Psychiatry 168:50–57

Burnett S, Sebastian C, Cohen Kadosh K, Blakemore SJ (2011) The social brain in adolescence: evidence from functional magnetic resonance imaging and behavioural studies. Neurosci Biobehav Rev 35:1654–1664

Cauce AM, Hannan K, Sargeant M (1992) Life stress, social support, and locus of control during early adolescence—interactive effects. Am J Community Psychol 20:787–798

Champagne FA, Francis DD, Mar A, Meaney MJ (2003) Variations in maternal care in the rat as a mediating influence for the effects of environment on development. Physiol Behav 79:359–371

Cole DA, Turner JE (1993) Models of cognitive mediation and moderation in child depression. J Abnorm Psychol 102:271–281

Compas BE, Davis GE, Forsythe CJ (1985) Characteristics of life events during adolescence. Am J Community Psychol 13:677–691

Conley CS, Haines BA, Hilt LM, Metalsky GI (2001) The children's attributional style interview: developmental tests of cognitive diathesis-stress theories of depression. J Abnorm Child Psychol 29:445–463

Davila J, Steinberg SJ, Kachadourian L, Cobb R, Fincham F (2004) Romantic involvement and depressive symptoms in early and late adolescence: the role of a preoccupied relational style. Pers Relatsh 11:161–178

Durston S, Davidson MC, Tottenham N, Galvan A, Spicer J, Fossella JA, Casey BJ (2006) A shift from diffuse to focal cortical activity with development. Dev Sci 9:1–8

Eley TC, Stevenson J (2000) Specific life events and chronic experiences differentially associated with depression and anxiety in young twins. J Abnorm Child Psychol 28:383–394

Fenzel LM (2000) Prospective study of changes in global self-worth and strain during the transition to middle school. J Early Adolesc 20:93–116

Flint J, Chen Y, Shi S, Kendler KS; CONVERGE consortium (2012) Epilogue: lessons from the CONVERGE study of major depressive disorder in China. J Affect Disord 140(1):1–5. [Epub 28 Sep 2011].

Forbes EE, Christopher May J, Siegle GJ, Ladouceur CD, Ryan ND, Carter CS, Birmaher B, Axelson DA, Dahl RE (2006) Reward-related decision-making in pediatric major depressive disorder: an fMRI study. J Child Psychol Psychiatry 47:1031–1040

Forbes EE, Hariri AR, Martin SL, Silk JS, Moyles DL, Fisher PM, Brown SM, Ryan ND, Birmaher B, Axelson DA, Dahl RE (2009) Altered striatal activation predicting real-world positive affect in adolescent major depressive disorder. Am J Psychiatry 166:64–73

Gardner M, Steinberg L (2005) Peer influence on risk taking, risk preference, and risky decision making in adolescence and adulthood: an experimental study. Dev Psychol 41:625–635

Gesell A (1932) The developmental morphology of infant behavior pattern. Proc Natl Acad Sci U S A 18:139–143

Giedd JN, Rapoport JL (2010) Structural MRI of pediatric brain development: what have we learned and where are we going? Neuron 67:728–734

Gilman SE, Kawachi I, Fitzmaurice GM, Buka L (2003) Socio-economic status, family disruption and residential stability in childhood: relation to onset, recurrence and remission of major depression. Psychol Med 33:1341–1355

Gladstone TR, Kaslow NJ, Seeley JR, Lewinsohn PM (1997) Sex differences, attributional style, and depressive symptoms among adolescents. J Abnorm Child Psychol 25:297–306

Gotlib IH, Joormann J (2010) Cognition and depression: current status and future directions. Annu Rev Clin Psychol 6:285–312

Gotlib IH, Hamilton JP, Cooney RE, Singh MK, Henry ML, Joormann J (2010) Neural processing of reward and loss in girls at risk for major depression. Arch Gen Psychiatr 67(4):380–387

Grant KE, Compas BE, Thurm AE, Mcmahon SD, Gipson PY (2004) Stressors and child and adolescent psychopathology: measurement issues and prospective effects. J Clin Child Adolesc Psychol 33:412–425

Gunther Moor B, Guroglu B, Op De Macks ZA, Rombouts SA, Van Der Molen MW, Crone EA (2012) Social exclusion and punishment of excluders: neural correlates and developmental trajectories. Neuroimage 59:708–717

Guyer AE, Lau JY, Mcclure-Tone EB, Parrish J, Shiffrin ND, Reynolds RC, Chen G, Blair RJ, Leibenluft E, Fox NA, Ernst M, Pine DS, Nelson EE (2008) Amygdala and ventrolateral prefrontal cortex function during anticipated peer evaluation in pediatric social anxiety. Arch Gen Psychiatry 65:1303–1312

Guyer AE, Mcclure-Tone EB, Shiffrin ND, Pine DS, Nelson EE (2009) Probing the neural correlates of anticipated peer evaluation in adolescence. Child Dev 80:1000–1015

Haeffel GJ, Gibb BE, Metalsky GI, Alloy LB, Abramson LY, Hankin BL, Joiner TE, Swendsen JD (2008) Measuring cognitive vulnerability to depression: development and validation of the cognitive style questionnaire. Clin Psychol Rev 28:824–836

Halari R, Simic M, Pariante CM, Papadopoulos A, Cleare A, Brammer M, Fombonne E, Rubia K (2009) Reduced activation in lateral prefrontal cortex and anterior cingulate during attention and cognitive control functions in medication-naive adolescents with depression compared to controls. J Child Psychol Psychiatry 50:307–316

Hankin BL, Abramson LY, Moffitt TE, Silva PA, Mcgee R, Angell KE (1998) Development of depression from preadolescence to young adulthood: emerging gender differences in a 10-year longitudinal study. J Abnorm Psychol 107:128–140

Hankin BL, Fraley RC, Abela JR (2005) Daily depression and cognitions about stress: evidence for a traitlike depressogenic cognitive style and the prediction of depressive symptoms in a prospective daily diary study. J Pers Soc Psychol 88:673–685

Hankin BL, Lakdawalla Z, Carter IL, Abela JRZ, Adams P (2007) Are neuroticism, cognitive vulnerabilities and self-esteem overlapping or distinct risks for depression? evidence from exploratory and confirmatory factor analyses. J Soc Clin Psychol 26:29–63

Hartley CA, Phelps EA (2010) Changing fear: the neurocircuitry of emotion regulation. Neuropsychopharmacology 35:136–146

Hawker DS, Boulton MJ (2000) Twenty years' research on peer victimization and psychosocial maladjustment: a meta-analytic review of cross-sectional studies. J Child Psychol Psychiatry 41:441–455

Hoffman MA, Levyshiff R, Sohlberg SC, Zarizki J (1992) The impact of stress and coping—developmental-changes in the transition to adolescence. J Youth Adolesc 21:451–469

Hoffmann JP, Cerbone FG (1999) Stressful life events and delinquency escalation in early adolescence. Criminology 37:343–373

Jacobs RH, Reinecke MA, Gollan JK, Kane P (2008) Empirical evidence of cognitive vulnerability for depression among children and adolescents: a cognitive science and developmental perspective. Clin Psychol Rev 28:759–782

Kendler KS, Hettema JM, Butera F, Gardner CO, Prescott CA (2003) Life event dimensions of loss, humiliation, entrapment, and danger in the prediction of onsets of major depression and generalized anxiety. Arch Gen Psychiatry 60:789–796

Kendler KS, Gardner CO, Lichtenstein P (2008) A developmental twin study of symptoms of anxiety and depression: evidence for genetic innovation and attenuation. Psychol Med 38:1567–1575

Lamb DJ, Middeldorp CM, Van Beijsterveldt CE, Bartels M, Van Der Aa N, Polderman TJ, Boomsma DI (2010) Heritability of anxious-depressive and withdrawn behavior: age-related changes during adolescence. J Am Acad Child Adolesc Psychiatry 49:248–255

Larson RW, Richards MH, Moneta G, Holmbeck G, Duckett E (1996) Changes in adolescents' daily interactions with their families from ages 10 to 18: disengagement and transformation. Dev Psychol 32:744–754

Lau JY, Eley TC (2006) Changes in genetic and environmental influences on depressive symptoms across adolescence and young adulthood. Br J Psychiatry 189:422–427

Lau JY, Eley TC (2008a) Attributional style as a risk marker of genetic effects for adolescent depressive symptoms. J Abnorm Psychol 117:849–859

Lau JY, Eley TC (2008b) Disentangling gene-environment correlations and interactions on adolescent depressive symptoms. J Child Psychol Psychiatry 49:142–150
Lau JY, Rijsdijk F, Eley TC (2006) I think, therefore I am: a twin study of attributional style in adolescents. J Child Psychol Psychiatry 47:696–703
Lau JY, Gregory AM, Goldwin MA, Pine DS, Eley TC (2007) Assessing gene-environment interactions on anxiety symptom subtypes across childhood and adolescence. Dev Psychopathol 19:1129–1146
Lau JY, Goldman D, Buzas B, Fromm SJ, Guyer AE, Hodgkinson C, Monk CS, Nelson EE, Shen PH, Pine DS, Ernst M (2009) Amygdala function and 5-HTT gene variants in adolescent anxiety and major depressive disorder. Biol Psychiatry 65:349–355
Lau JY, Britton JC, Nelson EE, Angold A, Ernst M, Goldwin M, Grillon C, Leibenluft E, Lissek S, Norcross M, Shiffrin N, Pine DS (2011a) Distinct neural signatures of threat learning in adolescents and adults. Proc Natl Acad Sci U S A 108:4500–4505
Lau JY, Molyneaux E, Telman MD, Belli S (2011b) The plasticity of adolescent cognitions: data from a novel cognitive bias modification training task. Child Psychiatry Hum Dev 42:679–693
Lau JY, Belli S, Chopra R, Cognitive bias modification training in adolescents reduces anxiety to a psychological challenge. Clin Child Psychol Psychiatry (in press-a)
Lau JY, Guyer AE, Mcclure-Tone EB, Parrish J, Shiffrin ND, Reynolds RC, Chen G, Ernst M, Pine DS, Nelson EE, Neural responses to social rejection and acceptance. Int J Behav Dev (in press-b)
Lester KJ, Field AP, Muris P (2010) Experimental modification of interpretation bias regarding social and animal fear in children. J Anxiety Disord 25:697–705
Levitan RD, Rector NA, Sheldon T, Goering P (2003) Childhood adversities associated with major depression and/or anxiety disorders in a community sample of Ontario: issues of co-morbidity and specificity. Depress Anxiety 17:34–42
Lothmann C, Holmes EA, Chan SWY, Lau JY (2011) Cognitive bias modification training in adolescents: effects on interpretation biases and mood. J Child Psychol Psychiatry Allied Disciplines 52:24–32
Masten CL, Eisenberger NI, Borofsky LA, Pfeifer JH, Mcnealy K, Mazziotta JC, Dapretto M (2009) Neural correlates of social exclusion during adolescence: understanding the distress of peer rejection. Soc Cogn Affect Neurosci 4:143–157
Mcclure EB, Monk CS, Nelson EE, Parrish JM, Adler A, Blair RJ, Fromm S, Charney DS, Leibenluft E, Ernst M, Pine DS (2007) Abnormal attention modulation of fear circuit function in pediatric generalized anxiety disorder. Arch Gen Psychiatry 64:97–106
Monk CS, Nelson EE, Mcclure EB, Mogg K, Bradley BP, Leibenluft E, Blair RJ, Chen G, Charney DS, Ernst M, Pine DS (2006) Ventrolateral prefrontal cortex activation and attentional bias in response to angry faces in adolescents with generalized anxiety disorder. Am J Psychiatry 163:1091–1097
Muris P, Huijding J, Mayer B, Hameetman M (2008) A space odyssey: experimental manipulation of threat perception and anxiety-related interpretation bias in children. Child Psychiatry Hum Dev 39:469–480
Muris P, Huijding J, Mayer B, Remmerswaal D, Vreden S (2009) Ground control to major Tom: experimental manipulation of anxiety-related interpretation bias by means of the "space odyssey" paradigm and effects on avoidance tendencies in children. J Anxiety Disord 23:333–340
Nolen-Hoeksema S, Girgus JS, Seligman ME (1992) Predictors and consequences of childhood depressive symptoms: a 5-year longitudinal study. J Abnorm Psychol 101:405–422
O'Connor TG, Mcguire S, Reiss D, Hetherington EM, Plomin R (1998a) Co-occurrence of depressive symptoms and antisocial behavior in adolescence: a common genetic liability. J Abnorm Psychol 107:27–37
O'Connor TG, Neiderhiser JM, Reiss D, Hetherington EM, Plomin R (1998b) Genetic contributions to continuity, change, and co-occurrence of antisocial and depressive symptoms in adolescence. J Child Psychol Psychiatry 39:323–336

Pine DS, Cohen E, Cohen P, Brook J (1999) Adolescent depressive symptoms as predictors of adult depression: moodiness or mood disorder? Am J Psychiatry 156:133–135

Plomin R, Defries JC, Mclearn GE, Mcguffin P (2008) Behavioral genetics. Worth Books, New York

Puskar KR, Rohay JM (1999) School relocation and stress in teens. J Sch Nurs 15:16–22

Quamma JP, Greenberg MT (1994) Childrens experience of life stress—the role of family social support and social problem-solving skills as protective factors. J Clin Child Psychol 23: 295–305

Rice F, Harold GT, Thapar A (2002) Assessing the effects of age, sex and shared environment on the genetic aetiology of depression in childhood and adolescence. J Child Psychol Psychiatry 43:1039–1051

Risch N, Herrell R, Lehner T, Liang KY, Eaves L, Hoh J, Griem A, Kovacs M, Ott J, Merikangas KR (2009) Interaction between the serotonin transporter gene (5-HTTLPR), stressful life events, and risk of depression: a meta-analysis. JAMA 301:2462–2471

Sallinen M, Ronka A, Kinnunen U, Kokko K (2007) Trajectories of depressive mood in adolescents: does parental work or parent-adolescent relationship matter? A follow-up study through junior high school in Finland. Int J Behav Dev 31:181–190

Scourfield J, Rice F, Thapar A, Harold GT, Martin N, Mcguffin P (2003) Depressive symptoms in children and adolescents: changing aetiological influences with development. J Child Psychol Psychiatry 44:968–976

Sebastian C, Viding E, Williams KD, Blakemore SJ (2010) Social brain development and the affective consequences of ostracism in adolescence. Brain Cogn 72:134–145

Sebastian CL, Fontaine NM, Bird G, Blakemore SJ, De Brito SA, Mccrory EJ, Viding E (2011a) Neural processing associated with cognitive and affective theory of mind in adolescents and adults. Soc Cogn Affect Neurosci 7(1):53–63

Sebastian CL, Tan GC, Roiser JP, Viding E, Dumontheil I, Blakemore SJ (2011b) Developmental influences on the neural bases of responses to social rejection: implications of social neuroscience for education. Neuroimage 57:686–694

Somerville LH, Jones RM, Casey BJ (2010) A time of change: behavioral and neural correlates of adolescent sensitivity to appetitive and aversive environmental cues. Brain Cogn 72:124–133

Steinberg L, Monahan KC (2007) Age differences in resistance to peer influence. Dev Psychol 43:1531–1543

Steinberg L, Morris AS (2001) Adolescent development. Annu Rev Psychol 52:83–110

Stroud LR, Foster E, Papandonatos GD, Handwerger K, Granger DA, Kivlighan KT, Niaura R (2009) Stress response and the adolescent transition: performance versus peer rejection stressors. Dev Psychopathol 21:47–68

Thapar A, Mcguffin P (1994) A twin study of depressive symptoms in childhood. Br J Psychiatry 165:259–265

Turner JE, Cole DA (1994) Developmental differences in cognitive diatheses for child depression. J Abnorm Child Psychol 22:15–32

Van Der Valk JC, Van Den Oord EJ, Verhulst FC, Boomsma DI (2003) Genetic and environmental contributions to stability and change in children's internalizing and externalizing problems. J Am Acad Child Adolesc Psychiatry 42:1212–1220

Zimmer-Gembeck MJ (2002) The development of romantic relationships and adaptations in the system of peer relationships. J Adolesc Health 31:216–225

Animal Models of Depression Vulnerability

Jaanus Harro

Abstract The rapid increase in the number of proposed animal models of depression reflects the dissatisfaction with our current state of knowledge on neurobiology of depression and unsuccessful drug development. Results obtained with even the best validated models can be difficult to compare. Because evidence from epidemiological studies suggests that depression occurs in biologically predisposed subjects under the impact of adverse life events, increasing attempts have been made to use the diathesis–stress concept in animal models. In this way, factors underpinning vulnerability to depression have been identified by measuring behavioural traits analogous to facets of human personality, or created by inducing neurochemical lesions. Stressful interventions administered prenatally, in early life or in adulthood have been combined with other vulnerability factors including genetic changes. As a result, several putative animal models of endophenotypes of depression or depression vulnerability have been proposed. Diathesis–stress models may aid in separating adaptive and maladaptive strategies in coping with stress, and understanding the relevant neurobiology. Studies comparing effects of stress on males and females should reveal to which extent the pathogenetic processes leading to depression can be specific to sex/gender.

Keywords Depression · Animal models · Predisposition · Resilience · Diathesis–stress · Endophenotypic · Acute and chronic stress · Maternal separation · Inter-individual variability

J. Harro (✉)
Unit of Psychophysiology, Department of Psychology,
Estonian Centre of Behavioural and Health Sciences,
University of Tartu, Ravila 14a, 50411 Tartu, Estonia
e-mail: Jaanus.Harro@ut.ee

Curr Topics Behav Neurosci (2013) 14: 29–54
DOI: 10.1007/7854_2012_221

Published Online: 7 August 2012

Contents

1 Introduction

In recent years, animal models of human psychopathology have been continuously growing in number, and this is particularly true for animal models of depression. Such a trend probably reflects a variety of developments in neuroscience, but one of these, pertinent to the present chapter, is the increased understanding that realistic and translational animal models of depression must incorporate the notion that not all people develop clinical depression even if being genetically similar or living at equally harsh conditions (Anisman et al. 2008). Furthermore, as one of the important aims of animal models is to aid in screening for novel drugs active as antidepressants, the limited success of recent psychotropic drug development is suggesting that there may be something wrong with the conventional approach. It would be unfair to lay all blame on insufficient animal models, as many other key shortfalls are evident in the process (Williams 2011). Nevertheless, rethinking the way, we consider the functions of animal models and their development, appears due.

While many available models are sensitive to existing antidepressants, few new leads have made it to the clinic, and many false positives have been reported. Indeed, if the drugs with antidepressant properties require large numbers of patients to be treated to detect efficacy compared to placebo, how can experiments based on small groups of laboratory animals, provided in distinct batches from a variety of breeding companies and uniformly treated, consistently yield such amazing efficacy of the known drugs? It appears likely that animal models are highly optimised for detecting the effect of the established antidepressants, and consequently the models often lead us to findings that are not at the core of the neurobiology of depression but somehow collateral to it. This may have been at least implicitly perceived by depression researchers who are increasingly looking for novel models or modifications that often are laboratory specific, hampering meaningful comparison of the data.

Regarding the often-used animal models of depression and antidepressant screening tests, there are many comprehensive reviews available that describe the mainstream models; these also describe several less well-characterised models and discuss conceptual issues and methodological caveats (e.g., Cryan et al. 2002; Cryan and Mombereau 2004; Cryan and Slattery 2007; El Yacoubi and Vaugeois 2007; Holmes 2003; Jacobson and Cryan 2007; Kalueff et al. 2007; McArthur and Borsini 2006; Nestler et al. 2002; O'Neill and Moore 2003; Refojo and Deussing 2012; Renoir et al. 2012; Samuels et al. 2011; Willner 1990, 1997). Several of these reviews also systematically address the issues of construct, face and predictive validity of animal models, as laid out by McKinney and Bunney (1969), Willner (1990), and Geyer and Markou (1995). Despite the fact that validity concerns always receive much attention, there is often disagreement between the authors on how best to model depression. The reasons for this can be found in the above cited papers. The present chapter is not aimed to provide another review of the well-known models or attempt to rectify the many controversies in the current literature. Instead, it is dedicated to some important conceptual issues revolving around the concept of depression vulnerability, and to views on the best approaches to model it.

2 What is Vulnerability to Depression?

Oxford Dictionary (2011) defines "vulnerability" as follows: "exposed to the possibility of being attacked or harmed, either physically or emotionally." The origin of the word has been traced to early seventeenth century, and comes from late Latin *vulnerabilis*, derived from *vulnerare* 'to wound,' and from *vulnus* 'wound.' According to the collective wisdom of Wikipedia (2011), vulnerability refers to the inability to withstand the effects of a hostile environment. Vulnerability is not thought of as a persistent state: a window of vulnerability is a time frame within which defensive measures are reduced, compromised or lacking. Hence, we ought to consider the harmful factors that reduce our capabilities to resist or compensate for the impact of such factors that can lead to depression. Factors causing vulnerability to depression and the clinical condition can be extracted from the epidemiological studies. It has to be acknowledged that individuals affected by depression display a wide variation in clinical symptoms and signs, and the diagnostic conventions are consensus driven, and thus to some extent arbitrary (Fava and Kendler 2000). Nevertheless, the large numbers of epidemiological studies in this area provide the most reliable starting point.

Epidemiological studies suggest that predisposing factors with a probable causal link to depression include sex/gender, stressful life events, adverse childhood experiences and certain personality traits (Fava and Kendler 2000). Of the latter, the best evidence is for neuroticism, a stable personality trait that reflects the predisposition to develop emotional upset under stress (Eysenck 1951). Many medical conditions can lead to clinical depression, and there are cases of depression inducing drugs as was recognised many years ago for the monoamine depleting alkaloid reserpine (Drevets and Furey 2009). Genetic factors contribute to the overall risk of depression, but also influence the sensitivity of individuals to the pathogenetic effects of environmental adversity (Kendler et al. 1995).

Altogether, this evidence suggests that the pathogenetic process of depression proceeds stepwise, building on blocks of previously emerged predisposition, and alterations brought about by adverse environment. How this could happen in terms of neurobiology has been elaborated on (e.g., Harro and Oreland 2001), but is mostly beyond the scope of this chapter. For the development of depression vulnerability models, the evidence that depression has both environmental and genetic origins provides a guide to approaches that are based on enhanced vulnerability of some animals as compared to other, more resilient counterparts.

3 How Do We Define Animal Models of Depression Vulnerability?

The first explicit attempt to define animal models of depression vulnerability, and to compare the validity of such models, was made 10 years ago. For this analysis, Willner and Mitchell (2002) suggested that some animal models introduced as models of depression should rather be considered models of vulnerability to depression. These authors defined models of depression vulnerability as models that "increase the ease with which an analogue of major depression may be evoked, or a presentation analogous to dysthymia (chronic mild depression)." According to the vulnerability approach, animals in the models of depressive diathesis would either express "depressive" features even in the absence of stress (corresponding to the DSM-IV condition of dysthymia or minor depression) or appear no different from their non-vulnerable counterparts if there is no manipulation present to precipitate depression. The authors went further by listing 11 animal models of vulnerability to depression, and assessed these models for their predictive, face and construct validity. The authors classified these models into genetic, genomic, developmental and lesion models; the models were congenital learned helplessness, Flinders Sensitive Line (FSL) rats (muscarinic hypersensitivity), Roman High Avoidance rats, Fawn Hooded rats, hypothalamus-pituitary-adrenal axis (HPA) transgenic mice, 5-HT transporter knockout mice, corticotropin-releasing factor (CRF) receptor subtype knockout mice, tachykinin receptor knockout mice, neonatal antidepressant treatment, prenatal/neonatal stress and olfactory bulbectomy.

Many readers might consider most of these models as models of depression per se, without thinking in terms of whether they might represent models of depression

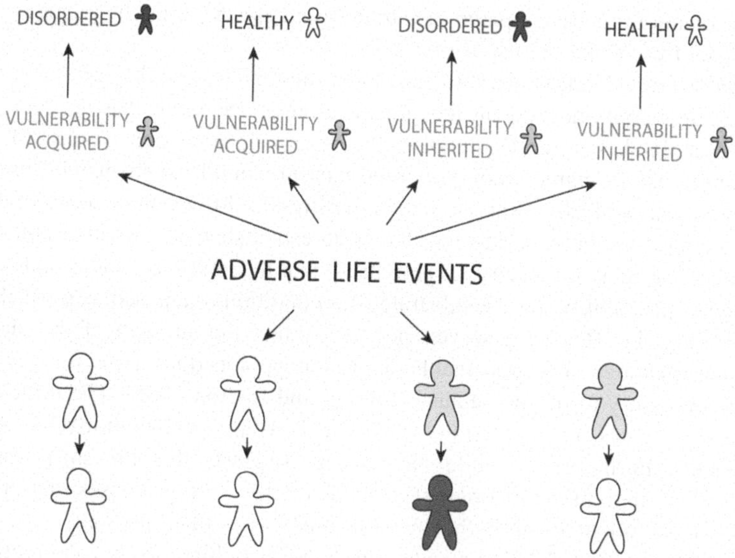

Fig. 1 *The diathesis–stress concept* Depression develops under strain brought about by adverse life events in those individuals who have respective predisposition. Such a predisposition can be either inherited or acquired during lifetime, and does not cause the disorder even not in all subjects who are subject to lifetime adversities owing to protective factors that for psychiatric disorders remain poorly defined

vulnerability. Indeed, it would be difficult to distinguish models of depression and depression vulnerability on the basis of assumptions on aspects of validity. This introduction of the vulnerability concept is, nevertheless, most valuable as it puts the focus on the salience of environmental factors in the pathogenesis of depression, and thus on the controllable aspects of construct validity of the models. In the following sections the approach to vulnerability models will be top down, departing from the known etiopathogenetic factors of depression that have guided us to the diathesis–stress concept.

4 The Diathesis–Stress Concept

The diathesis–stress or vulnerability–stress model of psychopathology has been used to explain behaviour as a result of two types of impact. A diathesis, or vulnerability, is a predispositional factor, or set of factors, that makes possible a disordered mental state (Ingram and Luxton 2005). The source of the disposition can be defined more or less broadly, and it could be either genetic or developmental, but it ought to be explainable in biological terms, and possess a relatively permanent, enduring trait. The stress from life experiences, superimposed on pre-existent biological differences conferring the vulnerability, will then be capable of eliciting psychopathology that would not be brought about in those individuals free of the predisposition.

It can be argued that vulnerabilities can be modifiable, and they probably are as the biological traits depend on the organisation of the brain that is meta stable rather than fixed. For model building the pragmatic approach would still keep vulnerability as the stable predispositional measure, and treat changeable states as the consequences of life events. It is then the stressful life event or series of events that challenges the homeostasis that has been established in the vulnerable brain, and shakes the state of the nervous activity so that the next meta stable state could, with higher probability, be the substrate of a condition diagnosed as a psychiatric disorder (Fig. 1). This is, however, not to say that the strength of the diathesis would never change: it is conceivable that at some points during life, relatively less stress is necessary to tip the balance (Ingram and Luxton 2005). The difficulty of uncovering "genes for depression" makes it likely that vulnerability has several components. Furthermore, while the concept suggests that the shift from the vulnerability stage to the disorder stage is qualitative, the processes that underlie the predisposition state can be balanced in more ways than just one.

How exactly stressful life events cause the switching from vulnerability to disorder is unclear. It is conceivable that greater exposure to life's adversities is pathogenetic, and it is also possible that some predispositions lead individuals to choose lifestyles that expose them to adverse life events that are either more serious or occur with higher probability. On the other hand, it has also been proposed that patients with depression have not encountered more severe life events, and it is their reactions that are exaggerated (Richardson 1991).

At this point, it has to be recalled that stressful life events, adverse childhood experiences and personality traits like neuroticism are not specifically associated with major depression, nor are there any gene products known to specifically cause depression. But neither is any symptom of affective disorders pathognomonic or even highly specific. Nevertheless, while the current diagnostic system has made psychiatric comorbidities rather a rule than an exception, each affected individual eventually is in a certain clinical state that must be the end product of the interaction of predisposition and environmental adversities. Thus, with the available knowledge, joint analysis of the effects of known risk factors will be helpful in disentangling the web of psychiatric nosology into more self-standing units. We cannot expect a greater level of understanding from the animal models in terms of symptoms, endophenotypes, biomarkers, aetiology or pathogenesis than that achieved from our level of understanding of the clinical pathology.

5 How Do We Learn Whether Vulnerability Is Truly Vulnerability?

One of the fundamental problems with modelling behavioural disorders has been the attempt to simulate changes in behaviour without really knowing the causes of the deviations (McArthur and Borsini 2006). Using simple unspecific behaviours of rodents as endpoints to measure human symptoms have led to endless discussions on, for example, what does immobility in the forced swimming test measure? This author believes that behaviours displayed in the forced swimming test can reliably measure a number of different psychological states depending upon the specific characteristics of the task and the neurobiological state of the animal (strain). Moreover, we can only find out the most likely construct for each well-defined condition by a laborious validation procedure (Harro 2004). Similarly, reduction in sucrose intake can probably reflect other psychological dimensions than just anhedonia, especially if it is recorded in food and water deprived animals.

An objective measure of vulnerability in an animal model should be congruent with a measure of human vulnerability to depression. This is very difficult to achieve for behavioural endpoints in rodents, because the adaptive value of simple behaviours in conditions highly artificial for the species remains a matter of interpretation. This, indeed, has allowed depression modelling to accept counter-intuitive approaches, as in the case of the olfactory bulbectomy model that displays a reliable behavioural readout in terms of hyperactivity in an open field. Nevertheless, this hyperactivity that hardly helps to meet the face validity requirement for a depression model is sensitive to a range of antidepressants, and the model is characterised by a number of neurobiological deviations reminiscent of depression (Song and Leonard 2005).

There appears to be two major approaches to increase our understanding of what we measure when taking behavioural readouts. First, to focus on ethologically relevant and situation-specific measures that reflect with high probability affective neurobiological states of the animal (Panksepp 1998). Second, to apply the diathesis–stress concept-derived models and pay special attention to behaviours that emerge under stress in vulnerable individuals. Subsequently, in either case, higher priority should be given to endophenotypic changes that can be directly translated to and from the pathogenesis of clinical depression.

The neurobiological vulnerability concept was introduced to the interpretation of animal models by Anisman and Zacharko (1990), who suggested that factors that "favor the provocation of amine depletions," or that diminish the intrinsic capacity of the nervous system to adapt to such depletions, increase vulnerability to depression. Monoamine-related measures still appear as legitimate targets, while a number of other candidate measures have been discovered. Since much work is going on to identify biomarkers of psychiatric conditions in humans, this approach should also yield the detection of vulnerability markers. Then, future research should aim at the development of animal models that display similar markers. As such objective markers of vulnerability are not yet available, the safest way forward appears to be to rely on the etiopathogenetic principle, that is, applying stress on animals with a biological background thought of as depression related.

6 A Note on Genetic Models and the Diathesis–Stress Concept

As this volume contains a chapter on genetic models of depression (see chapter by Barkus), these are not discussed in detail in the present chapter, but it should be mentioned that several advances have been recently made in using the diathesis–stress approach with vulnerable mouse strains. Indeed, some strains of mice have behavioural and physiological features that suggest construct and face validity as a model of depression, and some such models have been found to provide predictive validity. While many mouse models are used without including adverse life events, life events take place for the laboratory animals notwithstanding the experimental design, and the vulnerability factor could play a role even if there is formally no stress applied. Naturally, this is relevant not only for genetically modified mice but also for any animal stock with presumed predisposition to depression-like pathogenetic process. In addition to the minor life events that are inevitably happening in the animal house, vulnerability may be producing psychopathology via other routes such as internal, physical deviations building up stress from inside as well as behavioural tendencies that make the animals play a role in creating their own stresses, for example, by engaging in aggressive encounters with cage mates (Ingram and Luxton 2005; Harro 2010). However, each strain will establish over a few generations adaptive strategies to maintain homeostasis, such that the behaviours displayed by the vulnerable strains could indicate different coping styles rather than anything congruent with human clinical symptoms.

There is a large body of the literature on affect-related differences between rodent strains, and these data are not always easy to reconcile. For example, while C57BL/6 mice are considered the resilient strain in comparison with BALB/c mice, they have been used as a stress susceptibility strain in comparison with DBA/2 mice (Ventura et al. 2002). A recent review by Jacobson and Cryan (2007) compiled a large amount of information on the behavioural features of different mouse strains, and pointed out a number of complicating factors. However, one conclusion that emerges from analyses of models of anxiety, depression and antidepressant sensitivity is that the BALB/c strain has a reproducible predisposition to negative affect (Belzung and Griebel 2001; Jacobson and Cryan 2007). This is confirmed by the diathesis–stress approach: BALB/c mice were compared to C57BL/6 with regard to responsivity to chronic mild stress, and it was found that stress reduced learning specifically in the BALB/c mice (Palumbo et al. 2009). Interestingly, a differential alteration in levels of neuronal nitric oxide synthase (NOS) isoforms was also observed. NOS activity has been previously implicated in neurobiology of depression and may be attenuated during 5-HT hypofunction (Harvey et al. 2006).

Maternal separation has recently been reported to decrease the genetic expression of histone deacetylase in neocortex in adult BALB/c mice, an alteration that was not observed in C57BL/6 mice (Levine et al. 2012). These data suggest that epigenetic programming may be differentially recruited in mouse strains with different affective behaviour. Several genetically modified mice with suspected vulnerability to depression have also been submitted to chronic stress regimens. Examples include cannabinoid receptor $CB_1^{-/-}$ mice (Aso et al. 2011), vesicular glutamate transporter $vGLUT1^{+/-}$ mice (Garcia–Garcia et al. 2009), mice with impaired glucocorticoid receptor expression (GR-i mice) and 5-HT transporter $^{-/-}$ mice (Lanfumey et al. 2000). Importantly, all these studies have been supportive of the diathesis–stress concept, with the stress vulnerability of the genetically manipulated mice being greater than wild-type controls.

7 Adverse Early Life Events

A range of adversities in childhood including physical and sexual abuse, poor parent–child relationships and parental discord and divorce, almost certainly increase the risk for depression later in life (Fava and Kendler 2000). In animal studies, early life stress paradigms have utilised various manipulations, in particular handling, maternal separation, social deprivation and enriched versus impoverished environments. The validity of these paradigms as models of depression has been considered limited but in theory the programming effects elicited by the manipulations should, indeed, confer vulnerability rather than disease (Schmidt et al. 2011). Thus, early life stress paradigms could be very useful as part of the diathesis–stress approach.

Early life stress studies have revealed major changes in neuroplasticity and learning in adulthood that depends on environmental adversities. For example, in rats the adult male offspring of less caring mothers had reduced synaptic plasticity in stress-free conditions, but enhanced synaptic plasticity in stressful conditions (Champagne et al. 2008). These differences in neuroplasticity were mirrored in performance in learning tasks. Similar findings were obtained when rats were submitted to the presumably more stressful maternal separation paradigm; for example, while neurogenesis was reduced by severe early life stress and spatial learning was impaired, long-term potentiation was greater and fear conditioning was enhanced (Oomen et al. 2010). It is tempting to speculate that such neurobiological alterations underlie the selective attention to negative stimuli and rumination on dysfunctional thoughts in the development of depression.

7.1 Maternal Separation

Several animal models are based on the notion that disruption of the mother-infant bond is an important risk factor for disorders of affect. How the adult will emotionally respond depends on the quality of early environment, and this association appears to be mediated by early influences of the parents on brain development. The most robust approach to disrupt the mother-infant bond is daily separation of mothers from their offspring, and this has been an increasingly applied technique to mimic early life adversities brought about by inadequate maternal care. In monkey infants initial protests and the subsequent despair-like state induced by separation from their mothers was an early depression model in primates (McKinney and Bunney 1969), and the model can be applied in a variety of other species (e.g., Panksepp et al. 1978; Francis et al. 1999).

Probably, the most carefully studied model so far in the diathesis–stress approach is the FSL rat (originally bred for increased responses to an anticholinesterase drug) that exhibits several depression-like features when compared to the Flinders Resistant Line (FRL) (Overstreet et al. 2005). In FRL rats exposure to early life stress in the form of maternal separation did not reduce significantly the swimming behaviour in the modified Porsolt's test, but further reduced swimming in the vulnerability model, FSL rats (El Khoury et al. 2006). Whereas escitalopram was ineffective in FRL rats, the antidepressant brought swimming behaviour of maternally separated FSL rats to control levels (El Khoury et al. 2006). Swimming was recorded as the sum of all active behaviours in these experiments. In further studies, the stress resilience of the FRL rat was, however, less clear-cut. In one experiment, maternal separation increased immobility in both lines (non-significantly in FSL, but this could have been related to a ceiling effect), and the efficacy of escitalopram was notable in naïve FSL but much less potent in maternally separated FSL (Musazzi et al. 2010). In another experiment, maternal separation had no statistically significant effect on immobility, but a strong trend toward an increase was observed in both strains (Piubelli et al. 2011). Interestingly, in this experiment escitalopram was not effective in the maternally separated rats. Nevertheless, escitalopram attenuated the upregulation of hippocampal NMDA receptor subunit 1 immunoreactivity, which is an alteration observed selectively in FSL rats after maternal separation (Ryan et al. 2009). Further studies should reveal reliable endpoints in this diathesis–stress model, and whether these endpoints have relevance to clinical depression.

7.2 Prenatal Stress

An increasing number of prospective studies have revealed that if a mother has been under stress while pregnant, her child is substantially more likely to have

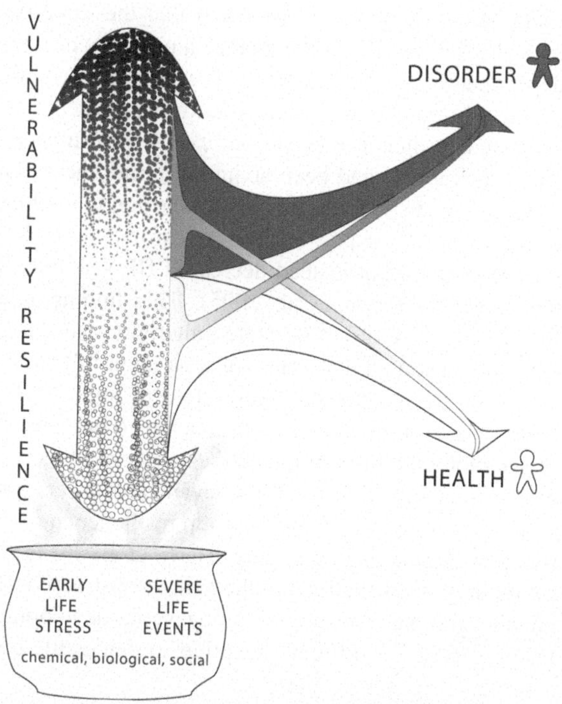

Fig. 2 *Dimensionality of the diathesis–stress concept* Both vulnerability and resilience to depression are probably quantitative in nature, and do not represent a single dimension. While the probability to respond to the variety of lifetime adversities with depression is increased by diathesis and reduced by resilience, mechanisms appear to exist that mediate an increase in resistance in subjects who would be considered as predisposed to the disorder

emotional or cognitive problems, independently of maternal postnatal affective state (Talge et al. 2007). Prenatal environment can have long lasting consequences that track into adulthood. For example, in animal studies, repeated exposure to severe stress during the second half of pregnancy increased anxiety and vulnerability to addictive drugs and led to a phase advance in the circadian rhythm and increased in corticosterone secretion and paradoxical sleep in adulthood of the offspring (Maccari et al. 2003). Thus, prenatal stress applied to animals with depression vulnerability represents a promising research strategy.

8 Stressful Life Events During Adulthood

Many environmental adversities, such as job loss, marital difficulties, major health problems and loss of close personal relationships, are linked with a substantial increase in risk for the onset of depression (Fava and Kendler 2000). In animal modelling, comparisons between studies can also be very approximate, for instance, because of the many types and schedules of stress applied. A simple classification can separate acute stress from chronic, chronic stress with a single stressor (often restraint, Rosecrans and De Feo 1965) from chronic variable stress (Katz et al. 1981; Willner et al. 1992) and perhaps chronic social stress from non-social stressors. It should be noted that the same type of stressor could serve as both the source of environmental adversity and signal of induced vulnerability, dependent on the paradigm; for example, early life stress applied to animals genetically predisposed to depression would serve as a stressor, while maternally separated animals can be further exposed to stress as adults, with the expectation that vulnerability had been acquired during the early life adversity.

One such an attempt, however, revealed that while maternal separation elicited exaggerated HPA axis responses, chronic variable stress in adulthood did not further increase but instead decreased HPA activity as well as CRF expression in the amygdala (Ladd et al. 2005). This finding leads to reconsideration of the functional value of stress-related alterations in the CNS (Fig. 2). Further studies have demonstrated that maternal separation followed by social defeat in adulthood can lead to heightened expression of the 5-HT transporter and tryptophan hydroxylase in the dorsal raphe nucleus (Gardner et al. 2009a, b), findings interpreted along the lines of the diathesis–stress concept. Social defeat stress may be a more severe but ethologically relevant stressor compared to chronic variable stress and less manageable by the built-up resilience, a notion that receives support from oxidative metabolism mapping (Harro et al. 2011). Studies that apply stress early in life may still benefit from the neuro-constructivist approach that emphasises the natural gradual process of ontogenetic development (Karmiloff-Smith 2006), leaving space for different adaptive strategies to adversities.

8.1 Acute Stress

While the opinion prevails that humans suffer from depression mostly after chronic stress, a single episode of stress can also precipitate depression, but it must be very intense or interpreted as such. When using a behavioural measure as the readout in the diathesis–stress model, it is preferable to select an independent test not related to performance during stress. For example, if the acute stressor is forced swimming, then immobility during stress may not be a reliable indicator. While we have little insight into the animal's subjective experiences of the severity of stress, it appears safer to make use of chronic stress paradigms. Humans are more likely to be subjected to periods of mild intermittent or prolonged stresses (McArthur and Borsini 2006), and behavioural alterations after such impacts seem less likely to reflect positively adaptive shifts than reactions to acute insults. Methodologically, it is also worth considering that the time course of the stress response to an acute stressor is an additional source of variation in comparisons across different stressors and laboratory conditions. For studies of chronic stress it is to be safer to assume a stabilised stress response than for an acute challenge. Nevertheless, additional adaptive changes to the stress response can occur at later stages (Matrov et al. 2011).

A few interesting studies have, however, attempted to precipitate a depression-like state in vulnerable animals by applying an acute but severe stressful stimulus.

Roman low- and high-avoidance (RLA/RHA) rat strains were bred for different performance in avoidance tasks. While the parameter for selection was learning, the segregating trait might rather be anxiety since the RLA-rats appear as more anxious (Broadhurst and Bignami 1965; Brush 2003). However, whilst a single 1-h session of social defeat elicited depression-like behaviour lasting for up to 10 days, it did not differentiate between the lines (Meerlo et al. 1997). Nevertheless, social stress may not be the optimal stressor for the RHA/RLA paradigm, as the anxiety tests that differentiate between the lines do not include a social interaction test but rather anxiety tests based on exploration of the environment. Furthermore, the strains differ at baseline regarding sensitivity to rewards (see Harro 2010).

In the case of the olfactory bulbectomy rat, arguably one of the more reliable models for detecting antidepressant drugs, animals were compared with regard to their acoustic startle reflex and shock-induced sensitisation of the startle reflex (McNish and Davis 1997). It was found that while bulbectomised animals were normal at baseline, they displayed a pronounced increase in responding during conditioning, and developed sensitisation of the startle reflex at shock intensities that did not produce sensitisation in control animals. This evidence supports the notion that olfactory bulbectomy leads to enhanced vulnerability to stressors.

The diathesis–stress approach has been applied with stressors for which it is not obvious whether they are (sub)chronic or acute, since the stressed state can last for days even though stress exposure time is relatively short. In the case of the FSL/FRL model, rats were submitted to 5 days of swimming stress of which four sessions were escapable but the final, fifth session inescapable (Wegener et al. 2010). Animals were sacrificed 2 h after the last session, at which point increased hippocampal constitutive NOS activity and neuronal NOS levels were found only in the FSL rats. Thus, no production was increased in hippocampus of the vulnerable strain. Since NOS activity has been implicated in neurobiology of depression and may be attenuated during 5-HT hypofunction (Harvey et al. 2006), this study suggests one of the possible pathways that could explain the diathesis–stress mechanism.

Withdrawal from addictive drugs, especially psychostimulants, can induce a number of behavioural symptoms reminiscent of depression, and a variety of associated neurochemical changes in monoaminergic and other transmitter systems (Renoir et al. 2012). Depression can also occur after serious somatic disease, and especially after challenges to the immune system (Anisman and Merali 2003). In the Fawn-Hooded rat model of depression (Rezvani et al. 2002), responses to interleukin were found to be distinct from the Sprague–Dawley rat (Simmons and Broderick 2005). Thus, these stressors also offer interesting possibilities to examine the neurobiology of predisposition to depression.

8.2 Chronic Stress

Long-term intermittent distress caused by adverse life events, either single or repeated, is an important contributor to the development of depression. In animals, this has been modelled by administering different types of stressors repeatedly over a period of time, often for many weeks. Different chronic stress protocols have been used, and it is increasingly recognised that inter-individual differences exist between animals in terms of their vulnerability to chronic stress. This variability provides the possibility to elucidate the neurobiology of depression vulnerability and resilience. Importantly, the depressed-like FSL rat was found to be more sensitive than the FRL rat with regard to the anhedonia inducing effect of chronic mild stress (Pucilowski et al. 1993). However, a more detailed knowledge of the specific behavioural traits and associated neurobiological systems that mediate this increased sensitivity to stress is desirable.

Regarding the neurochemistry of depression vulnerability, Curzon (1988) proposed that 5-HT abnormalities in depression vary between subgroups, and can be either state or trait dependent, and that disturbances in 5-HT function underlie the vulnerability to depression. However, we do not know the precise molecular mechanisms by which the 5-HT deficits emerge. Given the capacity of pre- and postsynaptic mechanisms to compensate for changes by means of shifts in 5-HT availability and signal transduction, it appears unlikely that there is a single mechanism responsible for the deficiency in 5-HT function. A number of studies have examined the effect of chronic variable stress after lesioning of the 5-HT system. Two main approaches have been made to produce a lesion of the 5-HT system as the vulnerability model, one based on 3,4-methylenedioxymethamphetamine (MDMA) and another using para-chloroamphetamine (PCA).

MDMA is a phenylethylamine derivative that elicits the release of monoamines, particularly 5-HT, and inhibits monoamine reuptake and monoamine oxidase (Green et al. 2003). Repeated administration of MDMA induces partial degeneration of 5-HT neurons, but without major behavioural changes that might be expected if the 5-HT system were severely dysfunctional. It has, therefore, been considered that MDMA treatment leads to a possible subclinical state that could correspond to a depression vulnerability model (Cunningham et al. 2009; van Donkelaar et al. 2010). When MDMA was administered four times at 2 h intervals in the dose of 7.5 mg/kg to Sprague–Dawley rats, 5-HT transporter immunoreactivity was reduced to about 30 % of control (Cunningham et al. 2009). In these animals anxiety-like behaviour in the elevated plus-maze test was immediately increased, but had returned to control levels after 10 days of chronic mild stress. While this finding may be difficult to interpret as a stress diathesis effect, chronic mild stress also led to a learning impairment in the Morris water maze in the MDMA-pretreated rats. In another study, administration of 20 mg/kg of MDMA twice daily for 4 days to Wistar rats significantly reduced paroxetine binding to 5-HT transporters, indicative of partial loss of 5-HT terminals (van Donkelaar et al. 2010). This treatment produced a decrease in blood flow relative to cerebral metabolism as measured by deoxyglucose uptake. In comparison, acute tryptophan depletion, a technique that also reduces 5-HT availability in the brain and may predict vulnerability to depression in humans (Ruhe et al., 2007), increased blood flow relative to cerebral metabolism in MDMA-pretreated rats. These findings suggest that the 5-HT regulation of cerebrovascular tone may be altered after such a partial lesion. It remains to be established whether this measure can be used in modelling depression vulnerability. However, information obtained from the model may be applicable to primates as MDMA exposure also alters behavioural and electrophysiological response to acute tryptophan depletion in rhesus monkeys (Taffe et al. 2003). Furthermore, a small study has suggested that former recreational MDMA users may indeed experience a higher level of depressive symptoms (MacInnes et al. 2001).

With the rationale that 5-HT deficits play a pivotal role at least in a subset of depressive syndromes (van Praag 2001), and that biological predispositions are likely to be complex and of quantitative rather than qualitative nature, a diathesis–stress model of applying chronic variable stress after administration of a low dose of PCA was introduced (Harro et al. 2001). PCA elicits a large release of 5-HT and temporary 5-HT depletion, followed by partial degeneration of 5-HT terminals and axons (Sanders-Bush et al. 1975). At variance with MDMA, a single dose of PCA is sufficient to elicit long lasting deficits in 5-HT. Our experience with monoamine neurotoxins is that large lesions are accompanied by compensatory changes that under stress lead to unexpected behavioural responses in conventional tests, and may thus reflect very different neurochemical balances which are hard to interpret in depression modelling (Kask et al. 1997; Harro et al., 1999). The size of lesion of the different neurotoxins depends on a number of factors including animal strain and housing conditions.

In our hands, PCA in a dose of 2 mg/kg consistently elicited a 20–30 % depletion of 5-HT (Häidkind et al. 2004). These PCA-induced partial 5-HT lesions, similar to chronic variable stress, increased anxiety in the social interaction test and reduced immobility at the first but not second session of forced swimming; the effect of combined PCA treatment and chronic variable stress was similar (Harro et al. 2001). Interestingly, sucrose intake in free-fed animals increased in the PCA-pretreated animals during chronic variable stress. This result is reminiscent of Wurtman's suggestion that "atypical" depression, that includes increased craving for carbohydrates, is specifically caused by a deficit of 5-HT in stressful conditions (Wurtman and Wurtman 1996). Both increased sucrose intake and increased struggling during swimming observed in animals treated with PCA plus stress were prevented by administration of citalopram (Tõnissaar et al. 2008). This suggests that the behavioural vulnerability was associated with a 5-HT deficiency.

9 Personality as a Vulnerability Factor

Of the four major risk factors of depression identified by epidemiological studies (Fava and Kendler 2000), temperament or personality represents the vulnerability side in the diathesis–stress approach. Personality traits, especially neuroticism, are related to affective disorders (e.g., Lesch 2007), and longitudinal twin studies suggest that neuroticism is strongly associated with a genetic risk of major depression (Kendler et al. 2006). Personality traits are defined as enduring patterns of perceiving, relating to, and thinking about the environment, and hence, as predispositions to react to it. What has been established on the structure of personality suggests that similar constructs can be applied across mammalian species (Davis and Panksepp 2011). Some of the animal models of depression vulnerability may, therefore, already implicitly incorporate the factor of neuroticism, or synonymously, low emotional stability. Unfortunately, most of the established vulnerability models have used other selection principles, and therefore reflect rather a constellation of traits. Nevertheless, the data available suggest that new vistas may be opened through research on temperamental differences in animal models.

One of the most extensively exploited models of inter-individual differences is the high responder/low responder (HR/LR) to novelty approach. HR/LR rats are classified on the basis of the level of spontaneous locomotor activity in a novel environment, and in its original formulation the model was proposed as a measure of vulnerability to drug abuse (Piazza et al. 1989). A variety of selection principles have been reported, and HR/LR rats have been recently selectively bred (Stead et al. 2006). Given that psychomotor retardation is an important symptom of depression, the LR rats possess some face validity as a depression vulnerability model; furthermore, the bred LR rats were found to be more anxious. Bred LR rats developed a reduction in sucrose preference by the second week of stress exposure, and a further dramatic decline by the third week (Stedenfeld et al. 2011), suggesting that the LR animals are indeed more predisposed to react to chronic stress. However, another study has reported opposite findings in a social defeat paradigm: 2 weeks after four daily consecutive social defeat exposures, HR rats exhibited not only higher anxiety levels, but also reduced body weight gain and sucrose preference, and marked social avoidance (Duclot et al. 2011). In a somewhat different investigation, chronic variable stress during adolescence elicited antidepressant-like effects (reduced immobility in the forced swimming test) in the HR rats, this being accompanied by epigenetic activation of the hippocampal BDNF (Oztan et al. 2011).

Since neuroticism is often expressed as heightened anxiety response, measurement of anxiety may offer a means to create a depression vulnerability model based around this trait construct. Inter-individual differences in anxiety-related activity as measured in elevated plus maze, open field or social interaction test have been established, and several interesting neurobiological leads for modelling depression vulnerability have been derived from these studies (see Harro 2010 for a review). Nevertheless, testing of an explicit diathesis–stress model remains rare as yet.

With the aim of clustering two landmark symptoms of depression, negative emotion and low motivation, we created the exploration box test that separates rats bimodally into high anxiety/low exploratory drive and low anxiety/high exploratory drive groups (Mällo et al. 2007a). These phenotypes are persistent, and the low exploring (LE) rats appear to be more responsive to acute stress as compared to the high exploring (HE) rats in a number of tests. When LE and HE rats were submitted to chronic variable stress, this did not alter the defining phenotype and led to largely similar effects in both groups (Matrov et al. 2011). In fact, in a number of studies comparing stress effects on animals with different expression of a trait, the defining phenotype survives chronic stress but other behavioural and neurochemical differences do not (see Harro 2010 for review).

Interestingly, there were many differences between the LE and HE rats in terms of gene expression patterns as measured on the Illumina microarray platform in cortex, hippocampus and raphe (Alttoa et al. 2010), with all of these differences linked with anxiety or depression in the expected direction. Notwithstanding this, LE and HE rats behaved similarly in a number of tasks, suggesting that higher expression of a significant number of "anxiety or depression genes" does not cumulatively produce general dysfunction. Furthermore, both LE and HE rats were able to adapt to 3 weeks of chronic mild stress, in terms of recovery of reduced sucrose intake (Matrov et al. 2011).

While cross-species affective neuroscience has recently cast doubt on the grouping together of several basic emotions under one umbrella measured as neuroticism, high negative emotionality or low emotional stability (Davis and Panksepp 2011), neuroticism has at least featured rather consistently in personality studies, and reliable scales are broadly used. Much less is known on the possible role of low positive emotionality, even though positive affect is not simply the other pole of negative affect but rather orthogonal to it, and diminished positive affect is a central feature to depression (Clark and Watson 1991; Forbes and Dahl 2005). Response to antidepressant treatment was found to be predicted better by an increase in experiencing positive emotions than a decrease in negative emotionality (Geschwind et al. 2011).

A positive affective state can be measured in rats by recording 50 kHz ultrasonic vocalisations in response to tickling-like manual stimulation (Burgdorf and Panksepp 2001). There is a large inter-individual difference in responsivity, and these differences are persistent (Mällo et al. 2007b). When male rats with higher and lower levels of 50 kHz chirping (HC and LC, respectively) were submitted to chronic variable stress in adulthood, it was found that compared to HC rats, the stressed LC rats exhibited increased levels of 22 kHz vocalisations (indicative of negative emotionality), an earlier and more stable reduction of weight gain, reduced sucrose intake and increased immobility in the forced swimming test (Mällo et al. 2009). This vulnerability to chronic stress in the LC rats was also reflected in changes in cerebral oxidative metabolism. Also, after chronic variable stress LC rats had a more reactive HPA response to stress, and altered regulation of extracellular levels of 5-HT (Raudkivi et al. 2012). This animal model, thus, supports the notion that low inherent positive affectivity can serve as a diathesis for affect-related disorders.

10 Vulnerability Based on Post hoc Assessment

The salience of understanding the neurobiology of inter-individual differences for biological psychiatry and psychopharmacology is increasingly recognised (Harro 2010). A major impact relevant to depression studies has been influential report of

Berton and colleagues (Berton et al. 2006) on the role of brain-derived neurotrophic factor (BDNF) in the mesolimbic dopamine pathway in response to social defeat stress. This study turned the fact that not all animals react equally to stressful stimuli into an advantage. A comparison of mice with reduced social engagement after repeated defeat with their resilient counterparts revealed a number of alterations associated with defeated behaviour, including reduced sucrose intake. Since then many studies have identified mice and rats as social stress sensitive or -resilient, using quite a variety of methods of stress induction combined with criteria for defining the vulnerability (e.g., Schmidt et al. 2010; Kanarik et al. 2011). Other researchers have classified rats into vulnerable and resilient groups based on sucrose intake in the chronic mild stress paradigm (Bergström et al. 2008). These studies have identified a number of neurobiological differences between stress sensitive and -resilient animals. In a full diathesis–stress model, Strekalova et al. (2004) examined C57BL/6 mice in a resident-intruder paradigm, and submitted both the submissive and non-submissive animals to a variant of chronic variable stress for 4 weeks. They found that only submissive mice developed anhedonia, increased immobility in the forced swimming test and reduced exploratory behaviour.

11 The Sex/Gender Factor

It is well-known that prevalence of depression is higher in females than males, and sex/gender has been specified as a major risk factor of depression in epidemiological studies (Fava and Kendler 2000). At the first glance it would appear less important for animal model development because the causal link is hard to delineate, and some of the variance in prevalence may be socially constructed (i.e. rather the gender than sex issue). Also, depression quite frequently occurs in males. Nevertheless, the sex/gender factor would become extremely important for model building if the neurobiological basis of depression in males and females were different. Evidence is accumulating that this may be the case (Shaffery et al. 2003), and therefore comparison of stress effects in vulnerable male and female animals appears an important avenue. Some studies that have compared effects of prenatal or chronic variable stress in males and females have rather unexpectedly found that routine behavioural tests indicate higher vulnerability in males (Goel and Bale 2009; Mällo et al. 2009; Sun and Alkon 2006). Considering that most of the studies on modelling depression have been carried out on male animals, there is a possibility that behavioural readouts that are especially sensitive to negative affect and anhedonia in males have become preferred by researchers.

One type of life event unique to females can lead to vulnerability to depression; the post-partum period is the time of greatest risk for women to develop major depression (Goodman 2007). Animal models of post-partum depression make use of knowledge on hormonal changes around delivery, or apply chronic stress during pregnancy to elicit behavioural or neurobiological changes observed in other models of depression (Brummelte and Galea 2010). However, in this respect it appears that the diathesis–stress concept has not been tested as yet.

12 Limitations and Caveats

Coming closer to the pathogenetic principles, diathesis–stress models should offer advantages in the discovery of depression endophenotypes, and also for drug discovery if they prove to be robust predictors of clinical efficacy. However, diathesis–stress models require an increased number of experimental groups and animals, and thus recruitment of additional resources. Furthermore in a true vulnerability model, at the statistical level interaction effects are expected and this requires greater statistical power to achieve significance.

Naturally, as with any animal model of human psychopathology, painstaking work is required to validate the diathesis–stress-based models. As noted above, stressful life events are likely to 'immunize' the organism towards further impacts. Recently, a series of experiments combining prenatal stress with chronic mild stress in adulthood revealed a number of changes in behaviour, as well as cortical structure, that were suggestive of protective adaptations on prenatally stressed rats (Michelsen et al. 2007; van den Hove 2006). Another factor to be considered in diathesis–stress research is familial aggregation, that genetic predisposition which can be moderated by environmental conditions in relation to the genetic similarity of parents and offspring. For instance, two inbred mouse strains, BALB/c and C57BL/6, have often been compared as genetically different strains with more versus less anxiety behaviour and stress sensitivity. In recent research, BALB/c mice have also been described as poor mothers with inferior nest building, less time spent on the nest and longer pup retrieval latencies in experiments. Moreover, this mothering behaviour has been suggested to affect gene methylation patterns (Anisman et al. 2008; Tarantino et al. 2011). Similarly, the FSL rats have been reported to exhibit relatively poor mothering style (Lavi-Avnon et al. 2005). Cross-fostering experiments have shed some light on the relative contribution of genotype and mothering style (Anisman et al. 2008), but this research area is still in its early stage.

A further issue in model development is the decision regarding which readouts to use. Until recently, it prevailed that a model of depression is appropriate if it has behavioural similarity with the human clinically diagnosable condition. This is now being challenged by the idea that rather than trying to recreate or mimic the complete spectrum of symptoms comprising the syndrome of depression, animal models should focus on major constructs such as negative affect, positive affect or physiological hyper arousal (Frazer and Morilak 2005). Another view is that major well-understood symptoms or symptom clusters can be targeted (Fernando and Robbins 2011). As a further refinement of symptom or symptom cluster-based models, endophenotype-specific models have been advocated (Cryan and Slattery

2007). This approach requires identification of biomarkers that relate to the etiopathogenetic process of depression development.

While the history of biological psychiatry has seen an abundance of putative biomarkers for depression come and go, the advance of in vivo neuroimaging methods and refinement of selection of subjects and experimental tasks, should increase the probability of discovery of more reliable indicators. For example using functional magnetic resonance imaging (fMRI) and reward processing tasks, it was found that childhood adversity with increased symptoms of anhedonia and depression is associated with blunted subjective responses to reward-predicting cues as well as a weaker response in the left globus pallidus (Dillon et al. 2009). Of course, the question remains whether it is any longer depression that is being modelled, if the focus shifts to symptoms. We should probably return in future to attempts to model the whole syndrome, but this requires better understanding of the human pathology. If the underlying molecular mechanisms are well known, it will be more feasible to consider animal models of the entire clinical disorder (Fernando and Robbins 2011).

In conclusion, the etiopathogenetic diathesis–stress approach to developing animal models of depression vulnerability appears promising and has yielded a deeper understanding of neurobiology of depression as a subject to inter-individual differences. Furthermore, systematic studies should clarify whether or not this approach can improve antidepressant drug screening, detect compounds with novel mechanisms of action with low false positive rate and advance more personalised medicine.

Acknowledgments Relevant research by the author has been supported by the Hope for Depression Research Foundation, Institute for the Study of Affective Neuroscience, the Estonian Ministry of Education and Science project 0180027 and the EU Framework 6 Integrated Project NEWMOOD (LSHM-CT-2004-503474).

References

Alttoa A, Kõiv K, Hinsley TA, Brass A, Harro J (2010) Differential gene expression in a rat model of depression based on persistent differences in exploratory activity. Eur Neuropsychopharmacol 20:288–300

Anisman H, Merali Z (2003) Cytokines, stress and depressive illness: brain-immune exposure. Ann Med 35:2–11

Anisman H, Zacharko RM (1990) Multiple neurochemical and behavioral consequences of stressors: implications for depression. Pharmacol Ther 46:119–136

Anisman H, Merali Z, Stead JDH (2008) Experiential and genetic contributions to depressive- and anxiety-like disorders: clinical and experimental studies. Neurosci Biobehav Rev 32:1185–1206

Aso E, Ozaita A, Serra M-A, Maldonado R (2011) Genes differentially expressed in CB_1 knockout mice: involvement in the depressive-like phenotype. Eur Neuropsychopharmacol 21:11–22

Belzung C, Griebel G (2001) Measuring normal and pathological anxiety-like behaviour in mice: a review. Behav Brain Res 125:141–149

Bergström A, Jayatissa MN, Mørk A, Wiborg O (2008) Stress sensitivity and resilience in the chronic mild stress rat model of depression: an in situ hybridization study. Brain Res 1196:41–52
Berton O, McClung CA, Dileone RJ, Krishnan V, Renthal W, Russo SJ, Graham D, Tsankova NM, Bolanos CA, Rios M, Monteggia LM, Self DW, Nestler EJ (2006) Essential role of BDNF in the mesolimbic dopamine pathway in social defeat stress. Science 311:864–868
Broadhurst PL, Bignami G (1965) Correlative effects of psychogenetic selection: a study of the Roman high and low avoidance strains of rats. Behav Res Ther 3:273–280
Brummelte S, Galea LAM (2010) Depression during pregnancy and postpartum: contribution of stress and ovarian hormones. Prog Neuro-Psychopharmacol Biol Psychiatry 34:766–776
Brush FR (2003) Selection for differences in avoidance learning: the Syracuse strains differ in anxiety, not learning ability. Behav Genet 33:677–696
Burgdorf J, Panksepp J (2001) Tickling induces reward in adolescent rats. Physiol Behav 72: 167–173
Champagne DL, Bagot RC, van Hasselt F, Ramakers G, Meaney MJ, de Kloet ER, Joels M, Krugers H (2008) Maternal care and hippocampal plasticity: evidence for experience-dependent structural plasticity, altered synaptic functioning, and differential responsiveness to glucocorticoids and stress. J Neurosci 28:6037–6045
Clark LA, Watson D (1991) Tripartite model of anxiety and depression: psychometric evidence and taxonomic implications. J Abnorm Psychol 100:316–336
Cryan JF, Mombereau C (2004) In search of a depressed mouse: utility of models for studying depression-related behavior in genetically modified mice. Mol Psychiatry 9:326–357
Cryan JF, Slattery DA (2007) Animal models of mood disorders: recent developments. Curr Opin Psychiatry 20:1–7
Cryan JF, Markou A, Lucki I (2002) Assessing antidepressant activity in rodents: recent developments and future needs. Trends Pharmacol Sci 23:238–245
Cunningham JI, Raudensky J, Tonkiss J, Yamamoto BK (2009) MDMA pretreatment leads to mild chronic unpredictable stress-induced impairments in spatial learning. Behav Neurosci 123:1076–1084
Curzon G (1988) Serotonergic mechanisms of depression. Clin Neuropharmacol 11(S2):S11–S20
Davis KL, Panksepp J (2011) The brain's emotional foundations of human personality and the affective neuroscience personality scales. Neurosci Biobehav Rev 35:1946–1958
Dillon DG, Holmes AJ, Birk JL, Brooks N, Lyons-Ruth K, Pizzagalli DA (2009) Childhood adversity is associated with left basal ganglia dysfunction during reward anticipation in adulthood. Biol Psychiatry 66:206–213
Drevets WC, Furey ML (2009) Depression and the brain. Encyclopedia of Neuroscience. Elsevier, New York, pp 459–470
Duclot F, Hollis F, Darcy MJ, Kabbaj M (2011) Individual differences in novelty-seeking behavior in rats as a model for psychosocial stress-related mood disorders. Physiol Behav 104:296–305
El Khoury A, Gruber SHM, Mørk A, Mathé AA (2006) Adult life behavioral consequences of early maternal separation are aleviated by escitalopram treatment in a rat model of depression. Prog Neuro-Psychopharmacol Biol Psychiatry 30:535–540
El Yacoubi M, Vaugeois J-M (2007) Genetic rodent models of depression. Curr Opin Pharmacol 7:3–7
Eysenck HJ (1951) The organization of personality. J Pers 20:101–117
Fava M, Kendler KS (2000) Major depressive disorder. Neuron 28:335–341
Fernando ABP, Robbins TW (2011) Animal models of neuropsychiatric disorders. Annu Rev Clin Psychol 7:39–61
Forbes EE, Dahl RE (2005) Neural systems of positive affect: relevance to understanding child and adolescent depression? Dev Psychopathol 17:827–850
Francis DD, Champagne FA, Liu D, Meaney MJ (1999) Maternal care, gene expression, and the development of individual differences in stress reactivity. Ann N Y Acad Sci 896:66–84

Frazer A, Morilak DA (2005) What should animal models of depression model? Neurosci Biobehav Rev 29:515–523
Garcia–Garcia AL, Elizalde N, Matrov D, Harro J, Wojcik SM, Venzala E, Ramirez MJ, Del Rio J, Tordera RM (2009) Increased vulnerability to depressive-like behavior of mice with decreased expression of VGLUT1. Biol Psychiatry 66:275–282
Gardner KL, Hale MW, Lightman SL, Plotsky PM, Lowry CA (2009a) Adverse early life experience and social stress during adulthood interact to increase serotonin transporter mRNA expression. Brain Res 1305: 47–63
Gardner KL, Hale MW, Oldfield S, Lightman SL, Plotsky PM, Lowry CA (2009b) Adverse experience during early life and adulthood interact to elevate tph2 mRNA expression in serotonergic neurons within the dorsal raphe nucleus. Neuroscience 163:991–1001
Geschwind N, Nicolson NA, Peeters F, van Os J, Barge-Schaapveld D, Wichers M (2011) Early improvement in positive rather than negative emotion predicts remission from depression after pharmacotherapy. Eur Neuropsychopharmacol 21:241–247
Geyer MA, Markou A (1995) Animal models of psychiatric disorders. In: Bloom FE, Kupfer DJ (eds) Psychopharmacology: the fourth generation of progress. Raven Press, New York, pp 787–798
Goel N, Bale TL (2009) Examining the intersection of sex and stress in modelling neuropsychiatric disorders. J Neuroendocrinol 21:415–420
Goodman SH (2007) Depression in mothers. Annu Rev Clin Psychol 3:107–135
Green AR, Mechan AO, Elliot JM, O'Shea E, Colado MI (2003) The pharmacology and clinical pharmacology of 3,4-methylenedioxymethamphetamine (MDMA, "ecstasy"). Pharmacol Rev 55:463–506
Häidkind R, Eller M, Kask A, Harro M, Rinken A, Oreland L, Harro J (2004) Increased behavioural activity of rats in forced swimming test after partial denervation of serotonergic system by parachloroamphetamine treatment. Neurochem Int 45:721–732
Harro J (2004) Animal models for better antidepressants: can pathogenetic approaches make a difference? Preclinica 2:402–407
Harro J (2010) Inter-individual differences in neurobiology as vulnerability factors for affective disorders: implications for psychopharmacology. Pharmacol Ther 125:402–422
Harro J, Oreland L (2001) Depression as a spreading adjustment disorder of monoaminergic neurons: a case for primary implication of the locus coeruleus. Brain Res Rev 38:79–128
Harro J, Häidkind R, Harro M, Modiri A-R, Gillberg P-G, Pähkla R, Matto V, Oreland L (1999) Chronic mild unpredictable stress after noradrenergic denervation: attenuation of behavioural and biochemical effects of DSP-4 treatment. Eur Neuropsychopharmacol 10:5–16
Harro J, Tõnissaar M, Eller M, Kask A, Oreland L (2001) Chronic variable stress and partial 5-HT denervation by parachloroamphetamine treatment in the rat: effects on behavior and monoamine neurochemistry. Brain Res 899:227–239
Harro J, Kanarik M, Matrov D, Panksepp J (2011) Mapping patterns of depression-related brain regions with cytochrome oxidase histochemistry: relevance of animal affective systems to human disorders, with a focus on resilience to adverse events. Neurosci Biobehav Rev 35:1876–1889
Harvey BH, Retief R, Korff A, Wegener G (2006) Increased hippocampal nitric oxide synthase activity and stress responsiveness after imipramine discontinuation: role of 5HT(2A/C)-receptors. Metab Brain Dis 21:211–220
Holmes PV (2003) Rodent models of depression: reexamining validity without anthropomorphic inference. Crit Rev Neurobiol 15:143–174
Ingram RE, Luxton DD (2005) Vulnerability-stress models. In: Hankin BL, Abela JRZ (eds) Development of psychopathology. A Vulnerability-Stress Perspective. Sage, Thousand Oaks, pp 32–46
Jacobson LH, Cryan JF (2007) Feeling strained? Influence of genetic background on depression-related behavior in mice: a review. Behav Genet 37:171–213
Kalueff AV, Wheaton M, Murphy DL (2007) What's wrong with my mouse model? Advances and strategies in animal modeling of anxiety and depression. Behav Brain Res 179:1–18

Kanarik M, Alttoa A, Matrov D, Kõiv K, Sharp T, Panksepp J, Harro J (2011) Brain responses to chronic social defeat stress: effects on regional oxidative metabolism as a function of a hedonic trait, and gene expression in susceptible and resilient rats. Eur Neuropsychopharmacol 21:92–107

Karmiloff-Smith A (2006) The tortuous route from genes to behavior: a neuroconstructivist approach. Cogn Affect Behav Neurosci 6:9–17

Kask A, Harro J, Tuomainen P, Rägo L, Männistö PT (1997) Overflow of noradrenaline and dopamine in frontal cortex after [N-(2-chloroethyl)-N-ethyl-2-bromobenzylamine] (DSP-4) treatment: in vivo microdialysis study in anaesthetized rats. Naunyn-Schmiedeberg's Arch Pharmacol 355:267–272

Katz RJ, Roth KA, Carroll BJ (1981) Acute and chronic stress effects on open field activity in the rat: implications for a model of depression. Neurosci Biobehav Rev 5:247–251

Kendler KS, Kessler RC, Walters EE, MacLean CJ, Sham PC, Neale MC, Heath AC, Eaves LJ (1995) Stressful life events, genetic liability and onset of an episode of major depression in women. Am J Psychiatry 152:833–842

Kendler KS, Gatz M, Gardner CO, Pedersen NL (2006) Personality and major depression. A Swedish longitudinal, population-based twin study. Arch Gen Psychiatry 63:1113–1120

Ladd CO, Thrivikraman KV, Huot RL, Plotsky PM (2005) Differential neuroendocrine responses to chronic variable stress in adult Long Evans rats exposed to handling-maternal separation as neonates. Psychoneuroendocrinology 30:520–533

Lanfumey L, Mannoury La Cour C, Froger N, Hamon M (2000) 5-HT-HPA interactions in two models of transgenic mice relevant to major depression. Neurochem Res 25:1199–1206

Lavi-Avnon Y, Yadid G, Overstreet DH, Weller A (2005) Abnormal patterns of maternal behavior in a genetic animal model of depression. Physiol Behav 84:607–615

Lesch K-P (2007) Linking emotion to the social brain. EMBO Reports 8:S24–S29

Levine A, Worrell TR, Zimnisky R, Schmauss C (2012) Early life stress triggers sustained changes in histone deacetylase expression and histone H4 modifications that alter responsiveness to adolescent antidepressant treatment. Neurobiol Dis 45:488–498

Maccari S, Darnaudery M, Morley-Fletcher S, Zuena AR, Cinque C, van Reeth O (2003) Prenatal stress and long-term consequences: implications of glucocorticoid hormones. Neurosci Biobehav Rev 27:119–127

MacInnes N, Handley SL, Harding GF (2001) Former chronic methylenedioxymethamphetamine (MDMA or ecstasy) users report mild depressive symptoms. J Psychopharmacol 15:181–186

Mällo T, Alttoa A, Kõiv K, Tõnissaar M, Eller M, Harro J (2007a) Rats with persistently low or high exploratory activity: behaviour in tests of anxiety and depression, and extracellular levels of dopamine. Behav Brain Res 177:269–281

Mällo T, Matrov D, Herm L, Kõiv K, Eller M, Rinken A, Harro J (2007b) Tickling-induced 50-kHz ultrasonic vocalization is individually stable and predicts behaviour in tests of anxiety and depression. Behav Brain Res 184:57–71

Mällo T, Matrov D, Kõiv K, Harro (2009) Effect of chronic stress on behavior and cerebral oxidative metabolism in rats with high or low positive affect. Neuroscience 164:963–974

Matrov D, Vonk A, Herm L, Rinken A, Harro J (2011) Activating effects of chronic variable stress in rats with different exploratory activity: association with dopamine D_1 receptor function in nucleus accumbens. Neuropsychobiology 64:110–122

McArthur R, Borsini F (2006) Animal models of depression in drug discovery: a historical perspective. Pharmacol Biochem Behav 84:436–452

McKinney WT Jr, Bunney WE Jr (1969) Animal models of depression. I: review of evidence: implications for research. Arch Gen Psychiatry 21:240–248

McNish KA, Davis M (1997) Olfactory bulbectomy enhances sensitization of the acoustic startle reflex produced by acute or repeated stress. Behav Neurosci 111:80–91

Meerlo P, Overkamp GJ, Koolhaas JM (1997) Behavioural and physiological consequences of a single social defeat in Roman high- and low-avoidance rats. Psychoneuroendocrinology 22:155–168

Michelsen KA, van den Hove DL, Schmitz C, Segers O, Prickaerts J, Steinbusch HW (2007) Prenatal stress and subsequent exposure to chronic mild stress influence dendritic spine density and morphology in the rat medial prefrontal cortex. BMC Neurosci 8:107
Musazzi L, Mallei A, Tardito D, Gruber SHM, El Khoury A, Racagni G, Mathe AA, Popoli M (2010) Early-life stress and antidepressant treatment involve synaptic signaling and Erk kinases in a gene-environment model of depression. J Psychiat Res 44:511–520
Nestler EJ, Gould E, Manji H, Bucan M, Duman RS, Gershenfeld HK, Hen R, Koester S, Lederhendler I, Meaney M, Robbins T, Winsly L, Zalcman S (2002) Preclinical models: status of basic research in depression. Biol Psychiatry 52:503–528
O'Neill MF, Moore NA (2003) Animal models of depression: are there any? Hum Psychopharmacol 18:239–254
Oomen CA, Soeters H, Audureau N, Vermunt L, van Hasselt FN, Manders EMM, Joels M, Lucassen PJ, Krugers H (2010) Severe early life stress hampers spatial learning and neurogenesis, but improves hippocampal synaptic plasticity and emotional learning under high-stress conditions in adulthood. J Neurosci 30:6635–6645
Overstreet DH, Friedman E, Mathe AA, Yadid G (2005) The Flinders Sensitive Line rat: a selectively bred putative animal model of depression. Neurosci Biobehav Rev 29:739–759
Oxford Dictionaries (2011) http://oxforddictionaries.com/definition/vulnerable-?q=vulnerability-#vulnerable__3. 31 Dec.2011
Oztan O, Aydin C, Isgor C (2011) Stressful environmental and social stimulation in adolescence causes antidepressant-like effects associated with epigenetic induction of the hippocampal BDNF and mossy fibre sprouting in the novelty-seeking phenotype. Neurosci Lett 501: 107–111
Palumbo ML, Zorrilla Zubilete MA, Cremaschi GA, Genaro AM (2009) Different effect of chronic stress on learning and memory in BALB/c and C57BL/6 inbred mice: involvement of hippocampal NO production and PKC activity. Stress 12:350–361
Panksepp J (1998) Affective Neuroscience: the foundations of human and animal emotions. Oxford University Press, New York
Panksepp J, Vilberg T, Bean NJ, Coy DJ, Kastin AJ (1978) Reduction of distress vocalization in chicks by opiate-like peptides. Brain Res Bull 3:663–667
Piazza PV, Deminiere JM, Le Moal M, Simon H (1989) Factors that predict individual vulnerability to amphetamine self-administration. Science 245:1511–1515
Piubelli C, Vighini M, Mathe AA, Domenici E, Carboni L (2011) Escitalopram modulates neuron-remodelling proteins in a rat gene-environment interaction model of depression as revealed by proteomics. Part I: genetic background. Int J Neuropsychopharmacol 14:796–833
Pucilowski O, Overstreet DH, Rezvani AH, Janowsky DS (1993) Chronic mild stress-induced anhedonia: greater effect in a genetic rat model of depression. Physiol Behav 54:1215–1220
Raudkivi K, Mällo T, Harro J (2012) Effect of chronic variable stress on corticosterone levels and hippocampal extracellular 5-HT in rats with persistent differences in positive affectivity. Acta Neuropsychiatrica (in press)
Refojo D, Deussing JM (2012) Animal models of depression. In: Lopez-Munos F, Alamo C (eds) Neurobiology of depression. Frontiers in neuroscience. CRC Press, Boca Raton, pp 1–25
Renoir T, Pang TY, Lanfumey L (2012) Drug withdrawal-induced depression: serotonergic and plasticity changes in animal models. Neurosci Biobehav Rev 36:696–726
Rezvani AH, Parsian A, Overstreet DH (2002) The Fawn-Hooded (FH/Wjd) rat: a genetic animal model of comorbid depression and alcoholism. Psychiatr Genet 12:1–16
Richardson JS (1991) Animal models of depression reflect changing views on the essence and etiology of depressive disorders in humans. Prog Neuro-Psychopharmacol Biol Psychiatry 15:199–204
Rosecrans JA, De Feo JJ (1965) The interrelationships between chronic restraint stress and reserpine sedation. Arch Int Pharmacodyn Ther 157:487–498
Ruhe HG, Mason NS, Schene AH (2007) Mood is indirectly related to serotonin, norepinephrine and dopamine levels in humans: a meta-analysis of monoamine depletion studies. Mol Psychiatry 12:331–359

Ryan B, Musazzi L, Mallei A, Tardito D, Gruber SHM, El Khoury A, Anwyl R, Racagni G, Mathe AA, Rowan MJ, Popoli M (2009) Remodelling by early-life stress of NMDA receptor-dependent synaptic plasticity in a gene-environment rat model of depression. Int J Neuropsychopharmacol 12:553–559

Samuels BA, Leonardo ED, Gadient R, Williams A, Zhou J, David DJ, Gardier AM, Wong EHF, Hen R (2011) Modeling treatment-resistant depression. Neuropharmacology 61:408–413

Sanders-Bush E, Bushing JA, Sulser F (1975) Long-term effects of *p*-chloroamphetamine and related drugs on central serotonergic mechanisms. J Pharmacol Exp Ther 192:33–41

Schmidt MV, Scharf SH, Sterlemann V, Ganea K, Liebl C, Holsboer F, Müller MB (2010) High susceptibility to chronic social stress is associated with a depression-like phenotype. Psychoneuroendocrinology 35:635–643

Schmidt MV, Wang X-D, Meijer OC (2011) Early life stress paradigms in rodents: potential animal models of depression? Psychopharmacology 214:131–140

Shaffery J, Hoffmann R, Armitage R (2003) The neurobiology of depression: perspectives from animal and human sleep studies. Neuroscientist 9:82

Simmons DA, Broderick PA (2005) Cytokines, stressors, and clinical depression: augmented adaptation responses underlie depression pathogenesis. Prog Neuro-Psychopharmacol Biol Psychiatry 29:793–807

Song C, Leonard BE (2005) The olfactory bulbectomized rat as a model of depression. Neurosci Biobehav Rev 29:627–647

Stead JD, Clinton S, Neal C, Schneider J, Jama A, Miller S, Vazguez DM, Watson SJ, Akil H (2006) Selective breeding for divergence in novelty-seeking traits: heritability and enrichment in spontaneous anxiety-related behaviours. Behav Genet 36:697–712

Stedenfeld KA, Clinton SM, Kerman IA, Akil H, Watson SJ, Sved AF (2011) Novelty-seeking behavior predicts vulnerability in a rodent model of depression. Physiol Behav 103:210–216

Strekalova T, Spanagel R, Bartsch D, Henn FA, Gass P (2004) Stress-induced anhedonia in mice is associated with deficits in forced swimming and exploration. Neuropsychopharmacology 29:2007–2017

Sun M-K, Alkon DL (2006) Differential gender-related vulnerability to depression induction and converging antidepressant responses in rats. J Pharmacol Exp Ther 316:926–932

Taffe MA, Huitron-Resendiz S, Schroeder R, Parsons LH, Henriksen SJ, Gold LH (2003) MDMA exposure alters cognitive and electrophysiological sensitivity to rapid tryptophan depletion in rhesus monkeys. Pharmacol Biochem Behav 76:141–152

Talge NM, Neal C, Glover V (2007) Antenatal maternal stress and long-term effects on child neurodevelopment: how and why? J Child Psychol Psychiatry 48:245–261

Tarantino LM, Sullivan PF, Meltzer-Brody S (2011) Using animal models to disentangle the role of genetic, epigenetic, and environmental influences on behavioral outcomes associated with maternal anxiety and depression. Front Psychiatry 2:44

Tõnissaar M, Mällo T, Eller M, Häidkind R, Kõiv K, Harro J (2008) Rat behavior after chronic variable stress and partial lesioning of 5-HT-ergic neurotransmission: effects of citalopram. Prog Neuro-Psychopharmacol Biol Psychiatry 32:164–177

van den Hove DLA (2006) Prenatal stress and adult psychopathology. University of Maastricht 2006

van Donkelaar EL, Kelly PAT, Dawson N, Blokland A, Prickaerts J, Steinbusch HWM, Ferrington L (2010) Acute tryptophan depletion potentiates 3,4-methylenedioxymethamphetamine-induced cerebrovascular hyperperfusion in adult male Wistar rats. J Neurosci Res 88:1557–1568

van Praag HM (2001) Anxiety/aggression driven depression: a paradigm of functionalization and verticalization of psychiatric diagnosing. Prog Neuro-Psychopharmacol Biol Psychiatry 12:28–39

Ventura R, Cabib S, Puglisi-Allegra S (2002) Genetic susceptibility of mesocortical dopamine to stress determines liability to inhibition of mesoaccumbens dopamine and to behavioural "despair" in a mouse model of depression. Neuroscience 115:999–1007

Wegener G, Harvey BH, Bonefeld B, Müller HK, Volke V, Overstreet DH, Elfving B (2010) Increased stress-evoked nitric oxide signalling in the Flinders sensitive line (FSL) rat: a genetic animal model of depression. Int J Neuropsychopharmacol 13:461–473

Wikipedia (2011) http://en.wikipedia.org/wiki/Vulnerability. 31 Febr 2011

Williams M (2011) Productivity shortfalls in drug discovery: contributions from the preclinical sciences? J Pharmacol Exp Ther 336:3–8

Willner P (1990) Animal models of depression: an overview. Pharmacol Ther 45:425–455

Willner P (1997) Validity, reliability and utility of the chronic mild stress model of depression: a 10-year review and evaluation. Psychopharmacology 134:319–329

Willner P, Mitchell PJ (2002) The validity of animal models of predisposition to depression. Behav Pharmacol 13:169–188

Willner P, Muscat R, Papp M (1992) Chronic mild stress-induced anhedonia: a realistic animal model of depression. Neurosci Biobehav Rev 16:525–534

Wurtman RJ, Wurtman JJ (1996) Brain serotonin, carbohydrate-craving, obesity and depression. Adv Exp Biol Med 398:35–41

Genetic Mouse Models of Depression

Christopher Barkus

Abstract This chapter focuses on the use of genetically modified mice in investigating the neurobiology of depressive behaviour. First, the behavioural tests commonly used as a model of depressive-like behaviour in rodents are described. These tests include those sensitive to antidepressant treatment such as the forced swim test and the tail suspension test, as well as other tests that encompass the wider symptomatology of a depressive episode. A selection of example mutant mouse lines is then presented to illustrate the use of these tests. As our understanding of depression increases, an expanding list of candidate genes is being investigated using mutant mice. Here, mice relevant to the monoamine and corticotrophin-releasing factor hypotheses of depression are covered as well as those relating to the more recent candidate, brain-derived neurotrophic factor. This selection provides interesting examples of the use of complimentary lines, such as those that have genetic removal or overexpression, and also opposing behavioural changes seen following manipulation of closely related genes. Finally, factors such as the issue of background strain and influence of environmental factors are reflected upon, before considering what can realistically be expected of a mouse model of this complex psychiatric disorder.

Keywords Knockout mice · Forced swim test · Serotonin transporter · Noradrenaline transporter · CRF · BDNF

C. Barkus (✉)
Department of Psychiatry, University of Oxford, Oxford, UK
e-mail: chris.barkus@psych.ox.ac.uk

Curr Topics Behav Neurosci (2013) 14: 55–78
DOI: 10.1007/7854_2012_224

Published Online: 15 August 2012

Contents

1 Introduction

Experimental animal models allow for methods of study into disease and treatment that would be unethical, impractical or otherwise impossible in humans. By using experimental animals, the molecular underpinnings of disorders and their treatment can be investigated, but modelling complex psychiatric diseases in rodents is problematic.

While there are other ways to model depressive-like phenotypes in rodents, such as olfactory bulbectomy (Song and Leonard 2005) and chronic stress paradigms (Willner 2005, but see also Forbes et al. 1996), this chapter will focus on the use of mutant genetic mouse lines in investigations of emotional behaviour as well as what constitutes a behavioural model of depression in rodents.

2 Depressed Mice?

A major depressive episode in humans comprises a number of different core symptoms and biological changes, but it is invariably a multifaceted event. Of the many symptoms that can comprise a depressive episode, some are simply impossible to model in rodents. These include the subjective feelings of fatigue and guilt, as well as suicidal ideation. However, for many core symptoms there are analogous models, which are reviewed in detail elsewhere (e.g., Crawley 2000). Here, some of these models will be noted in brief to contextualise a subsequent more detailed discussion of the work done using putative mutant mouse models of depression.

A good rodent model of depression would feature many of the depressive-like behaviours detailed below, as well as possibly some of the associated physiological changes commonly seen in depressed individuals, such as altered EEG sleep architecture. When the depressive-like phenotype detected in a mutant mouse model can also be reversed by clinically effective antidepressants, this adds to the validity of the model.

2.1 Depressed Mood

To avoid suggesting we can measure mood *per se* in rodents, the term "behavioural despair" is often applied to tests sensitive to antidepressant treatment. The two most commonly used tests for behavioural despair in mice are the forced swim test and the tail suspension test.

2.1.1 The Forced Swim Test

The forced swim test was first developed in the late 1970s as a test sensitive to antidepressant activity in both rats and mice (Porsolt et al. 1977a, b, 1978a, b). At its simplest, this test involves placing the rodent in a non-escapable cylinder of water and monitoring the resultant behaviour. Measurement of the time spent actively moving in an attempt to escape as well as the time spent passively immobile, are thought to provide a way of quantifying the rodent's propensity toward helplessness (summarised in Cryan et al. 2005b). An important indicator of the validity of the test is that the time spent immobile is typically decreased by clinically effective antidepressant drugs.

Activity during non-immobile phases is thought by some to have separate underlying neurobiological mechanisms. In particular, Denke and colleagues have proposed that noradrenergic mechanisms mediate increased climbing, whereas serotonergic mechanisms mediate increased swimming (Detke et al. 1997, 1995; Detke and Lucki 1996).

The specifics of how this test is carried out vary somewhat between studies and labs, but the typical duration of the forced swim test is 5–6 min. A day or so prior to the test, there may also be a pre-exposure to the non-escapable cylinder of water, which acts as a stressor and may improve the sensitivity of the test both to pharmacological and genetic manipulations. Importantly, the test ought to be paired with an assessment of general locomotor activity to exclude the potential confounding effect of hyperactivity induced by the genetic manipulation or pharmacological agent.

2.1.2 The Tail Suspension Test

Similar to the forced swim test, the tail suspension test involves recording immobility of mice exposed to an inescapable stressful situation, in this case suspension by the tail from a fixed structure, usually for 6 min (Steru et al. 1985). This test has been extensively validated pharmacologically (see Cryan et al. 2005a for review) and can be used alongside, or as an alternative to, the forced swim test.

As with the forced swim test, independent assessments of locomotor activity should be made to ensure the selectivity of any changes in immobility. A further issue with this test is the curious ability of mice on the C57Bl/6 background to climb up their own tails, therefore confounding assessment of the time spent immobile (Mayorga and Lucki 2001). Those mice that excessively display this acrobatic attempt to escape usually are excluded from further analysis, which can lead to sampling bias.

2.2 Anhedonia

Anhedonia is also a core symptom of depression that can be measured in rodents, and is sometimes overlooked in work with mutant mouse lines. Anhedonia is typically assessed in rodents by measuring preference for sucrose or saccharine over water (Papp et al. 1991). Reduced sucrose/saccharine preference is a typical characteristic of rodents exposed to chronic stress paradigms, and has also observed in genetic mouse models (El Yacoubi et al. 2003).

Decreased social interaction is often associated with anhedonia, as a reduction in socialising is specifically mentioned in the DSM criteria under anhedonia (DSM IV-TR 2000). In rodents, social behaviour can be quantified in a number of ways, the simplest being to place two unfamiliar mice in an open arena for a fixed period and measure the time spent engaged in active social interaction. However, the level of sociability of mice in such a paradigm is affected by other behavioural drives such as exploration of the environment or a reluctance to explore due to high anxiety. Social interaction may therefore not offer an absolute measure of anhedonia, but is, nonetheless, interesting as a naturalistic behaviour that encompasses certain behaviours relevant to depression.

2.3 Anxiety

Although anxiety disorders form a class of psychiatric illness in their own right, anxiety is also highly co-morbid with clinical depression (DSM IV-TR 2000; Zimmerman et al. 2000). Many tests of anxiety exist in rodents that have been developed and validated using clinically effective anxiolytic drugs such as benzodiazepines. These tests include the elevated plus maze (Handley and Mithani 1984; Lister 1987; Pellow et al. 1985), open field (Hall 1936; Treit and Fundytus 1988) and light/dark box (Blumstein and Crawley 1983; Crawley and Goodwin 1980; Crawley 1981). These tests typically rely on measurement of a conflict between competing behaviours in rodents; for example, on one hand to explore novel environments and on the other to avoid possibly threatening situations such as open, brightly lit areas which trigger an innate fear of predators.

Assessments of locomotor activity during these tests can provide a useful control to ensure that apparently altered levels of anxiety are not secondary to changes in general activity. Such assessments include distance moved across the arena, closed arm entries (elevated plus maze) or light/dark transitions. These assessments are best paired with an independent measurement of locomotor activity, to ascertain if changes in movement are a likely product of being placed in an anxious situation; a mouse may move less because it is anxious and thus make fewer transitions, for example.

2.4 Other Symptoms

In addition to the measures of depressive-like symptoms mentioned above, many other symptoms characteristic of a depressive episode can be modelled in rodents. For example, alterations in weight and feeding behaviour can easily be assessed, as can disrupted activity over the diurnal cycle. Further to this, sleep architecture, which is often disturbed in depressed patients (Steiger and Kimura 2010), can be assessed in rodents using implanted EEG electrodes.

There are many cognitive parameters that may be disrupted by a depressive episode, including decisiveness and concentration/attention. One task that is thought to give a measure of attention in rodents is the serial reaction time task (Robbins 2002), although this task is more commonly used in rats than mice. Tests of spatial memory are often substituted for measures of attention, although this seems to conflate quite different cognitive processes.

Further tests may also be helpful to probe specific questions about the nature of depressive-like state. For example, fear conditioning can be used as a model of aversive learning which may be relevant to the underpinnings of depression. There are also physiological disturbances often associated with depression in humans that can be easily measured in rodents such as hypothalamic-pituitary adrenal (HPA) axis functionality.

The remainder of this chapter will illustrate how the behavioural tasks detailed above can provide phenotypic measures of depressive-like behaviour in genetic mouse models.

3 Genetic Mouse Models of Depression

Many mutant mouse lines have been created on the basis of the growing knowledge of genes implicated in the pathogenesis of depression and its treatment. The majority of these lines involve the inactivation of a candidate gene for depression, resulting in an antidepressive-like phenotype. Examples of such models include 5-HT_{1A} receptor knockout mice (Heisler et al. 1998; Parks et al. 1998; Ramboz et al. 1998) and noradrenaline transporter knockout mice (Xu et al. 2000). The number of genetic models that result in depressive behaviours is much fewer, although there are still examples across several neurobiological systems.

A selection of these are discussed below along with some mutant mice that display antidepressive-like behaviour that provide interesting counterpoints to mice with a depressive-like phenotype.

New mutant mouse models and gene targets of interest are constantly being developed and investigated. The focus here will be on models relating to the established monoamine hypothesis of depression and also the stress hypothesis centred on corticotrophin-releasing factor (CRF). Also discussed is another hypothesis of depression set around brain-derived neurotrophic factor (BDNF), a more recent and contentious player in mood disorders. Each of these areas provides examples which illustrate the usefulness of the behavioural tests noted in determining the extent of the depressive phenotype of the model. Other reviews and perspectives of this field can also be found in the literature (e.g. Cryan and Mombereau 2004; Gardier et al. 2009).

3.1 5-HT Transporter

The serotonin (5-hydroxytryptamine or 5-HT) system has long been implicated in the pathogenesis of depression (Schildkraut 1965) as well as its treatment as exemplified by the current front-line antidepressant drugs being selective serotonin reuptake inhibitors (SSRIs). Since these agents, like many other antidepressant drugs, act by blocking the 5-HT transporter (Owens et al. 1997; Owens and Nemeroff 1994), it places the latter as an important candidate gene for depression. A further reason to study the 5-HT transporter gene is that there is a large, naturally occurring variation in expression levels of this gene in the human population. One driver of this variation is thought to be a common insertion/deletion polymorphism in the promoter region of the gene (5-HTTLPR polymorphism) that generates short (s) and long (l) alleles which are either low (s/s) or high (l/l) expressing, at least in cell-based models (Heils et al. 1996; Lesch et al. 1996). It is thought that individuals with the low expressing variant of the 5-HT transporter gene are predisposed to mood and anxiety disorders (Furlong et al. 1998; Lasky-Su et al. 2005; Lotrich and Pollock 2004), although this remains contentious (e.g. Anguelova et al. 2003).

There is some disagreement regarding the impact of the genetic polymorphisms and actual levels of functional 5-HT transporter protein in vivo (Mann et al. 2000; Naylor et al. 1998; Rhodes et al. 2007; Shioe et al. 2003), which complicates interpretation of the reported 5-HT transporter genotype–phenotype associations. However, a reliable alteration in 5-HT transporter levels between individuals is something that can be generated genetically in mice. Moreover, the high degree of genetic homogeneity, within an experimental cohort of transgenic mice, is a major help in reducing the confounds of a high level of genetic heterogeneity in a human population, in addition to variation in the gene of interest. Furthermore, gene x environment interactions can also be better controlled with experimental animals than with human participants.

For investigations of the role of variation in the 5-HT transporter on behaviour in mice both knockout and overexpressor lines exist, thereby enabling the study of a range of expression levels in an otherwise well-controlled genetic system. Genetic knockout of the 5-HT transporter might be predicted to have the same effect as pharmacological blockade of the 5-HT transporter; that is, 5-HT transporter knockout mice would be expected to behave similar to wild-type mice treated with an SSRI. However, if we consider the findings from human gene association studies that individuals with the low expressing variant of the 5-HT transporter gene are predisposed to mood and anxiety disorders, then we might expect 5-HT transporter knockout mice to show an anxious, depressive-like phenotype.

Several laboratories developed 5-HT transporter knockout mice (first produced by Bengel and colleagues 1998) but conflicting results were found in the early studies. When bred onto a 129S6/SvEv line 5-HT transporter knockout mice were found to have increased immobility during single exposure to the forced swim test (Lira et al. 2003) but, conversely, decreased immobility in the tail suspension test (Holmes et al. 2002; Lira et al. 2003). Additionally, these mice appeared more anxious in a novelty-suppressed feeding paradigm but did not show robust differences from wild-type mice on the elevated plus maze (Holmes et al. 2003b; Lira et al. 2003).

In comparison, 5-HT transporter knockout mice generated from mice on a C57Bl/6 J-129 Sv mixed background, showed a decrease in immobility in the tail suspension test but no change in a two exposure form of the forced swim test (Perona et al. 2008). The latter mice showed no change in sucrose preference. However, on a pure C57Bl/6 background, 5-HT transporter knockout mice showed a robust increase in anxiety across a number of tests (Holmes et al. 2003a, b; Kalueff et al. 2007; Zhao et al. 2006; Line et al. 2011; but see also Adamec et al. 2006). These mice also showed an increase in immobility in both the tail suspension test (Zhao et al. 2006; but see also Holmes et al. 2002) and forced swim test (Wellman et al. 2007), although the latter depressive-like phenotype required multiple exposures to the test. These mice did not demonstrate altered sucrose preference (Kalueff et al. 2006).

Whilst the contrasting effect of 5-HT transporter knockout in tests of behavioural despair seem confusing, a robust pattern of increased anxiety can be seen across strains and tests. The results from the forced swim test across all strains suggest a depressive-like phenotype, albeit a subtle one that is only detectable following repeated swim stress exposure. The findings from the tail suspension test, however, are less easy to explain with antidepressant effects being detected in mice on either a 129 or mixed.

5-HT transporter overexpressor mice have also been developed and characterised, and the phenotype of these mice add much strength to the outcome of the knockout studies, at least in terms of the anxiety data. Thus, the 5-HT transporter overexpressing mice were found to have reduced anxiety, the opposite of what is generally seen in the knockout lines (Jennings et al. 2006; Line et al. 2011). In studies of 5-HT transporter variation, these mice offer an interesting and valuable contrast to the knockout mice and allow for a wide range of expression levels to be explored.

Overall, a likely important factor, in some of discrepant findings regarding the 5-HT transporter knockout mice, is the background strain of the mouse used to construct the knockout. Ironically, this indicates that the genetic heterogeneity that may account for conflicting results in human studies, also influence the outcome of the genetic mouse studies. Further discussion of the issue of background strain in mutant mice can be found later in this chapter.

3.2 *Noradrenaline*

Noradrenaline has also long been associated with depression. There is evidence for alterations of noradrenaline and its receptors in both depression pathophysiology and in response to antidepressant administration (e.g., Ordway et al. 2003; Deupree et al. 2007; see also chapter by Sharp in this volume). Perhaps the most compelling evidence for a role for noradrenaline in depression is that the noradrenaline transporter is a major target for tricyclic antidepressant drugs as well as recently developed selective noradrenaline reuptake inhibitors (SNRIs). Noradrenaline is also heavily involved in stress, which is a well-known risk factor for depression.

3.2.1 Adrenoceptors

Receptors for noradrenaline (adrenoceptors) are found throughout both the central and autonomic nervous systems. Indeed, many of the adrenoceptor mutant mice were initially constructed for investigations into the cardiovascular system. While the focus here is on behavioural changes in these mice that are relevant to modelling depression, changes in the periphery should not be forgotten as possible confounds or, indeed, have direct relevance to changes emotionality and stress levels.

Selective pharmacological tools for several subtypes of adrenoceptor are yet to be developed and so much of what is known about the actions of some of these receptors are derived from studies on adrenoceptor mutant mice. For example, drugs non-selective for the different types of α_2 adrenoceptor have been found to have wide-ranging effects that include altered sedation and cardiovascular changes, which have complicated interpretation of many behavioural measures. These effects are largely the result of the α_{2A} adrenoceptor as shown through use of α_{2A} knockout mice (Lakhlani et al. 1997).

In tests of depressive-like behaviour, α_{2A} adrenoceptor knockout mice displayed increased immobility in the forced swim test, mediated by decreased climbing and not changes in swimming (Schramm et al. 2001). Also these mice appeared more anxious in both the elevated plus maze (Lahdesmaki et al. 2002)

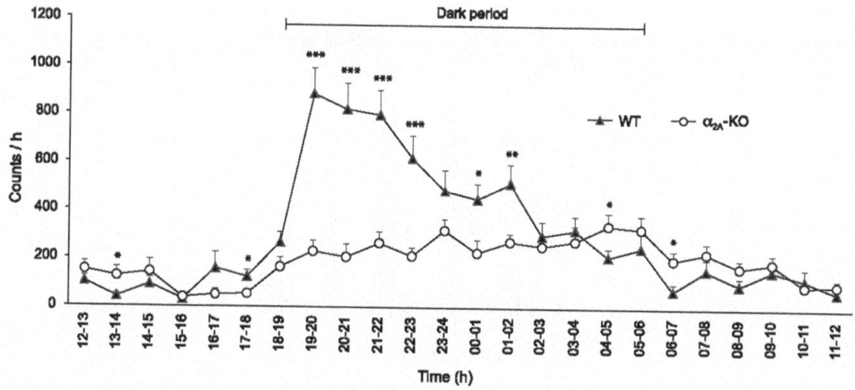

Fig. 1 The locomotor activity of wild type and α_{2A} adrenoceptor knockout mice over a 24 h period following a 2 h habituation period. Wild-type mice show a dramatic increase in locomotor activity in response to the onset of the dark period of the diurnal cycle but this is absent in α_{2A} adrenoceptor knockout mice. This may be relevant to the disruption of circadian rhythms seen during a depressive episode (from Lahdesmaki et al. 2002)

and light/dark box, but the latter only following injection stress (Schramm et al. 2001). The α_{2A} adrenoceptor knockout mice also displayed a flattened pattern of activity across the diurnal cycle (see Fig. 1), suggestive of disruptions to sleep/wake cycle, as seen in depression. It is suggested that altered diurnal pattern of activity may be due to the influence of the noradrenaline system on the synthesis of melatonin (Lahdesmaki et al. 2002).

The study by Schramm and colleagues provides an interesting example in which an antidepressant drug effect (imipramine) was seen in wild-type mice only when a two exposure form of the forced swim test was used and not after a single exposure (Schramm et al. 2001). This highlights the importance of how seemingly minor changes to the forced swim test protocol can markedly alter the sensitivity of the behavioural test.

In contrast to the α_{2A} adrenoceptor knockout mice, α_{2C} adrenoceptor knockout mice were found to display decreased immobility in the forced swim test (Sallinen et al. 1999). Again, the sensitivity of this test to detect this difference was dependent on pre-exposure to the swim test (see Fig. 2). To further validate the role of α_2 adrenoceptor subtypes in the forced swim test, α_{2C} adrenoceptor overexpressor mice were shown to have increased levels of immobility in this test (ibid.).

Overall, these data strongly implicate α_2 adrenoceptors in the forced swim test measure of depressive-like behaviour, with α_{2A} adrenoceptor-mediated activity appearing to be antidepressant and α_{2C} adrenoceptor-mediated activity being pro-depressive. While further testing needs to be done, the availability of both knockout and overexpressor lines of these adrenoceptor-specific mutant mice will allow for in-depth exploration of the role that these receptors play in emotional behaviour.

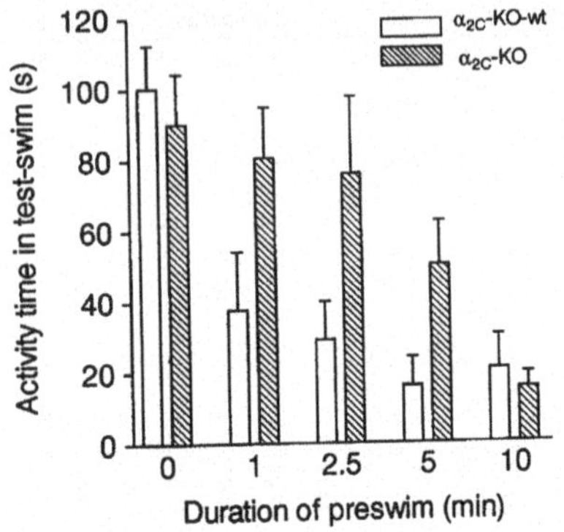

Fig. 2 The time spent active during a 5 min forced swim test following various lengths of prior exposure 24 h previously in wild type and α_{2C} adrenoceptor knockout mice. Prior exposure to forced swimming reduces activity in mice. In the case of this study, this effect was more dramatic in wild-type mice and so revealed an anti-depressive-like phenotype in α_{2C} adrenoceptor knockout mice. This suggests the use of a pre-swim session may be important in determining the sensitivity of the forced swim test. Taken from Sallinen et al. (1999)

It is worth noting, however, that other non-mood-related alterations in behaviour have been noted in some of these mouse lines including altered pre-pulse inhibition performance in both the α_{2C} adrenoceptor knockout and overexpressor mice (Sallinen et al. 1998). Nonetheless, these results have led to the α_{2C} adrenoceptor being proposed as a possible target for future therapeutic drugs that would lack the sedative effects of drugs that target α_{2A} adrenoceptors (Sallinen et al. 1999). Whether such agents would be sufficiently specific for the treatment of depression without having more wide-ranging effects remains to be seen.

The role of α_1 adrenoceptors in depressive behaviours has also been explored using mutant mice. In this case, instead of the receptor being knocked out, mice expressing a constitutively active form of the receptor have been developed. Such mutants have been constructed for both α_{1A} (Rorabaugh et al. 2005) and α_{1B} (Zuscik et al. 2000) adrenoceptors. The use of these mice has revealed complimentary alterations in behaviour. Thus, mice with constitutively active α_{1A} adrenoceptors having decreased immobility in both the forced swim test (Doze et al. 2009) and tail suspension test (Doze et al. 2009, 2011), whereas mice with constitutively active α_{1B} receptors showed increased immobility in these tests (Doze et al. 2009). The increased immobility in these tests of the latter mice contrasted with their general hyperactivity in an open field (with α_{1A} adrenoceptor constitutively active mutant mice having levels of locomotion similar to their wildtype controls). The anxiety profile of the α_{1A} adrenoceptor mutants, however, is unclear (Doze et al. 2009, 2011), the α_{1B} adrenoceptor mutant line do not appear to have altered levels of anxiety (Doze et al. 2009).

This work using sub-type specific manipulations of adrenoceptors reveals complex, opposing actions of these receptors in affective behaviour. The possibility of selectively targeting specific adrenoceptor populations using pharmacological agents is something yet to be taken advantage of in treatment of depression owing to the lack of appropriate drugs. While some current antidepressants act to block the reuptake of noradrenaline, and so affect activity at all adrenoceptors, targeting specific subtypes of the receptors for noradrenaline may produce more effective future therapeutics. This research provides an interesting example of how work with genetically altered mice has highlighted possible targets for future antidepressant drug development.

3.2.2 Noradrenaline Transporter

Noradrenaline transporter knockout mice display an antidepressant phenotype in both the forced swim test (Perona et al. 2008; Xu et al. 2000) and tail suspension test (Perona et al. 2008). This is, in spite of, also being found to be hypoactive (Perona et al. 2008; Xu et al. 2000). However, no alteration was seen to their sucrose preference at a range of sucrose concentrations (Perona et al. 2008). While the generally antidepressant-like phenotype is in line with the effects of tricyclic antidepressants and SNRIs that block the noradrenaline transporter, the data are in contrast to the findings from 5-HT transporter knockout mice which do not have an antidepressant phenotype (see above).

This difference between these noradrenaline and 5-HT transporter knockout lines is a curious one. The behavioural phenotype of the 5-HT transporter knockout mice, which is, in simple terms, the opposite to that seen following acute SSRI treatment, is often explained by a developmental origin of the phenotype, citing the role 5-HT has during development (Gingrich et al. 2003).

To examine the influence of the 5-HT and noradrenaline transporters in brain development, SSRIs and SNRIs have been administered to mice shortly after birth. Early SSRI treatment produced persistent behavioural changes including increased anxiety in adulthood (Ansorge et al. 2004, 2008). SNRI treatment, however, did not result in the same life-long changes (Ansorge et al. 2008). This finding may account for the contrasting phenotypes in the 5-HT transporter and noradrenaline transporter knockout mice; early 5-HT transporter blockade leads to (over-)compensatory changes that result in permanently altered emotionality, whereas early noradrenaline transporter blockade does not. This might help to explain why life-long 5-HT transporter blockade (5-HT transporter knockout) has the opposite effect to what would be expected from SSRI treatment, whereas life-long noradrenaline transporter blockade (noradrenaline transporter knockout) has effects similar to SNRI treatment.

The phenotypes of the transporter mutant mice illustrate the often complex relationship between pharmacological and genetic interventions, and also emphasises the importance of developmental effects of gene knockout on adult emotional behaviour.

3.3 CRF-Related Models

The role of corticotrophin-releasing factor (CRF) is traditionally seen as being imbedded in the hypothalamic-pituitary adrenal (HPA) axis, in that this neuropeptide is released from the hypothalamus to stimulate release of adrenocorticotrophic hormone (ACTH) from the pituitary that subsequently leads to the release of corticosteroid "stress hormones" from the adrenal glands. This process may in itself be important in depression, but receptors for CRF are expressed throughout the brain, not just in the hypothalamus. Therefore, extra-hypothalamic actions of CRF may also be important in the pathophysiology of depression.

CRF acting at CRF_1 receptors in limbic brain areas, such as the cerebral cortex, hippocampus and amygdala, is thought to lead to many of the behaviours associated with anxiety and depression (Holsboer 2000). This signal is modulated by a second class of CRF receptors, CRF_2 receptors, which bind CRF-like proteins known as urocortins. Thus far, three urocortins have been characterised; urocortin1 (Donaldson et al. 1996; Vaughan et al. 1995), urocortin2 (Reyes et al. 2001) and urocortin3 (Lewis et al. 2001). These peptides are expressed throughout the CNS in overlapping but unique patterns. Generally, activity at CRF_2 receptors is thought to dampen the effects of CRF_1 receptor activity and so is broadly anti-depressive and anxiolytic. Further modulation of this system is achieved by the CRF binding protein which sequesters all CRF-like proteins and prevents them binding to the receptors.

Mutant mouse lines with greatly increased CRF activity have been found to exhibit a Cushing's-like cluster of symptoms. For example, CRF overexpressor mice were found to have increased anxiety (Heinrichs et al. 1997; van Gaalen et al. 2002), but also thin skin, alopecia and altered fat and muscle deposition (Stenzel-Poore et al. 1992). Despite these many changes, CRF overexpressor mice provide an example of a mouse line that has been tested on some important but often ignored components of depressive-like behaviour. For instance, CRF mice have been shown to have a deficit in the serial reaction time task, a highly involved cognitive task that may be more relevant to the deficits seen in depression than assessments of, for example, spatial memory (van Gaalen et al. 2003).

CRF knockout when carried out during development has a profound effect on the viability of mouse offspring, but does not have this detrimental effect when carried out in adulthood (Muglia et al. 1995). For instance, mice with CRF genetically removed during adulthood show a blunted, but still intact stress response (Jacobson et al. 2000; Weninger et al. 1999).

Two independent lines of CRF_1 receptor knockout mice have been produced and both displayed decreased anxiety across several tests (Smith et al. 1998; Timpl et al. 1998). Interestingly, the decrease in anxiety following CRF_1 receptor knockout was also seen in a line of mice that have the CRF_1 receptor knockout restricted to the forebrain (Muller et al. 2003). These mice have a lack of CRF_1 receptors in limbic areas of the brain but still express these receptors in the pituitary. Because of this, measures of basal HPA axis activity are normal, but altered anxiety behaviours can still be detected. This is an important piece of evidence for the role of centrally acting CRF in mood in addition to its role in regulating corticosteroid release.

In contrast to the findings with CRF_1 receptor mutants, CRF_2 receptor knockout mice showed increased anxiety (Bale et al. 2000; Kishimoto et al. 2000; but see also Coste et al. 2000) as well as increased immobility in the forced swim test

(Bale and Vale 2003). This adds to the evidence that CRF_1 and CRF_2 receptors have broadly opposing roles in mediating emotional behaviour.

Further to this, weight and food intake were investigated in these mice. Body weight was not found to differ, and food intake was not altered at baseline. Following 24 h of food deprivation, however, CRF_2 receptor knockout mice were found to eat less than wild-type mice despite there still being no detectable difference in body weight (Bale et al. 2000). This may relate to the changes in feeding behaviour seen during a depressive episode, but are only noticeable after a mild stressor such as acute food deprivation is applied.

Receptors for the corticosteroids that are the product of HPA axis activity are also expressed centrally. Mineralocorticoid receptors have high affinity for corticosteroids, and several studies using mutant mice have helped in investigating the function of these receptors. Global knockout of these receptors proved to be perinatally lethal as corticosteroid signalling is vital to early lung development (Cole et al. 1995). Forebrain-specific deletion of mineralocorticoid receptors, however, produced viable offspring. These mice showed increased immobility in the forced swim and tail suspension tests (Boyle et al. 2005). However, overexpression of these receptors was also shown to result in increased immobility in the forced swim test, as well as increased anxiety-like behaviour in the elevated plus maze (Wei et al. 2004). This suggests an inverted U relationship between central corticosteroid signalling and emotionality.

The use of mutant mice has therefore shed some light on the roles of the CRF system in depression-related behaviours, and provided strong evidence for a key role for centrally expressed CRF receptors. This consideration of the wider CRF system and beyond the HPA axis has further avenues to be explored using mutant mice. For example, the testing of urocortin mutant mouse constructs in tests of depressive-like behaviour have begun (e.g., Chen et al. 2006; Neufeld-Cohen et al. 2010) although an extensive literature in this area is awaited. These models may provide further information on subtle but life-long changes to the CRF system.

3.4 BDNF

Recent theories regarding the mechanisms of depression and antidepressant action go beyond the role of neurotransmitter systems to emphasise the importance of neurotrophins (Hashimoto 2010). Trophic factors, such as BDNF, are integral to neuronal development and survival as well as neuroplasticity not only during early life but also in adulthood (Thoenen 1995). Moreover, expression of BDNF in the adult rodent brain was found to be reduced by corticosteroids and stress, and increased by antidepressants (Dwivedi et al. 2006). These findings and many others make BDNF an attractive addition to a unified neurobiological understanding of depression.

BDNF knockout mice were first produced in 1994 and found to have a plethora of detrimental phenotypes as well as perinatal lethality (Ernfors et al. 1994).

Behavioural work has therefore focused on heterozygote BDNF knockout mice. These mice, conversely, show relatively normal baseline behaviours including unaltered performance on the elevated plus maze and normal sucrose consumption (MacQueen et al. 2001). The mice also showed no change (MacQueen et al. 2001) or only subtle effects (Chourbaji et al. 2004) in various forms of the forced swim test. Overexpression of BDNF, however, was found to result in an antidepressant-like phenotype in the forced swim test, but also increased anxiety in the elevated plus maze (Govindarajan et al. 2006).

Conditional knockout mice that either have life-long or only adult reductions of BDNF have also been produced (Chan et al. 2006), allowing the dissection of the consequences of BDNF depletion during development from those during adulthood. The findings in these lines were surprisingly similar, with both showing a depressive-like phenotype in the tail suspension test but an antidepressant phenotype in a three exposure forced swim test paradigm (Chan et al. 2006). Neither line demonstrated changes in the elevated plus maze. These similarities between the two lines suggest that reductions in BDNF post-development are sufficient to produce the phenotypes, although once again, this phenotype is not consistently depressive-like.

Indeed, the role of BDNF in the pathogenesis of depression has been questioned (Groves 2007), although the case for a role of BDNF in antidepressant activity seems strong. The latter is supported by evidence that each of BDNF heterozygote mice, inducible BDNF knockout mice, and mice expressing non-functional versions of trkB, the high affinity receptor for BDNF, are resistance to the effects of antidepressant treatment in the forced swim test (Saarelainen et al. 2003; Adachi et al. 2008; Monteggia et al. 2007). It is worth noting that the baseline affective behaviours of the trkB mouse line are normal (Saarelainen et al. 2003; Zorner et al. 2003).

Overall, whilst the importance of BDNF in emotionality is still under investigation, current data suggest that at least in the hippocampus, BDNF gene expression is decreased by stress and corticosteroids and increased by antidepressant treatment. However, whilst data from genetically altered mice indicate that antidepressant effects require BDNF signalling, there does not appear to be such a strong link between BDNF signalling and depressive-like behaviours.

4 Concerns and Considerations

4.1 Background Strain

The issue of background strain has already been raised in this chapter, particularly in relation to the phenotype of 5-HT transporter knockout mice that differs across several lines. This is addressed and discussed extensively in Holmes et al. (2003a). The simplest explanation for the differences is that the various wild-type mice used

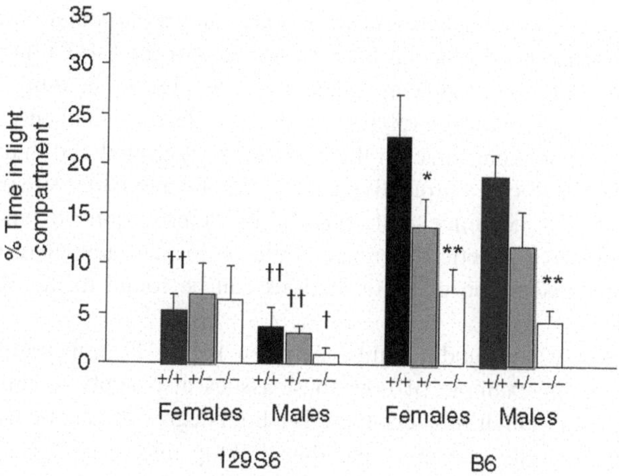

Fig. 3 Percentage of time spent in the light compartment of a 10 min exposure to the light/dark box for wild type and 5-HT transporter knockout mice on either a 129S6 or C57Bl/6 background. All mice on a 129S6 background spent very little time in the light, anxiogenic area of the light/dark box. In comparison, wild-type mice on a C57Bl/6 background spent a greater amount of time in the light compartment than their wild type 129S6 counterparts. The latter difference enables the anxiety phenotype of the 5-HT transporter knockout mice to be detected. Taken from Holmes et al. (2003a)

to generate the mutants do not provide a sufficiently sensitive baseline to detect differences. For instance, Holmes et al. (2003a) compare 5-HT transporter knockout mice on two different background strains, 129S6 and C57Bl/6, in the light/dark box (see Fig. 3). In this test, a decrease in time spent in the light compartment would reflect a high-anxiety phenotype. As can be seen, however, the 129S6 wild-type mice spend very little time in the light compartment, making a further decrease difficult to observe. In contrast, the C57Bl/6 wild-type mice spend more time in the light compartment, therefore giving a baseline from which a decrease can be observed. The behaviour of wild-type control mice, therefore, can affect the sensitivity of the tests. Further discussion of this issue can be found in Jacobson and Cryan (2007).

Another theory to explain the lack of consistent phenotype across different 5-HT transporter knockout lines is that genes flanking the mutant construct are incorporated into the genome of experimental mice during backcrossing onto a new background strain. In the case of the Holmes et al. (2003a) study, this would result in "129 genes" being inserted into C57Bl/6 mice. The observed decrease in time spent in the light compartment of the light/dark box may therefore be due to these flanking genes being inserted into the mutant mice and not because of loss of the 5-HT transporter gene *per se*. Sequencing of the site of insertion would reveal the likelihood of this confound.

Finally, genes that modify the effects of the mutant construct may exist at sites distal to its insertion, and these genes may only be present in particular background strains. For instance, one background strain may carry a particular allele of a gene that can compensate at a molecular level for the loss of the 5-HT transporter gene, whereas a different non-modifying allele may be present in other background strains. In this case, a possible explanation for the difference in phenotype between 5-HT transporter knockout mice on the C57Bl/6 background and that on a 129S6 background may be due to a protective genetic variant at another site present in the latter but not the former mice. This possibility would require extensive investigation as the putative genetic difference could be located anywhere in the entire genome. Further discussion of these theories can be found in the Holmes et al. (2003a) paper.

The controversy surrounding links between the 5-HT transporter and emotionality may be an example of how these issues also apply to human studies. When dealing with human subjects, there is a high degree of genetic heterogeneity. By genotyping for only one gene and then linking this to complex behavioural traits ignores the possibility that other genes may contribute, not only to the behaviour in question, but also to the expression level of the protein encoded by the gene in question. These gene × gene interactions may account for some of the "missing heritability" not yet accounted for by known genetic risk factors for various psychiatric conditions. While in vivo assessments of functional protein levels in humans can be expensive or impractical, histological measures of protein expression in experimental animals provide a way of ensuring that the genetic manipulation performed has the expected effects at a molecular level. While this may not always explain discrepancies in results, it is a useful validation step in using and producing mutant mouse lines.

4.2 Sex Effects

It is well established that females are more likely to develop clinical depression than males (DSM IV-TR; Weissman and Klerman 1977), a striking vulnerability difference for which the underlying mechanism is not yet known. In this regard, it would be valuable to recapitulate such sex difference in experimental animal models, but this is complicated by evidence that male mice and rats are more susceptible to learned helplessness than females (Caldarone et al. 2000; Steenbergen et al. 1990). This suggests that contrary to findings in humans, in rodents males are more predisposed to developing depressive-like phenotypes than females. What should be expected in terms of sexual dimorphism in depressive-like behaviours in mutant mice is therefore unclear.

As in the case of the baseline behaviour of mice with different background strains, male and female mice provide different baselines on a variety of behaviours. Thus, only investigating one sex may mask effects that would otherwise be apparent in the other. If striking differences are found following genetic manipulation, it may be that sex hormones are modulating the effects seen, in which case providing a further of avenue of study that could be highly translatable. Many basic studies with mutant mice use only males to avoid having to control for variation in the oestrus cycle during behavioural testing. Nevertheless, since sex hormones may impact on phenotypic differences between mutant and wild-type mice, investigations using both male and female mice could add further insight into the role that sex differences play in depression vulnerability.

4.3 Environmental Factors

As in humans, there is considerable evidence from animal studies that adverse life events such as stress, including that experienced during early life, increase depression vulnerability (e.g. Willner 2005; see chapter in this volume by Harro). Moreover, it seems highly likely that genetic and environmental risk factors interact in ways that are not yet fully understood to increase predisposition to depression (e.g., Kendler et al. 2001). A potentially powerful way to investigate this is through the combination of genetic and well-controlled environmental manipulations in mice. There are increasing examples of such studies emerging in the literature, which are reviewed elsewhere.

4.4 The Perfect Model?

So what can we hope for in attempting to model depression using mutant mice, and are we there yet?

Depression is a complex condition consisting of a variety of symptoms, only a small number of which are possible to investigate in rodents. However, as noted in this chapter, there are some core behavioural symptoms that are accessible to investigation in rodents in a highly relevant way, and these can be expanded to include measurement of physiological changes such as the sleep/wake cycles and food intake. Nonetheless, in the case of genetic mouse models of depression the majority of studies being reported at present often rely heavily on measures of behavioural despair such as the forced swim test and tail suspension test without exploring the wider symptomatology of depression. Often a more complex analysis of the current models would give greater insight into their validity. Hopefully, this will be a feature of future models.

While a genetic mouse model that recapitulates the fullness of the symptoms that typically feature in a clinical depressive episode seems overoptimistic and probably unachievable, a frank assessment of a wider variety of behaviours than is typically presented in the current literature may provide a clearer view of the likely contribution to depression pathophysiology of specific genes. Nevertheless, given that specific genes, and often genes with unclear functions, can be manipulated in an increasingly powerful and controlled way, key questions can now be asked about the contributions of a single gene to a specific behavioural processes that are highly relevant to depression vulnerability, in a way that is not possible in human gene association studies. This is not to deny, however, sometimes confounding effects that can be occur from targeting genes in mice, such as noted above with the example of the 5-HT transporter.

Acknowledgments The author wishes to thank Prof David Bannerman for his helpful discussions of the material in this chapter. This chapter was completed whilst working under an MRC Project Grant (G0700983).

References

Adachi M, Barrot M, Autry AE, Theobald D, Monteggia LM (2008) Selective loss of brain-derived neurotrophic factor in the dentate gyrus attenuates antidepressant efficacy. Biol Psychiatry 63:642–649

Adamec R, Burton P, Blundell J, Murphy DL, Holmes A (2006) Vulnerability to mild predator stress in serotonin transporter knockout mice. Behav Brain Res 170:126–140

Anguelova M, Benkelfat C, Turecki G (2003) A systematic review of association studies investigating genes coding for serotonin receptors and the serotonin transporter: i affective disorders. Mol Psychiatry 8:574–591

Ansorge MS, Zhou M, Lira A, Hen R, Gingrich JA (2004) Early-life blockade of the 5-HT transporter alters emotional behavior in adult mice. Science 306:879–881

Ansorge MS, Morelli E, Gingrich JA (2008) Inhibition of serotonin but not norepinephrine transport during development produces delayed, persistent perturbations of emotional behaviors in mice. J Neurosci 28:199–207

Bale TL, Vale WW (2003) Increased depression-like behaviors in corticotropin-releasing factor receptor-2-deficient mice: sexually dichotomous responses. J Neurosci 23:5295–5301

Bale TL, Contarino A, Smith GW, Chan R, Gold LH, Sawchenko PE, Koob GF, Vale WW, Lee KF (2000) Mice deficient for corticotropin-releasing hormone receptor-2 display anxiety-like behaviour and are hypersensitive to stress. Nat Genet 24:410–414

Bengel D, Murphy DL, Andrews AM, Wichems CH, Feltner D, Heils A, Mossner R, Westphal H, Lesch KP (1998) Altered brain serotonin homeostasis and locomotor insensitivity to 3, 4-Methylenedioxymethamphetamine ("Ecstasy") in serotonin transporter-deficient mice. Mol Pharmacol 53:649–655

Blumstein LK, Crawley JN (1983) Further characterization of a simple, automated exploratory model for the anxiolytic effects of benzodiazepines. Pharmacol Biochem Behav 18:37–40

Boyle MP, Brewer JA, Funatsu M, Wozniak DF, Tsien JZ, Izumi Y, Muglia LJ (2005) Acquired deficit of forebrain glucocorticoid receptor produces depression-like changes in adrenal axis regulation and behavior. Proc Natl Acad Sci U S A 102:473–478

Caldarone BJ, George TP, Zachariou V, Picciotto MR (2000) Gender differences in learned helplessness behavior are influenced by genetic background. Pharmacol Biochem Behav 66:811–817

Chan JP, Unger TJ, Byrnes J, Rios M (2006) Examination of behavioral deficits triggered by targeting Bdnf in fetal or postnatal brains of mice. Neuroscience 142:49–58

Chen A, Zorrilla E, Smith S, Rousso D, Levy C, Vaughan J, Donaldson C, Roberts A, Lee KF, Vale W (2006) Urocortin 2-deficient mice exhibit gender-specific alterations in circadian hypothalamus-pituitary-adrenal axis and depressive-like behavior. J Neurosci 26:5500–5510

Chourbaji S, Hellweg R, Brandis D, Zorner B, Zacher C, Lang UE, Henn FA, Hortnagl H, Gass P (2004) Mice with reduced brain-derived neurotrophic factor expression show decreased choline acetyltransferase activity, but regular brain monoamine levels and unaltered emotional behavior. Brain Res Mol Brain Res 121:28–36

Cole TJ, Blendy JA, Monaghan AP, Krieglstein K, Schmid W, Aguzzi A, Fantuzzi G, Hummler E, Unsicker K, Schutz G (1995) Targeted disruption of the glucocorticoid receptor gene blocks adrenergic chromaffin cell development and severely retards lung maturation. Genes Dev 9:1608–1621

Coste SC, Kesterson RA, Heldwein KA, Stevens SL, Heard AD, Hollis JH, Murray SE, Hill JK, Pantely GA, Hohimer AR, Hatton DC, Phillips TJ, Finn DA, Low MJ, Rittenberg MB, Stenzel P, Stenzel-Poore MP (2000) Abnormal adaptations to stress and impaired cardiovascular function in mice lacking corticotropin-releasing hormone receptor-2. Nat Genet 24:403–409

Crawley JN (1981) Neuropharmacologic specificity of a simple animal model for the behavioral actions of benzodiazepines. Pharmacol Biochem Behav 15:695–699

Crawley J (2000) What's wrong with my mouse: behavioral phenotyping of transgenic and knockout mice. Wiley-Liss, New York

Crawley J, Goodwin FK (1980) Preliminary report of a simple animal behavior model for the anxiolytic effects of benzodiazepines. Pharmacol Biochem Behav 13:167–170

Cryan JF, Mombereau C (2004) In search of a depressed mouse: utility of models for studying depression-related behavior in genetically modified mice. Mol Psychiatry 9:326–357

Cryan JF, Mombereau C, Vassout A (2005a) The tail suspension test as a model for assessing antidepressant activity: review of pharmacological and genetic studies in mice. Neurosci Biobehav Rev 29:571–625

Cryan JF, Valentino RJ, Lucki I (2005b) Assessing substrates underlying the behavioral effects of antidepressants using the modified rat forced swimming test. Neurosci Biobehav Rev 29: 547–569

Detke MJ, Lucki I (1996) Detection of serotonergic and noradrenergic antidepressants in the rat forced swimming test: the effects of water depth. Behav Brain Res 73:43–46

Detke MJ, Rickels M, Lucki I (1995) Active behaviors in the rat forced swimming test differentially produced by serotonergic and noradrenergic antidepressants. Psychopharmacology (Berl) 121:66–72

Detke MJ, Johnson J, Lucki I (1997) Acute and chronic antidepressant drug treatment in the rat forced swimming test model of depression. Exp Clin Psychopharmacol 5:107–112

Deupree JD, Reed AL, Bylund DB (2007) Differential effects of the tricyclic antidepressant desipramine on the density of adrenergic receptors in juvenile and adult rats. J Pharmacol Exp Ther 321:770–776

Diagnostic and statistical manual of mental disorders, IV-TR edn (2000). American Psychiatric Association, Washington

Donaldson CJ, Sutton SW, Perrin MH, Corrigan AZ, Lewis KA, Rivier JE, Vaughan JM, Vale WW (1996) Cloning and characterization of human urocortin. Endocrinology 137:3896

Doze VA, Handel EM, Jensen KA, Darsie B, Luger EJ, Haselton JR, Talbot JN, Rorabaugh BR (2009) alpha(1A)-and alpha(1B)-adrenergic receptors differentially modulate antidepressant-like behavior in the mouse. Brain Res 1285:148–157

Doze VA, Papay RS, Goldenstein BL, Gupta MK, Collette KM, Nelson BW, Lyons MJ, Davis BA, Luger EJ, Wood SG, Haselton JR, Simpson PC, Perez DM (2011) Long-term alpha1A-adrenergic receptor stimulation improves synaptic plasticity, cognitive function, mood, and longevity. Mol Pharmacol 80:747–758

Dwivedi Y, Rizavi HS, Pandey GN (2006) Antidepressants reverse corticosterone-mediated decrease in brain-derived neurotrophic factor expression: differential regulation of specific exons by antidepressants and corticosterone. Neuroscience 139:1017–1029

El Yacoubi M, Bouali S, Popa D, Naudon L, Leroux-Nicollet I, Hamon M, Costentin J, Adrien J, Vaugeois JM (2003) Behavioral, neurochemical, and electrophysiological characterization of a genetic mouse model of depression. Proc Natl Acad Sci U S A 100:6227–6232

Ernfors P, Lee KF, Jaenisch R (1994) Mice lacking brain-derived neurotrophic factor develop with sensory deficits. Nature 368:147–150

Forbes NF, Stewart CA, Matthews K, Reid IC (1996) Chronic mild stress and sucrose consumption: validity as a model of depression. Physiol Behav 60:1481–1484

Furlong RA, Ho L, Walsh C, Rubinsztein JS, Jain S, Paykel ES, Easton DF, Rubinsztein DC (1998) Analysis and meta-analysis of two serotonin transporter gene polymorphisms in bipolar and unipolar affective disorders. Am J Med Genet 81:58–63

Gardier AM, Guiard BP, Guilloux JP, Reperant C, Coudore F, David DJ (2009) Interest of using genetically manipulated mice as models of depression to evaluate antidepressant drugs activity: a review. Fundam Clin Pharmacol 23:23–42

Gingrich JA, Ansorge MS, Merker R, Weisstaub N, Zhou M (2003) New lessons from knockout mice: The role of serotonin during development and its possible contribution to the origins of neuropsychiatric disorders. CNS Spectr 8:572–577

Govindarajan A, Rao BS, Nair D, Trinh M, Mawjee N, Tonegawa S, Chattarji S (2006) Transgenic brain-derived neurotrophic factor expression causes both anxiogenic and antidepressant effects. Proc Natl Acad Sci U S A 103:13208–13213

Groves JO (2007) Is it time to reassess the BDNF hypothesis of depression? Mol Psychiatry 12:1079–1088

Hall CS (1936) Emotional behavior in the rat III. The relationship between emotionality and ambulatory activity. J Comp Psychol 22:345–352

Handley SL, Mithani S (1984) Effects of alpha-adrenoceptor agonists and antagonists in a maze-exploration model of 'fear'-motivated behaviour. Naunyn Schmiedebergs Arch Pharmacol 327:1–5

Hashimoto K (2010) Brain-derived neurotrophic factor as a biomarker for mood disorders: an historical overview and future directions. Psychiatry Clin Neurosci 64:341–357

Heils A, Teufel A, Petri S, Stober G, Riederer P, Bengel D, Lesch KP (1996) Allelic variation of human serotonin transporter gene expression. J Neurochem 66:2621–2624

Heinrichs SC, Min H, Tamraz S, Carmouche M, Boehme SA, Vale WW (1997) Anti-sexual and anxiogenic behavioral consequences of corticotropin-releasing factor overexpression are centrally mediated. Psychoneuroendocrinology 22:215–224

Heisler LK, Chu HM, Brennan TJ, Danao JA, Bajwa P, Parsons LH, Tecott LH (1998) Elevated anxiety and antidepressant-like responses in serotonin 5-HT1A receptor mutant mice. Proc Natl Acad Sci U S A 95:15049–15054

Holmes A, Yang RJ, Murphy DL, Crawley JN (2002) Evaluation of antidepressant-related behavioral responses in mice lacking the serotonin transporter. Neuropsychopharmacology 27:914–923

Holmes A, Lit Q, Murphy DL, Gold E, Crawley JN (2003a) Abnormal anxiety-related behavior in serotonin transporter null mutant mice: the influence of genetic background. Genes Brain Behav 2:365–380

Holmes A, Yang RJ, Lesch KP, Crawley JN, Murphy DL (2003b) Mice lacking the serotonin transporter exhibit 5-HT(1A) receptor-mediated abnormalities in tests for anxiety-like behavior. Neuropsychopharmacology 28:2077–2088

Holsboer F (2000) The corticosteroid receptor hypothesis of depression. Neuropsychopharmacology 23:477–501

Jacobson LH, Cryan JF (2007) Feeling strained? Influence of genetic background on depression-related behavior in mice: a review. Behav Genet 37:171–213

Jacobson L, Muglia LJ, Weninger SC, Pacak K, Majzoub JA (2000) CRH deficiency impairs but does not block pituitary-adrenal responses to diverse stressors. Neuroendocrinology 71:79–87

Jennings KA, Loder MK, Sheward WJ, Pei Q, Deacon RM, Benson MA, Olverman HJ, Hastie ND, Harmar AJ, Shen S, Sharp T (2006) Increased expression of the 5-HT transporter confers a low-anxiety phenotype linked to decreased 5-HT transmission. J Neurosci 26:8955–8964

Kalueff AV, Gallagher PS, Murphy DL (2006) Are serotonin transporter knockout mice 'depressed'?: hypoactivity but no anhedonia. NeuroReport 17:1347–1351

Kalueff AV, Jensen CL, Murphy DL (2007) Locomotory patterns, spatiotemporal organization of exploration and spatial memory in serotonin transporter knockout mice. Brain Res 1169:87–97

Kendler KS, Thornton LM, Gardner CO (2001) Genetic risk, number of previous depressive episodes, and stressful life events in predicting onset of major depression. Am J Psychiatry 158:582–586

Kishimoto T, Radulovic J, Radulovic M, Lin CR, Schrick C, Hooshmand F, Hermanson O, Rosenfeld MG, Spiess J (2000) Deletion of crhr2 reveals an anxiolytic role for corticotropin-releasing hormone receptor-2. Nat Genet 24:415–419
Lahdesmaki J, Sallinen J, MacDonald E, Kobilka BK, Fagerholm V, Scheinin M (2002) Behavioral and neurochemical characterization of alpha(2A)-adrenergic receptor knockout mice. Neuroscience 113:289–299
Lakhlani PP, MacMillan LB, Guo TZ, McCool BA, Lovinger DM, Maze M, Limbird LE (1997) Substitution of a mutant alpha2a-adrenergic receptor via "hit and run" gene targeting reveals the role of this subtype in sedative, analgesic, and anesthetic-sparing responses in vivo. Proc Natl Acad Sci U S A 94:9950–9955
Lasky-Su JA, Faraone SV, Glatt SJ, Tsuang MT (2005) Meta-analysis of the association between two polymorphisms in the serotonin transporter gene and affective disorders. Am J Med Genet B Neuropsychiatr Genet 133B:110–115
Lesch KP, Bengel D, Heils A, Sabol SZ, Greenberg BD, Petri S, Benjamin J, Muller CR, Hamer DH, Murphy DL (1996) Association of anxiety-related traits with a polymorphism in the serotonin transporter gene regulatory region. Science 274:1527–1531
Lewis K, Li C, Perrin MH, Blount A, Kunitake K, Donaldson C, Vaughan J, Reyes TM, Gulyas J, Fischer W, Bilezikjian L, Rivier J, Sawchenko PE, Vale WW (2001) Identification of urocortin III, an additional member of the corticotropin-releasing factor (CRF) family with high affinity for the CRF2 receptor. Proc Natl Acad Sci U S A 98:7570–7575
Line SJ, Barkus C, Coyle C, Jennings KA, Deacon RM, Lesch KP, Sharp T, Bannerman DM (2011) Opposing alterations in anxiety and species-typical behaviours in serotonin transporter overexpressor and knockout mice. Eur Neuropsychopharmacol 21:108–116
Lira A, Zhou M, Castanon N, Ansorge MS, Gordon JA, Francis JH, Bradley-Moore M, Lira J, Underwood MD, Arango V, Kung HF, Hofer MA, Hen R, Gingrich JA (2003) Altered depression-related behaviors and functional changes in the dorsal raphe nucleus of serotonin transporter-deficient mice. Biol Psychiatry 54:960–971
Lister RG (1987) The use of a plus-maze to measure anxiety in the mouse. Psychopharmacology (Berl) 92:180–185
Lotrich FE, Pollock BG (2004) Meta-analysis of serotonin transporter polymorphisms and affective disorders. Psychiatr Genet 14:121–129
MacQueen GM, Ramakrishnan K, Croll SD, Siuciak JA, Yu G, Young LT, Fahnestock M (2001) Performance of heterozygous brain-derived neurotrophic factor knockout mice on behavioral analogues of anxiety, nociception, and depression. Behav Neurosci 115:1145–1153
Mann JJ, Huang YY, Underwood MD, Kassir SA, Oppenheim S, Kelly TM, Dwork AJ, Arango V (2000) A serotonin transporter gene promoter polymorphism (5-HTTLPR) and prefrontal cortical binding in major depression and suicide. Arch Gen Psychiatry 57:729–738
Mayorga AJ, Lucki I (2001) Limitations on the use of the C57BL/6 mouse in the tail suspension test. Psychopharmacology (Berl) 155:110–112
Monteggia LM, Luikart B, Barrot M, Theobold D, Malkovska I, Nef S, Parada LF, Nestler EJ (2007) Brain-derived neurotrophic factor conditional knockouts show gender differences in depression-related behaviors. Biol Psychiatry 61:187–197
Muglia L, Jacobson L, Dikkes P, Majzoub JA (1995) Corticotropin-releasing hormone deficiency reveals major fetal but not adult glucocorticoid need. Nature 373:427–432
Muller MB, Zimmermann S, Sillaber I, Hagemeyer TP, Deussing JM, Timpl P, Kormann MS, Droste SK, Kuhn R, Reul JM, Holsboer F, Wurst W (2003) Limbic corticotropin-releasing hormone receptor 1 mediates anxiety-related behavior and hormonal adaptation to stress. Nat Neurosci 6:1100–1107
Naylor L, Dean B, Pereira A, Mackinnon A, Kouzmenko A, Copolov D (1998) No association between the serotonin transporter-linked promoter region polymorphism and either schizophrenia or density of the serotonin transporter in human hippocampus. Mol Med 4:671–674
Neufeld-Cohen A, Evans AK, Getselter D, Spyroglou A, Hill A, Gil S, Tsoory M, Beuschlein F, Lowry CA, Vale W, Chen A (2010) Urocortin-1 and -2 double-deficient mice show robust

anxiolytic phenotype and modified serotonergic activity in anxiety circuits. Mol Psychiatry 15(339):426–441
Ordway GA, Schenk J, Stockmeier CA, May W, Klimek V (2003) Elevated agonist binding to alpha2-adrenoceptors in the locus coeruleus in major depression. Biol Psychiatry 53:315–323
Owens MJ, Nemeroff CB (1994) Role of serotonin in the pathophysiology of depression: focus on the serotonin transporter. Clin Chem 40:288–295
Owens MJ, Morgan WN, Plott SJ, Nemeroff CB (1997) Neurotransmitter receptor and transporter binding profile of antidepressants and their metabolites. J Pharmacol Exp Ther 283:1305–1322
Papp M, Willner P, Muscat R (1991) An animal model of anhedonia: attenuation of sucrose consumption and place preference conditioning by chronic unpredictable mild stress. Psychopharmacology (Berl) 104:255–259
Parks CL, Robinson PS, Sibille E, Shenk T, Toth M (1998) Increased anxiety of mice lacking the serotonin1A receptor. Proc Natl Acad Sci U S A 95:10734–10739
Pellow S, Chopin P, File SE, Briley M (1985) Validation of open: closed arm entries in an elevated plus-maze as a measure of anxiety in the rat. J Neurosci Methods 14:149–167
Perona MT, Waters S, Hall FS, Sora I, Lesch KP, Murphy DL, Caron M, Uhl GR (2008) Animal models of depression in dopamine, serotonin, and norepinephrine transporter knockout mice: prominent effects of dopamine transporter deletions. Behav Pharmacol 19:566–574
Porsolt RD, Bertin A, Jalfre M (1977a) Behavioral despair in mice: a primary screening test for antidepressants. Arch Int Pharmacodyn Ther 229:327–336
Porsolt RD, Le Pichon M, Jalfre M (1977b) Depression: a new animal model sensitive to antidepressant treatments. Nature 266:730–732
Porsolt RD, Anton G, Blavet N, Jalfre M (1978a) Behavioural despair in rats: a new model sensitive to antidepressant treatments. Eur J Pharmacol 47:379–391
Porsolt RD, Bertin A, Jalfre M (1978b) "Behavioural despair" in rats and mice: strain differences and the effects of imipramine. Eur J Pharmacol 51:291–294
Ramboz S, Oosting R, Amara DA, Kung HF, Blier P, Mendelsohn M, Mann JJ, Brunner D, Hen R (1998) Serotonin receptor 1A knockout: an animal model of anxiety-related disorder. Proc Natl Acad Sci U S A 95:14476–14481
Reyes TM, Lewis K, Perrin MH, Kunitake KS, Vaughan J, Arias CA, Hogenesch JB, Gulyas J, Rivier J, Vale WW, Sawchenko PE (2001) Urocortin II: a member of the corticotropin-releasing factor (CRF) neuropeptide family that is selectively bound by type 2 CRF receptors. Proc Natl Acad Sci U S A 98:2843–2848
Rhodes RA, Murthy NV, Dresner MA, Selvaraj S, Stavrakakis N, Babar S, Cowen PJ, Grasby PM (2007) Human 5-HT transporter availability predicts amygdala reactivity in vivo. J Neurosci 27:9233–9237
Robbins TW (2002) The 5-choice serial reaction time task: behavioural pharmacology and functional neurochemistry. Psychopharmacology (Berl) 163:362–380
Rorabaugh BR, Ross SA, Gaivin RJ, Papay RS, McCune DF, Simpson PC, Perez DM (2005) alpha1A-but not alpha1B-adrenergic receptors precondition the ischemic heart by a staurosporine-sensitive, chelerythrine-insensitive mechanism. Cardiovasc Res 65:436–445
Saarelainen T, Hendolin P, Lucas G, Koponen E, Sairanen M, MacDonald E, Agerman K, Haapasalo A, Nawa H, Aloyz R, Ernfors P, Castren E (2003) Activation of the TrkB neurotrophin receptor is induced by antidepressant drugs and is required for antidepressant-induced behavioral effects. J Neurosci 23:349–357
Sallinen J, Haapalinna A, Viitamaa T, Kobilka BK, Scheinin M (1998) Adrenergic alpha2C-receptors modulate the acoustic startle reflex, prepulse inhibition, and aggression in mice. J Neurosci 18:3035–3042
Sallinen J, Haapalinna A, MacDonald E, Viitamaa T, Lahdesmaki J, Rybnikova E, Pelto-Huikko M, Kobilka BK, Scheinin M (1999) Genetic alteration of the alpha2-adrenoceptor subtype c in mice affects the development of behavioral despair and stress-induced increases in plasma corticosterone levels. Mol Psychiatry 4:443–452

Schildkraut JJ (1965) The catecholamine hypothesis of affective disorders: a review of supporting evidence. Am J Psychiatry 122:509–522
Schramm NL, McDonald MP, Limbird LE (2001) The alpha(2a)-adrenergic receptor plays a protective role in mouse behavioral models of depression and anxiety. J Neurosci 21: 4875–4882
Shioe K, Ichimiya T, Suhara T, Takano A, Sudo Y, Yasuno F, Hirano M, Shinohara M, Kagami M, Okubo Y, Nankai M, Kanba S (2003) No association between genotype of the promoter region of serotonin transporter gene and serotonin transporter binding in human brain measured by PET. Synapse 48:184–188
Smith GW, Aubry JM, Dellu F, Contarino A, Bilezikjian LM, Gold LH, Chen R, Marchuk Y, Hauser C, Bentley CA, Sawchenko PE, Koob GF, Vale W, Lee KF (1998) Corticotropin releasing factor receptor 1-deficient mice display decreased anxiety, impaired stress response, and aberrant neuroendocrine development. Neuron 20:1093–1102
Song C, Leonard BE (2005) The olfactory bulbectomised rat as a model of depression. Neurosci Biobehav Rev 29:627–647
Steenbergen HL, Heinsbroek RP, Van Hest A, Van de Poll NE (1990) Sex-dependent effects of inescapable shock administration on shuttlebox-escape performance and elevated plus-maze behavior. Physiol Behav 48:571–576
Steiger A, Kimura M (2010) Wake and sleep EEG provide biomarkers in depression. J Psychiatr Res 44:242–252
Stenzel-Poore MP, Cameron VA, Vaughan J, Sawchenko PE, Vale W (1992) Development of Cushing's syndrome in corticotropin-releasing factor transgenic mice. Endocrinology 130:3378–3386
Steru L, Chermat R, Thierry B, Simon P (1985) The tail suspension test: a new method for screening antidepressants in mice. Psychopharmacology (Berl) 85:367–370
Thoenen H (1995) Neurotrophins and neuronal plasticity. Science 270:593–598
Timpl P, Spanagel R, Sillaber I, Kresse A, Reul JM, Stalla GK, Blanquet V, Steckler T, Holsboer F, Wurst W (1998) Impaired stress response and reduced anxiety in mice lacking a functional corticotropin-releasing hormone receptor 1. Nat Genet 19:162–166
Treit D, Fundytus M (1988) Thigmotaxis as a test for anxiolytic activity in rats. Pharmacol Biochem Behav 31:959–962
van Gaalen MM, Stenzel-Poore MP, Holsboer F, Steckler T (2002) Effects of transgenic overproduction of CRH on anxiety-like behaviour. Eur J Neurosci 15:2007–2015
van Gaalen MM, Stenzel-Poore M, Holsboer F, Steckler T (2003) Reduced attention in mice overproducing corticotropin-releasing hormone. Behav Brain Res 142:69–79
Vaughan J, Donaldson C, Bittencourt J, Perrin MH, Lewis K, Sutton S, Chan R, Turnbull AV, Lovejoy D, Rivier C et al (1995) Urocortin, a mammalian neuropeptide related to fish urotensin I and to corticotropin-releasing factor. Nature 378:287–292
Wei Q, Lu XY, Liu L, Schafer G, Shieh KR, Burke S, Robinson TE, Watson SJ, Seasholtz AF, Akil H (2004) Glucocorticoid receptor overexpression in forebrain: a mouse model of increased emotional lability. Proc Natl Acad Sci U S A 101:11851–11856
Weissman MM, Klerman GL (1977) Sex differences and the epidemiology of depression. Arch Gen Psychiatry 34:98–111
Wellman CL, Izquierdo A, Garrett JE, Martin KP, Carroll J, Millstein R, Lesch KP, Murphy DL, Holmes A (2007) Impaired stress-coping and fear extinction and abnormal corticolimbic morphology in serotonin transporter knock-out mice. J Neurosci 27:684–691
Weninger SC, Dunn AJ, Muglia LJ, Dikkes P, Miczek KA, Swiergiel AH, Berridge CW, Majzoub JA (1999) Stress-induced behaviors require the corticotropin-releasing hormone (CRH) receptor, but not CRH. Proc Natl Acad Sci U S A 96:8283–8288
Willner P (2005) Chronic mild stress (CMS) revisited: consistency and behavioural-neurobiological concordance in the effects of CMS. Neuropsychobiology 52:90–110
Xu F, Gainetdinov RR, Wetsel WC, Jones SR, Bohn LM, Miller GW, Wang YM, Caron MG (2000) Mice lacking the norepinephrine transporter are supersensitive to psychostimulants. Nat Neurosci 3:465–471

Zhao S, Edwards J, Carroll J, Wiedholz L, Millstein RA, Jaing C, Murphy DL, Lanthorn TH, Holmes A (2006) Insertion mutation at the C-terminus of the serotonin transporter disrupts brain serotonin function and emotion-related behaviors in mice. Neuroscience 140:321–334

Zimmerman M, McDermut W, Mattia JI (2000) Frequency of anxiety disorders in psychiatric outpatients with major depressive disorder. Am J Psychiatry 157:1337–1340

Zorner B, Wolfer DP, Brandis D, Kretz O, Zacher C, Madani R, Grunwald I, Lipp HP, Klein R, Henn FA, Gass P (2003) Forebrain-specific trkB-receptor knockout mice: behaviorally more hyperactive than "depressive". Biol Psychiatry 54:972–982

Zuscik MJ, Sands S, Ross SA, Waugh DJ, Gaivin RJ, Morilak D, Perez DM (2000) Overexpression of the alpha1B-adrenergic receptor causes apoptotic neurodegeneration: multiple system atrophy. Nat Med 6:1388–1394

Part II
Cellular and Molecular Basis

The Genetic Basis of Depression

Karen Hodgson and Peter McGuffin

Abstract Since the publication of the working draft of the human genome just over a decade ago, there have been dramatic advances in our understanding of the role genetics play in both normal human functioning as well as in disease. The identification of genes, which influence an individual's susceptibility to depression, is not only an intriguing scientific endeavour in its own right, but further, if a gene can be confidently implicated in depression, then this could shed light on the aetiological processes involved in the disease. Moreover, a genetic association with depression may identify targets for consideration in the development of novel treatments for the illness. This chapter will summarise the current research into the genetic basis of depression. A number of genes of interest have been highlighted, although a genetic variant, that is unequivocally associated with increased risk for the disease, is yet to be identified. However, technologies and methodologies are evolving rapidly, and genetic approaches have helped shape how we conceptualise depression as an illness.

Keywords Candidate genes · Copy number variants · Epistasis · Gene-environment interactions · Genetic association · Genome-wide association studies GWAS · Heritability · Linkage · Quantitative genetics · Sequencing · Twin studies

Contents

K. Hodgson · P. McGuffin (✉)
MRC SGDP Centre, Institute of Psychiatry, King's College London, London, UK
e-mail: Peter.Mcguffin@kcl.ac.uk

Curr Topics Behav Neurosci (2013) 14: 81–99
DOI: 10.1007/7854_2012_225

Published Online: 24 August 2012

1 Identifying the Role of Genetics in the Aetiology of Depression

The first stage in genetic research is to establish that the trait of interest is under genetic influence. Major depressive disorder (MDD) has long been known to run in families and studies have demonstrated that first degree relatives of patients suffering from depression have a significantly increased risk of the disease relative to the general population. Looking across studies, the average odds ratio for a relative of a depressed proband to suffer from the illness, as compared with an unaffected control, has been estimated to be 2.84 (Sullivan et al. 2000), but in one study, it has been shown to be as high as 10 (Farmer et al. 2000). Whilst a pattern of heightened risk amongst family members is consistent with a genetic influence on disease liability, families not only share genes, but also many environmental factors (often termed as the familial or shared environment).

1.1 Quantitative Genetic Studies of Depression

Twin studies can be used to disentangle the relative contribution of genetics, familial (or shared) and non-shared environmental influences on variability in a trait such as depression liability, by comparing monozygotic (identical) and dizygotic (non-identical) twins. Monozygotic twins are genetically identical, and if brought up together, also share a familial environment. Dizygotic twins are assumed to share familial environments to the same extent as monozygotic twins; however, as they share only 50 % of their segregating genes, the differences in within-pair similarity between monozygotic and dizygotic twins can be used to estimate the heritability of a trait (the proportion of trait variance that is due to genetic variation between individuals). Once genetic factors have been considered, any remaining similarities between twins are due to shared familial environments, whilst within-pair differences between monozygotic twins can be attributed to non-shared environmental influences.

In a large meta-analysis of twin studies, Sullivan et al. (2000) estimated the heritability of depression to be 37 % (95 % CI 31–42 %). Shared environmental factors appeared to have very limited influence (with a point estimate of 0 %, 95 % CI 0–5 %), whilst 63 % of population variance in depression liability (95 % CI 58–67 %) could be attributed to non-shared environmental factors.

However, errors in trait measurement are incorporated into estimates of the influence of non-shared environments, and this may be an important factor when interpreting the results of twin studies into depression. Kendler et al. (1993) demonstrated that, when lifetime history of depression is assessed with a single interview, heritability estimates of the trait are around 40 %, but when two assessments were used to improve the reliability of diagnosis, heritability estimates rose to approximately 70 %. As many twin studies have relied upon a single psychiatric assessment in order to determine lifetime history of depression, this may mean that these studies underestimate the role of genetics in depression. Heritability estimates for depression also increase when more recurrent forms of the illness are considered (Kendler et al. 2007b; Sullivan et al. 2000). The pattern of depression in biological and adoptive relatives of adult adoptees supports the findings of twin studies, also indicating that genes play a key role in the aetiology of depression, but non-shared environmental influences are also important (Wender et al. 1986).

1.2 The Genetic Relationship Between Depression and Other Disorders

Quantitative genetic studies not only give an insight into the relative importance of genetics in the aetiology of depression, but can also indicate how the illness might relate to other disorders. For example, there is a very high degree of symptom overlap between depression and anxiety, as well as depression and bipolar disorder. There have been several multivariate quantitative genetic studies examining the relationship between these disorders, and high genetic overlap between anxiety and depression has been observed (Kendler et al. 2007a, 1992). Indeed, Kendler et al. (1992) propose that these two disorders are genetically very similar; the differences between anxiety and depression are driven by non-shared environmental factors. Similarly, work in bipolar affective disorder indicates significant genetic overlap with depression. The relationship between the two disorders is best explored using a correlated liability model, where patients can be classified as having depression, mania or both. Using this model, it has been shown that mania shares both genetic and non-shared environmental influences with depression. The genetic correlation between mania and depression was estimated at 0.65 (95 % CI, 0.58–0.75); however, there are also mania-specific genetic influences (McGuffin et al. 2003). These observed relationships between depression and other disorders indicate that the genetic causes of depression are less likely to be unique and suggest that there may be overlapping causal pathways involved in these disorders.

2 Identifying Which Genes are Associated with Depression

Whilst quantitative genetic methods are informative as to the relative importance of genes in depression, this does not tell us which variants are involved in the aetiology of depression. Depression is a complex disorder, where there are likely to be many genes of small effect which act to influence an individual's liability to the illness, as well as a number of non-genetic factors. This is indicated by two key factors. Firstly, whilst the disease is more common amongst relatives, there is no clear pattern of inheritance between generations. Furthermore, there are many cases of depression where no family history of the illness is seen. Thus, it is unlikely that a single gene drives the presence or absence of disease. Secondly, the symptoms of depression can occur along a spectrum of severity. Indeed, the boundary between "healthy" and "depressed", despite being operationalised in DSM-IV (American Psychiatric Association 1994) and ICD-10 (World Health Organisation 1992), is far from clear cut. Instead, depression appears to be a quantitative trait that varies in degree rather than kind. The quantitative nature of symptoms strongly suggests that the genetic influence on depression encompasses many genes, each of small effect. This is in contrast to some much rarer neuropsychiatric disorders with categorical phenotypes (for example Huntington's disease), which are driven by single genes with large genetic effects.

In order to identify the genetic variants involved in depression, molecular genetic methods are required. Two main approaches are used in gene finding, linkage and association.

2.1 *Molecular Genetic Methods: Linkage*

Linkage techniques aim to identify regions of the genome that contain disease-related genes, by examining the co-segregation of disease and genetic markers in families containing multiple affected members. If affected relatives share the same allele (alternative form) of a marker more frequently than expected by chance, then the genomic region surrounding this marker is likely to contain a genetic variant that is causally involved in the disorder.

Zubenko et al. (2003) published the first genome-wide linkage study of depression, reporting a linkage signal in the chromosomal region 2q33-34. However, the study was open to a number of methodological criticisms, including small sample size, and failure to correct for multiple hypothesis testing (Levinson 2006).

Several subsequent studies in depression have looked at genetic markers across the entire genome and four main genomic regions have been implicated at chromosomes 3p25-26, 12q22-24, 15q25-26 and 18q21-22, although findings have not been consistent across reports. In addition to variability in analytic approaches, studies have also examined a number of different phenotypes, such as recurrent and early onset MDD, as well as including genetically related disorders such as bipolar disorder and anxiety. Despite evidence of overlapping genetic aetiologies, these differences may contribute to inconsistencies between reports. For example, two studies selecting individuals from the same Utah sample using different phenotypic definitions reported divergent results. Abkevitch et al. (2003) examined individuals with MDD or bipolar depression and reported a signal within the 12q22-23.2 region, which became highly significant with fine mapping of the area. In contrast, Camp et al. (2005) found no evidence of linkage here when looking at those with MDD or anxiety (excluding those with bipolar depression). This later study did, however, find suggestive evidence of linkage at 3centr, 7p and 18q21.33-q22.2, the last of which was a region with suggestive support in the report from Zubenko et al. (2003). Sex-specific signals were also found in chromosomes 4q and 15q.

The signal on chromosome 15q received additional supporting evidence from the genetics of recurrent early onset depression (GenRED) study. In this independent large cohort, first wave findings reported linkage at 15q25-26 at a genome-wide significance level (Holmans et al. 2004).The second wave of the study including additional 359 families, found this signal had attenuated, reaching only suggestive levels of significance (Holmans et al. 2007). Nonetheless, Levinson et al. (2007) investigated the region in more detail, and found that with fine mapping, the signal became highly significant at a genome-wide level. The authors propose this area contains either one single locus that increases sibling MDD risk by around 20 %, or multiple loci of smaller effect.

McGuffin et al. (2005) also reported a suggestive signal in this 15q region in the depression network (DeNT) sample. In this first wave report, suggestive or modest significant linkage signals were also reported at 1p, 12q and 13q locations. The region in 12q was adjacent to that implicated by Abkevich et al. (2003). However, as seen in the GenRED study, in a final report from the DeNT sample (combining wave I and wave II data, to give a total of 971 affected sibling pairs), these regions showed no strong evidence of linkage (Breen et al. 2011b). Nonetheless, when the analysis was restricted to patients with more severe forms of recurrent MDD, a genome-wide significant association in the 3p25-26 region was observed. Although association mapping in the implicated region in an unrelated case–control did not identify a source for the signal, replication of the genome-wide signal in a closely overlapping location was reported by Pergadia et al. (2011) in an independent sample.

Middeldorp et al. (2009) did not detect any of these signals in their Australian and Dutch samples, instead finding suggestive linkage signals in chromosomes 17, 8p and a novel region of chromosome 2. The authors noted that the signal in chromosome 8p is in the same region as previously implicated in personality traits linked to depression (Neuroticism, Fullerton et al. 2003; Harm avoidance, Cloninger et al. 1998).

As can be seen, replication of linkage has been inconsistent. This may be due to a lack of statistical power to detect signals. It has been estimated that around 1,000 affected sibling pairs are required to detect a single locus which causes a 25–30 % MDD risk increase for siblings (Hauser et al. 1996; Levinson et al. 2003). Only two studies, GenRED and DeNT, have come close to this sample size and consequently a majority of studies may have been underpowered to detect true linkage signals.

Linkage is well-suited to address diseases resulting from a small number of genetic variants, each of large effect, but is less successful in probing complex diseases, as the presence or absence of a single variant is unlikely to drive the presence or absence of disease for an individual within a high-risk family. Nonetheless, in some families with more highly penetrant forms of the disorder, depression may be driven by rare mutations of large effect. The use of linkage methods to identify the genomic location of these variants involved in atypical and rare forms of depression would be invaluable in providing a window into the aetiological processes that can result in the symptoms of the disorder.

2.2 Molecular Genetic Methods: Candidate Gene Associations

Whilst linkage is better suited to detecting strong genetic effects, association techniques are more appropriate when looking for multiple genes each with a modest impact on a complex trait such as depression. Although within family methods can be used to test for association the simpler and more powerful approach involves examining genetic polymorphisms in both unrelated patients and disease-free controls and looking for differences in the frequency of these polymorphisms between the two groups. If a variant has a small influence on depression liability, then in order to obtain adequate statistical power to detect this small effect size, larger samples of cases and controls are required.

Genetic association studies can be undertaken either using a candidate gene approach or in a genome-wide hypothesis free manner. Candidate gene approaches involve selecting particular genes and polymorphisms based upon a prior hypothesis of association with the phenotype of interest. Key pathways that have been implicated in depression include monoaminergic (Hirschfeld 2000), neurotrophic (Hashimoto 2010; Castren et al. 2007) and stress response (Holsboer 2000) pathways. Candidate genes related to monoaminergic signalling include the serotonin transporter (*SLC6A4*), which is targeted by antidepressants and the serotonin 2A receptor (*HTR2A*). Neurotrophic genes include the neuroprotective protein brain-derived neurotrophic factor (*BDNF*) and its receptor Trk-B (*NTRK2*) whilst corticotrophin-releasing hormone receptor 1 (*CRHR1*) and FK506 binding protein 5 (*FKBP5*) have both been linked to regulation of the hypothalamic-pituitary-adrenal axis stress response pathway.

Whilst there is an extensive body of research looking at the association between candidate genes and depression, sample sizes are often limited and findings are frequently inconsistent. Meta analytic approaches are one method by which to tackle this issue, pulling together the available data from a number of studies. Lopez-Leon et al. (2008) performed meta-analyses for all polymorphisms examined in relation to MDD where three or more independent studies had been performed (listed in Table 1).

Table 1 Genes included in meta-analysis by Lopez-Leon et al. (2008)

Gene	Gene name	Variant	Significant association in meta-analysis
Angiotensin I-converting enzyme	ACE	Ins/Del-Intron 16	–
Apolipoprotein E	APOE	ε2/ε3/ε4	Yes
Brain-derived neurotrophic factor	BDNF	Val66Met	–
Catechol-O-methyltransferase	COMT	Val158Met	–
Dopamine D3 receptor	DRD3	Ser9Gly	–
Gamma-aminobutyric acid receptor, subunit alpha-3	GABRA3	CA repeat-Intron 8	–
Guanine nucleotide binding protein	GNB3	C825T	Yes
Serotonin 1A receptor	HTR1A	C-1019G	–
Serotonin 1B receptor	HTR1B	G861C	–
Serotonin 2A receptor	HTR2A	A-1438G	–
		T102C	–
Serotonin 2C receptor	HTR2C	Cys23Ser	–
Monoamine oxidase A	MAOA	VNTR-Promoter	–
Methylene tetrahydrofolate reductase	MTHFR	C677T	Yes
Norepinephrine transporter	SLC6A2	T-182C	–
Dopamine transporter	SLC6A3	40 bp VNTR-3′untranslated region	Yes
Serotonin transporter	SLC6A4	44 bp Ins/Del-Promoter	Yes
		VNTR-Intron 2	–
Tryptophan hydroxylase 1	TPH1	A218C	–

VNTR variable number tandem repeat

The results of this meta-analysis indicate that genetic effect sizes are often small; the estimated odds ratio for the most frequently investigated polymorphism (the *SLC6A4*-linked polymorphic repeat) was estimated to be 1.11 per "S" allele.

Of the 20 polymorphisms considered by three or more independent studies, five were shown to be significantly associated with depression; the *SLC6A4*-linked polymorphic repeat as well as *APOE* ε2, *GNB3* C825T, *MTHFR* C677T and *SLC6A3* VNTR.

Nonetheless, an attempt to replicate the *MTHFR* association in an independent large sample (which exceeded the number of cases and controls included in the meta-analyses) found no evidence of an association with depression (Gaysina et al. 2008), highlighting the need for further work before any associations can be considered accepted findings.

Furthermore, Lopez-Leon et al. found that 393 polymorphisms (in 102 genes) had been investigated in depression, but there were three or more replication attempts for only 6 % of these. Thus, the candidate gene literature can become focussed on a few "favourite" genes, such as the serotonin transporter. Many candidate gene analyses also make little attempt to capture all of the genetic variation that is present across the implicated gene, instead focussing on only one particular polymorphism (such as the serotonin-transporter-linked polymorphic region, 5HTTLPR). This means that the role of many other genetic variants in depression are yet to be fully examined with appropriate replication attempts.

Critically, candidate gene studies are reliant upon our existing understanding of the pathophysiology of depression to guide the selection of appropriate candidates. However, Shyn and Hamilton (2010) propose that several of the polymorphisms reaching significance in Lopez-Leon et al.'s meta analysis were those with a less obvious mechanistic connection to the neurobiology of depression as we currently understand it. This suggests that our current ability to select appropriate candidate genes is constrained by our lack of understanding of the pathological processes involved in depression.

2.3 Molecular Genetic Methods: Genome-Wide Association

In contrast, genome-wide association studies (GWAS) offer a hypothesis-free exploration of genetic variation across the entire genome, with no requirement to preselect genes of interest. If previously unsuspected genes are found to be associated with the disorder through these systematic analyses, genome-wide studies have the potential to push forward scientific theories of depression.

GWAS have been made possible by the development of microarray technology, where genotyping is performed at 500,000 to upwards of 1 million single nucleotide polymorphisms (SNPs), which result from single base changes in the DNA sequence. It is estimated that the SNP microarrays used in GWAS capture around 80 % of common genetic variation in European populations (Kruglyak 2008). Yet, performing over 500,000 tests of association confers a massive multiple-hypothesis testing burden. In order to protect against type I errors, it is necessary to impose a stringent significance threshold; conventionally $p < 5 \times 10^{-8}$ is considered a statistically significant result in a GWAS study (Dudbridge and Gusnanto 2008; Pe'er et al. 2008), whilst the threshold of $p < 5 \times 10^{-6}$ has been used to indicate a "suggestive" hit. These thresholds consider not only the number of SNPs tested, but also the local correlation (linkage disequilibrium) that is seen between SNPs. The stringent threshold of genome-wide significance, together with the small genetic effect sizes predicted in depression, mean that statistical power to detect association is a critical issue in GWAS, and so the collection of very large cohorts is imperative.

Additionally, whilst GWAS examining other phentoypes have often used unselected controls (as exemplified by the Wellcome Trust Case Control Consortium), the high lifetime prevalence of depression (estimated at around 15 %; Kessler et al. 2003) means that many individuals within an unscreened control group would have suffered from the disorder. This reduces the power of the analysis to detect a genetic association with the disease (Owen et al. 1997;

McCarthy et al. 2008) and so all genome-wide studies looking at depression, thus far, have used some form of screening for controls.

However, in any cohort there will be population stratification; systematic differences in allele frequencies between subpopulations arising from non-random mating (often due to physical separation). Population stratification must be accounted for in an association study to ensure that any genetic differences observed between cases and controls are due to a disease-associated gene locus, rather than simply a result of some underlying population structure. This is most frequently done with EIGENSTRAT, which uses principal component analysis to explicitly model and control for any ancestral differences between cases and controls in the sample (Price et al. 2006).

The first genome-wide study into depression was part of the genetic association information network (GAIN) initiative, looking at patients with a lifetime history of MDD. Sullivan et al. (2009) found a cluster of 11 SNPs within the *PCLO* gene were amongst the top 200 performing SNPs. *PCLO* encodes piccolo, a presynaptic protein that is involved in monoaminergic neurotransmission, although this protein had not previously been implicated in depression. However, the authors found no strong *PCLO* signal in an independent replication.

The genome-wide association analyses undertaken as part of the genetics of recurrent early onset depression (GenRED) project (Shi et al. 2011) implicated a location on chromosome 18q22.1 ($p = 1.83 \times 10^{-7}$) which was also implicated in linkage studies of depression (Zubenko et al. 2003; Camp et al. 2005). In a companion paper, Shyn et al. (2011) reported an initial GWAS on MDD patients recruited as part of the sequenced treatment alternatives to relieve depression (STAR*D) study, where no strong signals were found. They then combined the STAR*D, GenRED and GAIN MDD samples to perform a large meta-analysis on 3,956 cases and 3,428 controls. The authors drew attention to suggestive SNP associations within three genes; *ATP6V1B2*, *SP4* and *GRM7*. *ATP6V1B2* encodes a subunit of a vacuolar proton pump ATPase, whilst *SP4* encodes the brain-specific Sp4 zinc finger transcription factor, which has been previously associated with bipolar depression (Zhou et al. 2009). *GRM7* (encoding a metabotropic glutamate receptor subunit) has been proposed as a candidate in the pathophysiology of depression. In addition, Shyn et al. further examined a "narrow" phenotype of recurrent early onset depression. Their most significant finding in this "narrow" phenotypic analysis was the 18q22.1 region implicated in the GenRED analysis.

Muglia et al. (2010) also combined data from multiple studies, examining two European cohorts with recurrent MDD, both individually and in a meta-analysis. No genome-wide signals of interest were reported, although *GRM7* was the best performing gene from a list of candidates in their combined analysis. Whilst this association does not survive correction for multiple-hypothesis testing, it is consistent with the findings of Shyn et al.

Recurrent MDD was also the phenotypic definition employed by Lewis et al. (2010) in their UK based GWAS. Suggestive association was observed within the *BICC1* gene (an RNA binding protein), a novel gene in depression. However, in a meta analysis combining the UK sample with the two studies examined by Muglia et al. (2010), no support for the *BICC1* association was observed. There was some evidence in the meta analysis for association with *NLGN1*; this gene (which encodes neuroligin-1) was the third strongest association in the meta analysis with consistent patterns of association across all three studies when examined separately.

Wray et al. (2010) examined the MDD2000 + sample, which comprises of 2,431 MDD patients and 3,673 controls, taken from a number of different studies (thus requiring more stringent quality control criteria, which, to some extent, limited the power of the study). The authors also performed a meta analysis combining the data from the three largest samples; MDD2000 + , GAIN-MDD and the UK sample from Lewis et al. (2010), giving a total of 5,763 cases and 6,901 controls. No strong associations were found, although the authors highlight that *ADCY3* (encoding the G-protein signalling cascade enzyme adenylate cyclase 3) and *GAL* (which encodes the neuropeptide galanin) both perform well in gene-based analyses and have also been implicated in depression previously (Weiss et al. 1998; Hines and Tabakoff 2005, respectively).

The paucity of definitive associations emerging from GWAS studies and meta-analyses in depression may be because, despite efforts to combine data from a number of sources, studies are still underpowered to detect small genetic main effects. The most recent attempt to tackle this issue was undertaken by the Psychiatric GWAS Consortium (PGC). The consortium was able to perform a mega analysis (which, in contrast to a meta analysis, combines individual genotypic and phenotypic data rather than summary results) of nine GWAS studies in the discovery phase, assessing a total of 18,759 subjects; 9,240 cases and 9,519 controls (major depressive disorder working group of the psychiatric GWAS Consortium 2012). They then went on to evaluate the top hits in a replication phase drawing from seven independent samples (including 6,783 cases), as well as considering secondary analyses looking at various phenotypic subgroups. However no results reached genome-wide significance. The authors highlight that, given the high prevalence of depression, this sample may still be underpowered to detect genetic main effects, despite being the largest GWAS to date. Nonetheless, they also draw attention to the fact that genome-wide significant hits have been found in the vast majority of other phenotypes where sample sizes of comparable size have been collected (Hindorff et al. 2009).

Potentially, "polygenic heterogeneity" may contribute to the difficulty in finding genetic associations, where multiple different combinations of small genes of small effect, plus or minus genes of moderate effect (of the type indicated by positive linkage signals), may result in the same phenotype (Rucker and McGuffin 2010).

In an attempt to gain insight into the genetic relationships spanning psychiatric diagnostic boundaries, the PGC also conducted a MDD-bipolar cross-disorder GWAS analysis. A cluster of genome-wide significant associations was observed in a region of high linkage disequilibrium, in the 3p21.1 region. This is consistent with an earlier cross-disorder meta analysis by McMahon et al. (2010); however, Breen et al. (2011a) contended that the association was specific to bipolar disorder.

The PGC suggest similar caution in interpreting this more recent result, as the most significant SNP did not replicate in pure MDD samples.

Despite the disappointing lack of robust associations in MDD, the evidence from each of the GWAS to date has implicated a number of suggestive genes of interest and further work is needed to follow these up. However, it is notable that very few of the genes identified as likely depression candidates are amongst those reaching suggestive levels of significance in GWAS. This raised doubt regarding the role of currently popular candidates and underscores the value of genome-wide approaches in identifying novel genetic variants, and expanding our understanding of the processes involved in the disease. Furthermore, there are a number of developing technologies that will enable genetic research to progress further.

3 Developing Technologies

3.1 Copy Number Variants

Whilst the microarray chips used in GWAS assay common single nucleotide polymorphisms, this is not the only form of variation found in the genome; structural variants are also prevalent. For example, copy number variants (CNVs) are deletions or duplications of DNA greater than 1 kb in size and it has been estimated that copy number variable regions cover approximately 12 % of the genome. This means that CNVs account for more nucleotide variation within the genome than SNPs (Redon et al. 2006). Whilst some CNVs are common, inherited forms of variation, others occur de novo. SNP microarray data can be used to assess CNVs, by examining relative hybridisation intensities, although other forms of structural variation, such as inversions and translocations, remain more difficult to detect.

Recent evidence has associated CNVs with increased risk for a number of psychiatric phenotypes, including schizophrenia (Walsh et al. 2008; Xu et al. 2008), ADHD (Williams et al. 2010) and autism (Sebat et al. 2007). Two papers employing genome-wide analysis of CNVs in relation to depression have been published to date.

Glessner et al. (2010) looked for specific CNVs that were significantly associated with depression. A rare duplication in chromosome 5q35.1 was significantly associated with depression, in a total sample of 1,693 cases and 4,506 controls. The CNV was exclusive to MDD, occurring in five unrelated cases, and encompassed the *SLIT3* gene, which is involved in axonal guidance. As with SNP associations, this finding requires independent replication.

Rucker et al. (2011) took a different approach in their analysis of CNVs in depression, examining the pattern of CNVs across the genome. Whilst there was no significant relationship between depression and duplications, deletions were more common in recurrent MDD. Interestingly, a pattern of increasing burden of deletions was seen, with the lowest rates of occurrence in controls that were screened for the absence of depression, higher rates in unscreened controls and the highest numbers of CNV deletions seen in recurrent MDD cases. This trend was driven by deletions within protein-coding regions of the genome, and hence the authors suggest that the absence of exonic CNV deletions may be associated with resilience to depression. This finding is particularly interesting given that, in the general population, deletions tend not to be within coding regions (Conrad et al. 2006).

3.2 Next Generation Sequencing

Another new and rapidly developing technology for detecting other forms of genetic variation is so-called next-generation sequencing. Whilst the microarrays used in GWAS capture around 80 % of common SNP variation (where common variants are defined as those that have a frequency of over 5 % of the population), rare forms of variation, not captured by microarrays, are also likely to play a role in many disorders (Pritchard 2001). Instead, DNA sequencing can be used to detect these rare variants. Given the high cost of sequencing, initial targeted analyses focussing on previously implicated genomic regions and protein-coding sections of the genome seem sensible first steps towards comprehensive genome-wide sequencing analyses. This approach has been adopted with success in other complex phenotypes; for example, looking at blood pressure variation, Ji et al. (2008) sequenced exonic and flanking intronic regions in three candidate genes and found a number of rare variants which were associated with blood pressure. Nonetheless, further progress is needed to improve both cost-effectiveness and reliability of the technology, as well as to develop bioinformatic tools in order to analyse larger proportions of the genome.

Hence, it appears that the role of genetics in depression is complex, with a number of different forms of variation likely to each have a small effect in disease liability. This fits with evolutionary theory; any single genetic variant which confers a large increase in risk for the disorder would be rapidly removed from the population though negative selection, given the reproductive disadvantage associated with depression.

4 Interactions

The gene-hunting strategies outlined so far have all assumed genetic main effects on disease risk, where a detectable variant will act to increase or decrease vulnerability to depression across a population. Given that each implicated gene will act within complex biological systems, it seems likely that there are interactions between genes (epistasis), where the effect of one variant is dependent upon, or modified by another. For example, it has been proposed that there may be a biological interaction between *BDNF* and *SLC6A4*, two key candidates in depression (Martinowich and Lu 2008).

Additionally, as quantitative genetic methods have demonstrated, environmental factors also make an important contribution to an individual's risk of depression. Consequently, it has also been hypothesised that the effect of genetic risk variants may be conditional on environmental exposure (or conversely, the influence of environmental factors is modified by an individual's genotype).>

In a seminal paper, Caspi et al. (2003) tested this theory, examining whether the prominent candidate genetic variant in depression, 5HTTLPR, moderated the well-documented relationship between the experience of stressful life events and depression. The authors found that when 5HTTLPR genotype was considered in combination with an individual's exposure to stressful life events, a significant interaction was observed; stressful life events predicted depression amongst individuals with the "s" allele of the 5HTTLPR, but not for "l/l" homozygotes. Whilst there has been a degree of controversy surrounding the finding, Uher and McGuffin (2008, 2010) performed a comprehensive review and concluded that with accurate and objective measurement of stressful life events this interaction was consistently replicated in the literature.

Caspi et al. (2003) observed no main effect of 5HTTLPR in their sample, and gene-environment interactions in the absence of genetic main effects make evolutionary sense; if a genetic variant confers increased risk in the presence of certain environments, but its average effect across the whole population is neutral, it will be maintained in the population (Uher 2009). This means that the effects of genotype could be masked by variation in exposure to environmental risks within a sample, potentially explaining why efforts to identify genetic main effects in depression have had limited success to date. Moreover, inconsistent findings may arise if there is variation in environmental exposure between samples.

Further research is needed to clarify the precise nature of the 5HTTLPR × stressful life events interaction, and identify other important gene-environment interactions in depression. Researchers have considered key depression candidate genes such as *CRHR1* in relation to stressful life events (Bradley et al. 2008; Polanczyk et al. 2009), although to date findings have been mixed. The selection of appropriate genetic candidates should be guided by evidence linking the variant to variability in response to the environmental factor: for example, 5HTTLPR has been associated with stress response in animal models (e.g. Champoux et al. 2002) as well as in a human imaging study (Hariri et al. 2002), and CRHR1 has also been implicated in stress reactivity (De Kloet 2004).

If gene-environment interactions play an important role in depression aetiology, this indicates that environmental exposure should be considered in the design of genetic studies into the disorder, although the accurate measurement of these factors poses a practical challenge for large cohorts. Indeed, research into gene-environment interplay requires meticulous methodological care in both environmental and genotypic measurement and analysis. Nonetheless, given the reduction in statistical power for tests of interaction, sample size remains a critical consideration. One novel strategy that has been proposed is to sample individuals known to have been exposed to a key environmental risk factor. If all individuals have been exposed to the risk factor of interest, but only some have developed depression, it is possible to compare allele frequencies in cases and controls and identify genes which are associated with increased vulnerability to the environmental risk factor. In depression, one identified environmental risk factor that seems to lend itself well to this approach is childbirth. The incidence of depression in the first month after giving birth has been estimated to be three times higher than in controls (Cox et al. 1993), and it is hoped that if a sufficiently sized sample of women who have recently given birth can be collected, genetic variants which increase an individual's vulnerability to postpartum depression can be identified. Nevertheless, whether postpartum depression is genetically similar to other forms of depression is not well established.

Understanding gene-environment interactions also offers valuable guidance in our attempts to determine the neurobiology of depression. Whilst the nature of the relationship between the implicated genetic variant and the environmental risk factor may be unclear, an interaction indicates that there is some aetiological pathway to depression which is connected to the two factors. This triad of factors offers a useful tool by which to probe the processes resulting in depression.

5 Conclusions

Quantitative genetic methods have established that genetics plays an important role in determining an individual's liability to depression, but, critically, genes do not determine absolutely whether an individual suffers with this complex disease; environmental factors also play an important role.

In attempting to identify which genetic variants may contribute to depression risk, a range of molecular genetic methodologies have been employed. Linkage studies attempt to identify regions of the genome which segregate with disease in affected families and have implicated regions on chromosomes 3p, 12q, 15q and 18q; however, the interpretation of linkage findings is more difficult in complex, rather than single-gene, genetic disorders.

In terms of association studies, using current theories of the neurobiology of depression as a starting point, candidate genes within monoaminergic and stress-response systems have been investigated. Furthermore, technological developments have enabled hypothesis-free, systematic analyses and interrogation of common variants across the entire genome. However, clear associations between genotype and depression are yet to emerge. Whilst newer approaches such as CNV analysis and sequencing promise a more thorough consideration of all types of genetic variability, it appears increasingly likely that if genetic variants do exert a main effect on depression liability, the size of these effects is small. This fits with evolutionary theories; if a single variant confers a significant increase in depression risk, given the reproductive disadvantage associated with the illness, it would be under negative selection and removed from the population.

Instead, given the high prevalence of depression in the population and the observed role of environmental factors, research is increasingly focussing on how a number of genetic variants may act in combination with each other and environmental factors in order to alter the vulnerability of an individual to depression. These findings can then be fed into other methodological approaches to understand the aetiology of depression, and used to drive forward drug development strategies.

References

Abkevich V, Camp NJ, Hensel CH, Neff CD, Russell DL, Hughes DC, Plenk AM, Lowry MR, Richards RL, Carter C et al (2003) Predisposition locus for major depression at chromosome 12q22-12q23.2. Am J Hum Genet 73(6):1271–1281. doi:10.1086/379978

American Psychiatric Association (1994) Diagnostic and Statistical Manual of Mental Disorders, 4th edn. American Psychiatric Association, Washington DC

Bradley RG, Binder EB, Epstein MP, Tang Y, Nair HP, Liu W, Gillespie CF, Berg T, Evces M, Newport DJ et al (2008) Influence of child abuse on adult depression: moderation by the corticotropin-releasing hormone receptor gene. Arch Gen Psychiatry 65(2):190–200. doi:10.1001/archgenpsychiatry.2007.26

Breen G, Lewis CM, Vassos E et al (2011a) Replication of association of 3p21.1 with susceptibility to bipolar disorder but not major depression. Nat Genet 43 (1):3–5. doi:10.1038/ng0111-3

Breen G, Webb BT, Butler AW, van den Oord EJ, Tozzi F, Craddock N, Gill M, Korszun A, Maier W, Middleton L et al (2011b) A genome-wide significant linkage for severe depression on chromosome 3: the depression network study. Am J Psychiatry 168(8):840–847. doi:10.1176/appi.ajp.2011.10091342

Camp NJ, Lowry MR, Richards RL, Plenk AM, Carter C, Hensel CH, Abkevich V, Skolnick MH, Shattuck D, Rowe KG, Hughes DC, Cannon-Albright LA (2005) Genome-wide linkage analyses of extended Utah pedigrees identifies loci that influence recurrent, early-onset major depression and anxiety disorders. Am J Med Genet B Neuropsychiatr Genet 135B(1):85–93. doi:10.1002/ajmg.b.30177

Caspi A, Sugden K, Moffitt TE, Taylor A, Craig IW, Harrington H, McClay J, Mill J, Martin J, Braithwaite A, Poulton R (2003) Influence of life stress on depression: moderation by a polymorphism in the 5-HTT gene. Science 301(5631):386–389. doi:10.1126/science.1083968

Castren E, Voikar V, Rantamaki T (2007) Role of neurotrophic factors in depression. Curr Opin Pharmacol 7(1):18–21. doi:10.1016/j.coph.2006.08.009

Champoux M, Bennett A, Shannon C, Higley JD, Lesch KP, Suomi SJ (2002) Serotonin transporter gene polymorphism, differential early rearing, and behavior in rhesus monkey neonates. Mol Psychiatry 7(10):1058–1063. doi:10.1038/sj.mp.4001157

Cloninger CR, Van Eerdewegh P, Goate A, Edenberg HJ, Blangero J, Hesselbrock V, Reich T, Nurnberger J Jr, Schuckit M, Porjesz B et al (1998) Anxiety proneness linked to epistatic loci in genome scan of human personality traits. Am J Med Genet 81(4):313–317. doi:10.1002/(SICI)1096-8628(19980710)81:4<313:AID-AJMG7>3.0.CO;2-U

Conrad DF, Andrews TD, Carter NP, Hurles ME, Pritchard JK (2006) A high-resolution survey of deletion polymorphism in the human genome. Nat Genet 38(1):75–81. doi:10.1038/ng1697

Cox JL, Murray D, Chapman G (1993) A controlled study of the onset, duration and prevalence of postnatal depression. Br J Psychiatry 163:27–31. doi:10.1192/bjp.163.1.27

De Kloet ER (2004) Hormones and the stressed brain. Ann N Y Acad Sci 1018:1–15. doi:10.1196/annals.1296.001

Dudbridge F, Gusnanto A (2008) Estimation of significance thresholds for genomewide association scans. Genet Epidemiol 32(3):227–234. doi:10.1002/gepi.20297

Farmer A, Harris T, Redman K, Sadler S, Mahmood A, McGuffin P (2000) Cardiff depression study: a sib-pair study of life events and familiality in major depression. Br J Psychiatry 176:150–155. doi:10.1192/bjp.176.2.150

Fullerton J, Cubin M, Tiwari H, Wang C, Bomhra A, Davidson S, Miller S, Fairburn C, Goodwin G, Neale MC, Fiddy S, Mott R, Allison DB, Flint J (2003) Linkage analysis of extremely discordant and concordant sibling pairs identifies quantitative-trait loci that influence variation in the human personality trait neuroticism. Am J Hum Genet 72(4):879–890. doi:10.1086/374178

Gaysina D, Cohen S, Craddock N, Farmer A, Hoda F, Korszun A, Owen MJ, Craig IW, McGuffin P (2008) No association with the 5,10-methylenetetrahydrofolate reductase gene and major depressive disorder: results of the depression case control (DeCC) study and a meta-analysis. Am J Med Genet B Neuropsychiatr Genet 147B(6):699–706. doi:10.1002/ajmg.b.30665

Glessner JT, Wang K, Sleiman PM, Zhang H, Kim CE, Flory JH, Bradfield JP, Imielinski M, Frackelton EC, Qiu H, Mentch F, Grant SF, Hakonarson H (2010) Duplication of the SLIT3 locus on 5q35.1 predisposes to major depressive disorder. PLoS One 5(12):e15463. doi:10.1371/journal.pone.0015463

Hariri AR, Mattay VS, Tessitore A, Kolachana B, Fera F, Goldman D, Egan MF, Weinberger DR (2002) Serotonin transporter genetic variation and the response of the human amygdala. Science 297(5580):400–403. doi:10.1126/science.1071829

Hashimoto K (2010) Brain-derived neurotrophic factor as a biomarker for mood disorders: an historical overview and future directions. Psychiatry Clin Neurosci 64(4):341–357. doi:10.1111/j.1440-1819.2010.02113.x

Hauser ER, Boehnke M, Guo SW, Risch N (1996) Affected-sib-pair interval mapping and exclusion for complex genetic traits: sampling considerations. Genet Epidemiol 13(2): 117–137. doi:10.1002/(SICI)1098-2272(1996)13:2<117:AID-GEPI1>3.0.CO;2-5

Hindorff LA, Sethupathy P, Junkins HA, Ramos EM, Mehta JP, Collins FS, Manolio TA (2009) Potential etiologic and functional implications of genome-wide association loci for human diseases and traits. Proc Nat Acad Sci 106(23):9362–9367. doi:10.1073/pnas.0903103106

Hines LM, Tabakoff B (2005) Platelet adenylyl cyclase activity: a biological marker for major depression and recent drug use. Biol Psychiatry 58(12):955–962. doi:10.1016/j.biopsych.2005.05.040

Hirschfeld RM (2000) History and evolution of the monoamine hypothesis of depression. J Clin Psychiatry 61(Suppl 6):4–6

Holmans P, Zubenko GS, Crowe RR, DePaulo JR Jr, Scheftner WA, Weissman MM, Zubenko WN, Boutelle S, Murphy-Eberenz K, MacKinnon D et al (2004) Genomewide significant linkage to recurrent, early-onset major depressive disorder on chromosome 15q. Am J Hum Genet 74(6):1154–1167. doi:10.1086/421333

Holmans P, Weissman MM, Zubenko GS, Scheftner WA, Crowe RR, Depaulo JR Jr, Knowles JA, Zubenko WN, Murphy-Eberenz K, Marta DH et al (2007) Genetics of recurrent early-onset major depression (GenRED): final genome scan report. Am J Psychiatry 164(2):248–258. doi:10.1176/appi.ajp.164.2.248

Holsboer F (2000) The corticosteroid receptor hypothesis of depression. Neuropsychopharmacology 23(5):477–501. doi:10.1016/S0893-133X(00)00159-7

Ji W, Foo JN, O'Roak BJ, Zhao H, Larson MG, Simon DB, Newton-Cheh C, State MW, Levy D, Lifton RP (2008) Rare independent mutations in renal salt handling genes contribute to blood pressure variation. Nat Genet 40(5):592–599. doi:10.1038/ng.118

Kendler KS, Neale MC, Kessler RC, Heath AC, Eaves LJ (1992) Major depression and generalized anxiety disorder. Same genes, (partly) different environments? Arch Gen Psychiatry 49(9):716–722

Kendler KS, Neale MC, Kessler RC, Heath AC, Eaves LJ (1993) The lifetime history of major depression in women: reliability of diagnosis and heritability. Arch Gen Psychiatry 50(11):863–870. doi:10.1001/archpsyc.1993.01820230054003

Kendler KS, Gardner CO, Gatz M, Pedersen NL (2007a) The sources of co-morbidity between major depression and generalized anxiety disorder in a Swedish national twin sample. Psychol Med 37(3):453–462. doi:10.1017/S0033291706009135

Kendler KS, Gatz M, Gardner CO, Pedersen NL (2007b) Clinical indices of familial depression in the Swedish Twin Registry. Acta Psychiatr Scand 115(3):214–220. doi:10.1111/j.1600-0447.2006.00863.x

Kessler RC, Berglund P, Demler O, Jin R, Koretz D, Merikangas KR, Rush AJ, Walters EE, Wang PS (2003) The epidemiology of major depressive disorder: results from the National Comorbidity Survey Replication (NCS-R). JAMA 289(23):3095–3105. doi:10.1001/jama.289.23.3095

Kruglyak L (2008) The road to genome-wide association studies. Nat Rev Genet 9(4):314–318. doi:10.1038/nrg2316

Levinson DF (2006) The genetics of depression: a review. Biol Psychiatry 60(2):84–92. doi:10.1016/j.biopsych.2005.08.024

Levinson DF, Zubenko GS, Crowe RR, DePaulo RJ, Scheftner WS, Weissman MM, Holmans P, Zubenko WN, Boutelle S, Murphy-Eberenz K et al (2003) Genetics of recurrent early-onset depression (GenRED): design and preliminary clinical characteristics of a repository sample for genetic linkage studies. Am J Med Genet B Neuropsychiatr Genet 119B(1):118–130. doi:10.1002/ajmg.b.20009

Levinson DF, Evgrafov OV, Knowles JA, Potash JB, Weissman MM, Scheftner WA, Depaulo JR Jr, Crowe RR, Murphy-Eberenz K, Marta DH et al (2007) Genetics of recurrent early-onset major depression (GenRED): significant linkage on chromosome 15q25-q26 after fine mapping with single nucleotide polymorphism markers. Am J Psychiatry 164(2):259–264. doi:10.1176/appi.ajp.164.2.259

Lewis CM, Ng MY, Butler AW, Cohen-Woods S, Uher R, Pirlo K, Weale ME, Schosser A, Paredes UM, Rivera M et al. (2010) Genome-wide association study of major recurrent depression in the U.K. population. Am J Psychiatry 167 (8):949–957. doi:10.1176/appi.ajp.2010.09091380

Lopez-Leon S, Janssens AC, Gonzalez-Zuloeta Ladd AM, Del-Favero J, Claes SJ, Oostra BA, van Duijn CM (2008) Meta-analyses of genetic studies on major depressive disorder. Mol Psychiatry 13(8):772–785. doi:10.1038/sj.mp.4002088

Major Depressive Disorder Working Group of the Psychiatric GWAS Consortium (2012) A mega-analysis of genome-wide association studies for major depressive disorder. Mol Psychiatry. doi: 10.1038/mp.2012.21

Martinowich K, Lu B (2008) Interaction between BDNF and serotonin: role in mood disorders. Neuropsychopharmacology 33(1):73–83. doi:10.1038/sj.npp.1301571

McCarthy MI, Abecasis GR, Cardon LR, Goldstein DB, Little J, Ioannidis JP, Hirschhorn JN (2008) Genome-wide association studies for complex traits: consensus, uncertainty and challenges. Nat Rev Genet 9(5):356–369. doi:10.1038/nrg2344

McGuffin P, Rijsdijk F, Andrew M, Sham P, Katz R, Cardno A (2003) The heritability of bipolar affective disorder and the genetic relationship to unipolar depression. Arch Gen Psychiatry 60(5):497–502. doi:10.1001/archpsyc.60.5.497

McGuffin P, Knight J, Breen G, Brewster S, Boyd PR, Craddock N, Gill M, Korszun A, Maier W, Middleton L et al (2005) Whole genome linkage scan of recurrent depressive disorder from the depression network study. Hum Mol Genet 14(22):3337–3345. doi:10.1093/hmg/ddi363

McMahon FJ, Akula N, Schulze TG, Muglia P, Tozzi F, Detera-Wadleigh SD, Steele CJ, Breuer R, Strohmaier J, Wendland JR et al (2010) Meta-analysis of genome-wide association data identifies a risk locus for major mood disorders on 3p21.1. Nat Genet 42(2):128–131. doi:10.1038/ng.523

Middeldorp CM, Sullivan PF, Wray NR, Hottenga JJ, de Geus EJ, van den Berg M, Montgomery GW, Coventry WL, Statham DJ, Andrews G, Slagboom PE, Boomsma DI, Martin NG (2009) Suggestive linkage on chromosome 2, 8, and 17 for lifetime major depression. Am J Med Genet B Neuropsychiatr Genet 150B(3):352–358. doi:10.1002/ajmg.b.30817

Muglia P, Tozzi F, Galwey NW, Francks C, Upmanyu R, Kong XQ, Antoniades A, Domenici E, Perry J, Rothen S et al (2010) Genome-wide association study of recurrent major depressive disorder in two European case-control cohorts. Mol Psychiatry 15(6):589–601. doi:10.1038/mp.2008.131

Owen MJ, Holmans P, McGuffin P (1997) Association studies in psychiatric genetics. Mol Psychiatry 2(4):270–273. doi:10.1038/sj.mp.4000292

Pe'er I, Yelensky R, Altshuler D, Daly MJ (2008) Estimation of the multiple testing burden for genomewide association studies of nearly all common variants. Genet Epidemiol 32(4):381–385. doi:10.1002/gepi.20303

Pergadia ML, Glowinski AL, Wray NR, Agrawal A, Saccone SF, Loukola A, Broms U, Korhonen T, Penninx BW, Grant JD et al (2011) A 3p26-3p25 genetic linkage finding for DSM-IV major depression in heavy smoking families. Am J Psychiatry 168(8):848–852. doi:10.1176/appi.ajp.2011.10091319

Polanczyk G, Caspi A, Williams B, Price TS, Danese A, Sugden K, Uher R, Poulton R, Moffitt TE (2009) Protective effect of CRHR1 gene variants on the development of adult depression following childhood maltreatment: replication and extension. Arch Gen Psychiatry 66(9):978–985. doi:10.1001/archgenpsychiatry.2009.114

Price AL, Patterson NJ, Plenge RM, Weinblatt ME, Shadick NA, Reich D (2006) Principal components analysis corrects for stratification in genome-wide association studies. Nat Genet 38(8):904–909. doi:http://www.nature.com/ng/journal/v38/n8/suppinfo/ng1847_S1.html

Pritchard JK (2001) Are rare variants responsible for susceptibility to complex diseases? Am J Hum Genet 69(1):124–137. doi:10.1086/321272

Redon R, Ishikawa S, Fitch KR, Feuk L, Perry GH, Andrews TD, Fiegler H, Shapero MH, Carson AR, Chen W et al (2006) Global variation in copy number in the human genome. Nature 444(7118):444–454. doi:10.1038/nature05329

Rucker JJ, McGuffin P (2010) Polygenic heterogeneity: a complex model of genetic inheritance in psychiatric disorders. Biol Psychiatry 68(4):312–313

Rucker JJ, Breen G, Pinto D, Pedroso I, Lewis CM, Cohen-WoodsS, Uher R, Schosser A, Rivera M, Aitchison KJ et al (2011) Genome-wide association analysis of copy number variation in recurrent depressive disorder. Mol Psychiatry. doi: 10.1038/mp.2011.144

Sebat J, Lakshmi B, Malhotra D, Troge J, Lese-Martin C, Walsh T, Yamrom B, Yoon S, Krasnitz A, Kendall J et al (2007) Strong association of de novo copy number mutations with autism. Science 316(5823):445–449. doi:10.1126/science.1138659

Shi J, Potash JB, Knowles JA, Weissman MM, Coryell W, Scheftner WA, Lawson WB, DePaulo JR Jr, Gejman PV, Sanders AR et al (2011) Genome-wide association study of recurrent early-onset major depressive disorder. Mol Psychiatry 16(2):193–201. doi:10.1038/mp.2009.124

Shyn SI, Hamilton SP (2010) The genetics of major depression: moving beyond the monoamine hypothesis. Psychiatr Clin North Am 33(1):125–140. doi:10.1016/j.psc.2009.10.004

Shyn SI, Shi J, Kraft JB, Potash JB, Knowles JA, Weissman MM, Garriock HA, Yokoyama JS, McGrath PJ, Peters EJ et al (2011) Novel loci for major depression identified by genome-wide association study of sequenced treatment alternatives to relieve depression and meta-analysis of three studies. Mol Psychiatry 16(2):202–215. doi:10.1038/mp.2009.125

Sullivan PF, Neale MC, Kendler KS (2000) Genetic epidemiology of major depression: review and meta-analysis. Am J Psychiatry 157(10):1552–1562. doi:10.1176/appi.ajp.157.10.1552

Sullivan PF, de Geus EJ, Willemsen G, James MR, Smit JH, Zandbelt T, Arolt V, Baune BT, Blackwood D, Cichon S et al (2009) Genome-wide association for major depressive disorder: a possible role for the presynaptic protein piccolo. Mol Psychiatry 14(4):359–375. doi:10.1038/mp.2008.125

Uher R (2009) The role of genetic variation in the causation of mental illness: an evolution-informed framework. Mol Psychiatry 14(12):1072–1082. doi:10.1038/mp.2009.85

Uher R, McGuffin P (2008) The moderation by the serotonin transporter gene of environmental adversity in the aetiology of mental illness: review and methodological analysis. Mol Psychiatry 13(2):131–146. doi:10.1038/sj.mp.4002067

Uher R, McGuffin P (2010) The moderation by the serotonin transporter gene of environmental adversity in the etiology of depression: 2009 update. Mol Psychiatry 15(1):18–22. doi:10.1038/mp.2009.123

Walsh T, McClellan JM, McCarthy SE, Addington AM, Pierce SB, Cooper GM, Nord AS, Kusenda M, Malhotra D, Bhandari A et al (2008) Rare structural variants disrupt multiple genes in neurodevelopmental pathways in schizophrenia. Science 320(5875):539–543. doi:10.1126/science.1155174

Weiss JM, Bonsall RW, Demetrikopoulos MK, Emery MS, West CH (1998) Galanin: a significant role in depression? Ann N Y Acad Sci 863:364–382. doi:10.1111/j.1749-6632.1998.tb10707.x

Wender PH, Kety SS, Rosenthal D, Schulsinger F, Ortmann J, Lunde I (1986) Psychiatric disorders in the biological and adoptive families of adopted individuals with affective disorders. Arch Gen Psychiatry 43(10):923–929

Williams NM, Zaharieva I, Martin A, Langley K, Mantripragada K, Fossdal R, Stefansson H, Stefansson K, Magnusson P, Gudmundsson OO et al (2010) Rare chromosomal deletions and duplications in attention-deficit hyperactivity disorder: a genome-wide analysis. Lancet 376(9750):1401–1408. doi:10.1016/S0140-6736(10)61109-9

World Health Organisation (1992) ICD-10 Classifications of mental and behavioural disorder: clinical descriptions and diagnostic guidelines. World Health Organisation, Geneva

Wray NR, Pergadia ML, Blackwood DH, Penninx BW, Gordon SD, Nyholt DR, Ripke S, Macintyre DJ, McGhee KA, Maclean AW et al (2010) Genome-wide association study of major depressive disorder: new results, meta-analysis, and lessons learned. Mol Psychiatry. doi:10.1038/mp.2010.109

Xu B, Roos JL, Levy S, van Rensburg EJ, Gogos JA, Karayiorgou M (2008) Strong association of de novo copy number mutations with sporadic schizophrenia. Nat Genet 40(7):880–885. doi:10.1038/ng.162

Zhou X, Tang W, Greenwood TA, Guo S, He L, Geyer MA, Kelsoe JR (2009) Transcription factor SP4 is a susceptibility gene for bipolar disorder. PLoS One 4(4):e5196. doi:10.1371/journal.pone.0005196

Zubenko GS, Maher B, Hughes HB 3rd, Zubenko WN, Stiffler JS, Kaplan BB, Marazita ML (2003) Genome-wide linkage survey for genetic loci that influence the development of depressive disorders in families with recurrent, early-onset, major depression. Am J Med Genet B Neuropsychiatr Genet 123B(1):1–18. doi:10.1002/ajmg.b.20073

Neurochemical Imaging and Depressive Behaviours

Jeffrey H. Meyer

Abstract Neurochemical imaging is frequently applied to measure markers of pathological change so as to understand mechanisms that create symptoms of major depressive disorder. For example, indices of greater monoamine oxidase A(MAO-A) level, particularly in the prefrontal and anterior cingulate cortex, are associated with depressed mood states, and high-risk states for onset of major depressive episodes. MAO-A metabolises monoamines, and greater metabolism of monoamines occurs when MAO-A is elevated in brain. Lower extracellular serotonin is associated with greater pessimism in humans and chronic serotonin deficiency is associated with upregulation of 5-HT_{2A} (serotonin$_{2A}$) receptors in cortex. During major depressive episodes when pessimism is more severe, greater 5-HT_{2A} BP_{ND}, an index of density occurs in prefrontal and anterior cingulate cortex. These results argue for a mechanism of lowering extracellular serotonin in the prefrontal and anterior cingulate cortex, consequent to elevated MAO-A level. The relationship between elevated 5-HTT BP_{ND} and greater pessimism during major depressive episodes suggests that greater 5-HTT density in the context of elevated MAO-A level further contributes to serotonin deficiency in these brain regions. A similar mechanism may explain the association between neuroimaging indices of greater dorsal striatal D_2 density, DAT density and symptoms of motor retardation: Greater MAO-A level and relatively greater DAT density lower extracellular dopamine in the dorsal striatum, leading to motor retardation. Indices of greater 5-HT_{1A} density,

J. H. Meyer (✉)
Department of Psychiatry, University of Toronto, Toronto, Canada
e-mail: jeff.meyer@camhpet.ca

J. H. Meyer
Canada Research Chair in Neurochemistry of Depression Head,
Neurochemical Imaging Program in Mood Disorders Centre for Addiction
and Mental Health Research Imaging Center, Toronto, Canada

Curr Topics Behav Neurosci (2013) 14: 101–134
DOI: 10.1007/7854_2012_219

Published Online: 28 September 2012

particularly in the cingulate cortex, have been associated with major depressive disorder, and well as anxiety disorders, suggesting that this abnormality is mechanistically related to presence of anxiety symptoms. To date, abnormalities of Glx a measure reflecting glutamate and glutamine levels have been most strongly associated with presence of major depressive episodes, with greater levels in occipital cortex, and reduced levels in prefrontal cortex. Ultimately, the future for neurochemical imaging is to better understand the mechanisms that predispose toward onset of MDE so as to create biologically informed, novel, methods of prevention, and superior, more symptom-targeted treatments.

Keywords Monoamine oxidase A · Serotonin$_{2A}$ receptors · Serotonin transporter · Serotonin$_{1A}$ receptors · Dopamine$_2$ receptors · Dopamine transporter Glutamate · Major depressive disorder

Contents

1 Introduction

Major depressive disorder affects 2–5 % of the population at any time and is the third leading cause of death and disability, according to the world health organisation (World Health Organization 2008). Given its high prevalence an important issue is to better understand the aetiology of symptoms so as to develop new strategies for prevention and more targeted approach for treatment.

Neurochemical imaging techniques offer an opportunity to measure markers of neurophysiological change to better understand mechanisms that contribute toward onset of symptoms during major depressive disorder. By identifying neurochemical abnormalities of MAO-A level, 5-HT_{2A} (serotonin$_{2A}$), 5-HTT (serotonin transporter), 5-HT_{1A} (serotonin$_{1A}$), D_2, DAT as well as glutamate levels in functionally important regions, then relating these findings to symptoms, and illness state, neuroimaging investigations are making important contributions toward understanding the pathophysiology of major depressive disorder.

2 Monoamine Oxidase A and Depressed Mood

2.1 Monoamine Oxidase A, Monoamines and Mood

Monoamine oxidase-A (MAO-A) is an enzyme with diverse functions in the brain, some of which have important roles for influencing mood; for most brain tissues, MAO-A activity is the main route for serotonin metabolism, and a significant route of metabolism for other monoamines, including norepinephrine and dopamine (Youdim et al. 2006). MAO-A is detectable in cells that release these monoamines, with the highest levels in norepinephrine releasing neurons (Konradi et al. 1988, 1989; Luque et al. 1995; Moll et al. 1990; Saura et al. 1996); however, MAO-A is also present in cells that do not release monoamines, such as astrocytes and glia (Youdim et al. 2006). Within cells, MAO-A is mainly located on outer mitochondria membranes. MAO-A has a high density in brain regions that influence mood (Saura et al. 1992): While MAO-A density is highest in brainstem (within the locus coeruleus), it is moderately high in the cortex, hippocampus, striatum, much lower in cerebellar cortex and minimal in white matter tissue. In addition, MAO-A has pro-oxidant effect, via the production of hydrogen peroxide, and is functionally linked to apoptosis. The latter is based on the observations that MAO-A inhibitors reduce apoptosis and MAO-A expression is increased in cell lines that are in a pro-apoptotic state (Ou et al. 2006). Among these roles, it is the effect of MAO-A upon monoamine metabolism that is highly implicated in influencing mood state.

There is a considerable amount of data linking loss of extracellular serotonin, norepinephrine and dopamine in humans with onset of depressed mood and/or major depressive episodes. Overall, two temporal mechanisms have been observed, one acute and one chronic. For example, acute monoamine depletion, through either tryptophan depletion to lower brain serotonin or alpha-methyl-para-tyrosine to lower brain dopamine and norepinephrine, is associated with depressed mood (Freis 1954; Hasler et al. 2008; Neumeister et al. 2004b; Verhoeff et al. 2001; Young et al. 1985). The second type of mechanism observed is that long periods of monoamine depletion are associated with onset of MDE in humans as demonstrated by chronic reserpine administration (Freis 1954).

In brain tissue, the density of MAO-A correlates highly with the level of its metabolic activity. Thus, given the functional link between monoamine loss and low mood and/or MDE, and the role of MAO-A in metabolising multiple monoamines,

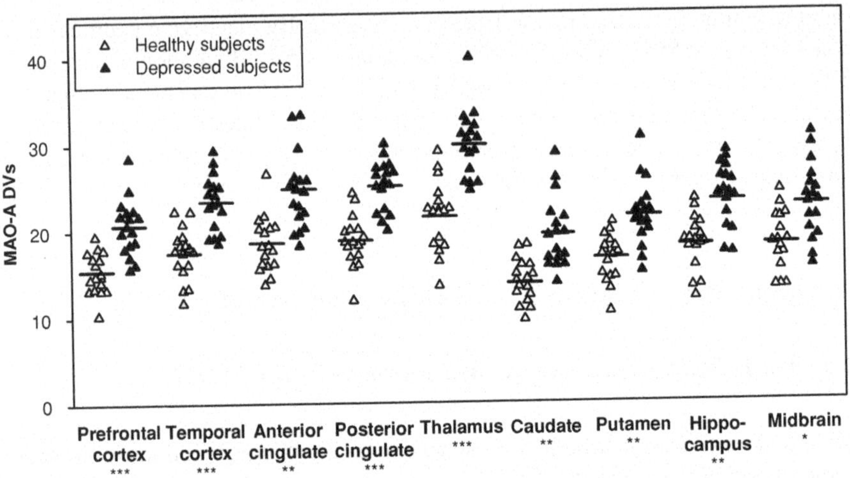

Fig. 1 Comparison of MAO-A DV_s between depressed and healthy subjects. On average MAO-A DV_s was elevated by 34 percent, or two standard deviations, in depressed individuals. Differences between groups were highly significant in each region: * p=0.001, ** p<0.0001, *** p<0.00001. Reprinted from Meyer et al. (2006a)

states that raise of MAO-A level in affect modulating brain regions would be expected to influence overall mood toward a depression. Additional evidence for a relationship between MAO-A and mood has been that one longstanding class of antidepressants, MAO inhibitors, has been effective treatments; this provides another reason to investigate this target in depressed mood states (Youdim et al. 2006). The relationship between MAO-A density and depressed mood may now be tested: recent advances in positron emission tomography (PET) neuroimaging such as $[^{11}C]$harmine PET enable optimal quantitation of MAO-A V_T, a measurement proportional to MAO-A density.

2.2 *Prefrontal and Anterior Cingulate Cortex Monoamine Oxidase A Binding, Depressed Mood and Major Depressive Disorder*

A key advantage of neuroimaging for investigating major depressive episodes (MDE) is that the in vivo measurement of MAO-A V_T may be conducted in medication free subjects and in the specific grouping of early onset MDD. Most MDD is related to early onset, and it vital to differentiate early from late onset for investigations of MDD, because late onset MDD is associated with neurodegenerative disease such as Parkinson's disease or Alzheimer's disease, which can be viewed as different pathologies. The issue of selecting early onset medication free subjects was not readily addressable in the first three postmortem studies of MAO-A levels and/or activity, hence prior to 2006 it was unknown whether MAO-A density or activity was increased in early onset major depressive disorder. In 2006, MAO-A V_S, an index of MAO-A density, was measured using [^{11}C] harmine PET in medication free MDE secondary to early onset MDD (Meyer et al. 2006a). Subjects with MDE were drug free for at least 5 months and most were antidepressant naive. All MDE subjects and controls were otherwise healthy. The MAO-A V_S was highly significantly elevated ($p < 0.001$ each region, average magnitude 34 % (or two standard deviations)) during MDE (see Fig. 1). This was the first definitive study of MAO-A binding in MDE because the clinical sample was carefully defined to focus upon MDD, the effect size was large and the PET radiotracer [^{11}C] harmine has outstanding qualities for measuring MAO-A binding.

Later studies have been highly consistent in support of this finding: Barton et al. (2008) reported elevated brain serotonin turnover in unmedicated depressed patients, a phenomenon which could be explained by greater brain MAO-A level (Barton et al. 2008). In 2009, the finding of greater MAO-A binding in MDE was replicated with [^{11}C] harmine PET in the same laboratory, and in 2011 the finding was replicated in antidepressant free MDE subjects in postmortem study of orbitofrontal cortex in a different laboratory applying Western blot (Johnson et al. 2011; Meyer et al. 2009).

Monoamine lowering processes may lead to lowered mood, hence, in recovery from MDD, it might be expected that MAO-A levels would normalise with euthymic mood. However, the recovered state of MDD is also a state of high risk for another MDE. The risk for a recurrent MDE over 2 years is 20–50 % depending upon treatment conditions. Elevated MAO-A binding may be considered an index of a monoamine lowering process and in the 1950s during treatment with reserpine-based antihypertensives, it was discovered that chronic monoamine lowering is associated with subsequent onset of MDEs which typically occurred 2 weeks to 4 months later (Meyer et al. 2009). The in vivo nature of neuroimaging enabled measurement of MAO-A V_T, an index of MAO-A density, in a study of recovered MDD. In this study, MAO-A binding was significantly elevated in prefrontal cortex, anterior cingulate cortex, striatum, hippocampus, thalamus and midbrain in a sample of 18 medication free recovered MDD subjects compared to 28 healthy controls (Meyer et al. 2009). Recovered MDD subjects who had recurrence of their MDE in the subsequent 6 months had the highest levels of MAO-A binding in the prefrontal and anterior cingulate cortex at the time of scanning (Meyer et al. 2009). The prefrontal cortex and anterior cingulate cortex were prioritised because these regions (and/or subregions of these structures) are often activated in mood induction studies (reflecting processes that generate sad mood) (Liotti et al. 2002) and these regions participate in cognitive functions like pessimism which create sad mood (Sharot et al. 2007; Tom et al. 2007). In this MDD sample (Meyer et al. 2009), other factors related to recurrence were accounted for: Subjects were medication free for at least 1 year, had no cognitive behavioural therapy within 3 years, were currently asymptomatic, and had no comorbid medical, psychiatric or substance abuse illnesses. Given the link between elevated MAO-A binding in prefrontal and anterior cingulate cortex and subsequent MDE, a stronger case can be made that new therapeutics are needed to decrease MAO-A levels in these brain regions, ideally with persistence beyond the duration of administration of the therapeutic to prevent recurrence.

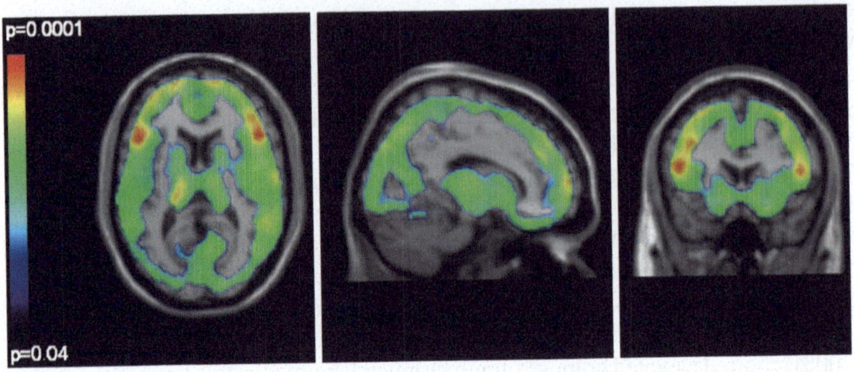

Fig. 2 Regional distribution of p-values reflecting elevated MAO-A binding in immediate postpartum period. Parametric maps of elevated monoamine oxidase A binding in the postpartum group vs the control group. Maps are superimposed on a T1-weighted magnetic resonance image that is normalized to the T1-weighted template (SPM2; Department of Cognitive Neurology, Wellcome Trust Centre for Neuroimaging, London, UK). **a** Tránsverse. **b** Sagittal. **c** Coronal. Individual voxel threshold was set at P<0.05; 86 412 voxels comprised a single cluster, which had a cluster-corrected significance of P=0.03. Mean regional difference was 43%

2.3 *Monoamine Oxidase A and Depressed Mood During Postpartum*

Depressed mood is common during early postpartum and can be categorised into three different syndromes. One common syndrome, occurring up to 75 % of the time, and considered within the healthy range of experience, is the "baby blues" or maternity blues. This consists of sad mood typically peaking on day 5 and ending with the first week postpartum accompanied by some irritability, desire to be alone, insomnia and trouble concentrating (O'Hara and Swain 1996). The second syndrome, which is also common with a prevalence rate of 13 %, is a full, clinical level MDE occurring within the first 1–3 months after delivery (O'Hara and Swain 1996; American Psychiatric Association 1994). The third syndrome, postpartum psychosis, is relatively rare with a prevalence rate of 0.1–0.2 %, and includes a combination of a MDE with hallucinations or delusions (Brockington et al. 1981; Kendell et al. 1987).

The first investigation of MAO-A in the postpartum period occurred in 2010: A [^{11}C] harmine PET study measured MAO-A V_T during postpartum blues (with scanning done between days 4 and 6 postpartum). A highly significant elevation of MAO-A binding was found, which, on average, was 43 % greater across the brain regions assayed (prefrontal cortex, anterior cingulate cortex, striatum, thalamus, hippocampus and midbrain) as compared to women not recently pregnant (Sacher et al. 2010). A voxel-based analysis demonstrated that the elevation in MAO-A binding was present throughout the grey matter of the brain (see Fig. 2) (Sacher et al. 2010).

This discovery led to a new neurobiological explanation for postpartum blues, involving a rapid decline in estrogen, followed by a rapid rise in MAO-A levels in affect-modulating brain regions, and finally subsequent sad mood with postpartum blues. During pregnancy, estradiol and estriol levels rise more than 100-fold and during the first week postpartum, with loss of the placenta, there is an enormous reduction in 17β-estradiol and estriol levels (O'Hara et al. 1991). Most of the decline in estrogens occur in the first 4 days followed by a more modest decline thereafter (O'Hara et al. 1991). Although MAO-A had never been previously investigated in the postpartum, it is known that estrogen decline is associated with a rise in either MAO-A density, activity or mRNA (see review (Sacher et al. 2010)). The 43 % elevation in MAO-A V_T in early postpartum confirms that this inverse relationship is applicable to the early postpartum period, after the early estrogen decline. MAO-A V_T can be viewed as an index of MAO-A levels and MAO-A levels correlate highly with MAO-A activity in brain tissue (Saura et al. 1992). Hence, the acute rise in MAO-A V_T in the early postpartum period represents a monoamine-lowering process and acute monoamine lowering processes are associated with sad mood (Ruhe et al. 2007).

2.4 Cigarette Withdrawal, Depressed Mood and MAO-A

Sad mood is an important problem for people who smoke cigarettes. First, during early withdrawal, sad mood frequently occurs, and when sad mood is prominent during withdrawal, it is associated with greater likelihood of relapse during quit attempts (Carey et al. 1993; Kenford et al. 2002). Second, cigarette smoking predisposes to MDD and vice versa; consequently, there is a very high comorbidity between cigarette smoking and major depressive disorder with 50 % of people with MDD also smoking cigarettes (Anda et al. 1990; Breslau et al. 1998).

The initial impression of the neuroimaging field was that MAO-A binding is reduced in those who smoke cigarettes in the active smoking state (Fowler et al. 1996). Given that, the plasma half-life of the key MAO-A binding substances found in cigarette smoke (harman and norharman) is only an hour (Rommelspacher et al. 2002), there was reason to specifically assess MAO-A V_T during both active smoking and withdrawal conditions. When MAO-A V_T was assessed in both conditions, it was discovered that prefrontal and anterior cingulate cortex MAO-A V_T rose during withdrawal in those who smoke heavily, that is, more than one pack of cigarettes per day, but not in those who smoke more moderately at less than one pack per day (see Fig. 3 from Bacher et al. (2011)). Interestingly, it is the heavy cigarette smoking group who are known to be at much greater risk for MDD (Pratt and Brody 2010), and it was this group that had 25 % elevated MAO-A V_T in the prefrontal and anterior cingulate cortex during withdrawal as compared to healthy controls, arguing for a process of elevated MAO-A level during withdrawal as a mechanism to create risk for MDD. Hence repeated exposure of elevated MAO-A level, in the prefrontal and anterior cingulate cortex, a mechanism associated MDD,

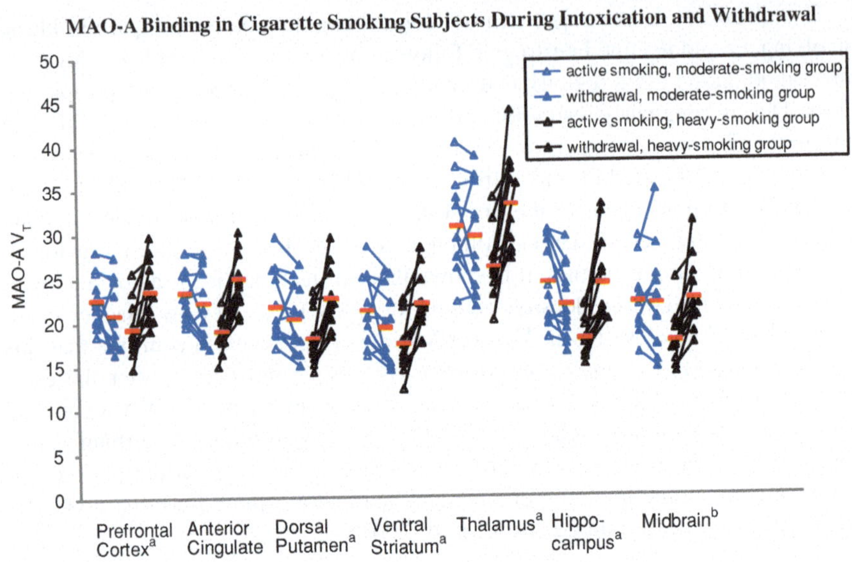

Fig. 3 Monoamine Oxidase-A binding in cigarette smoking subjects during active smoking and withdrawal. Repeated measures MANOVA found a highly significant effect between smoking severity and change in MAO-A V_T in the prefrontal cortex and anterior cingulate cortex regions ($F_{1,\,22} = 25.58$, $p<0.001$). Repeated measures MANOVA also found a significant effect between smoking severity and change in MAO-A V_T for all the regions assayed ($F_{1,\,22} = 28.24$, $p<0.001$). The effect was confirmed in each region with a repeated measures ANOVA (interaction between smoking severity and change in MAO-A V_T, repeated measures ANOVA, $F_{1,\,22} = 11.16$ to 28.87, $p = 0.003$ to $p<0.001$). [a] $p<0.001$ [b] $p = 0.003$. Greater MAO-A binding during acute cigarette withdrawal in heavy smoking group compared with healthy (MAO-A VT within groups were compared with an independent t-test. a $p \leq 0.001$ b $p \leq 0.005$ c $p \leq 0.01$ d $p \leq 0.05$ e $p \leq 0.1$ Arch Gen Psych (in press))

occurring after a short period of 8 h withdrawal could explain the predisposition to MDD in people who smoke cigarettes heavily.

As mentioned earlier, a second problem with mood in people who smoke cigarettes is the depressed mood of acute withdrawal (Carey et al. 1993; Kenford et al. 2002). In the imaging study of MAO-A during cigarette withdrawal, those who smoked heavily, the magnitude of rise in MAO-A V_T in prefrontal and anterior cingulate cortex during withdrawal was significantly correlated with the shift in visual analogue scales toward depressed mood (Bacher et al. 2011). This rise in MAO-A V_T also correlated with the decline in the MAO-A binding substance harman in those who smoke heavily. These results have significant implications for quitting heavy smoking. They suggest that rapid removal of harman from occupying MAO-A sites leaves a high level of available MAO-A for metabolising monoamines in prefrontal and anterior cingulate cortex resulting in depressed mood. This argues for testing of MAO-A inhibitor treatments in people who experience sad mood during early cigarette withdrawal as a strategy to assist in quitting, an important issue since 50 % of people tend to relapse in the first few days of trying to quit cigarette smoking (Garvey et al. 1992; Law and Tang 1995).

In conclusion, greater MAO-A V_T occurs, particularly in the prefrontal and anterior cingulate cortex, in a number of low mood states that are also associated with predisposition to major depressive disorder (Bacher et al. 2011; Meyer et al. 2009; Sacher et al. 2010). This data has major implications for preventing major depressive disorder as it suggests that better understanding and regulation of MAO-A may be helpful in avoiding the onset of major depressive episodes.

3 Pessimistic Perspective, 5-HT_{2A} Receptor and 5-HTT Receptor Imaging

3.1 5-HT_{2A} Receptor Imaging in Prefrontal and Anterior Cingulate Cortex

Negativistic thinking often occurs during major depressive episodes and it is important because high levels of hopelessness, a key component of pessimism, are associated with greater risk of suicide. Pessimism during major depressive episodes has also been captured by the concept of 'dysfunctional attitudes'. While a modest level of dysfunctional attitude can viewed as adaptive, dysfunctional attitudes increase significantly during major depressive episodes (Simons et al. 1986; Weissman 1979). Greater pessimism during major depressive episodes contributes to negative thoughts, and subsequent sad mood and this underlying pessimism is targeted by cognitive therapy (Simons et al. 1986; Weissman 1979). Dysfunctional attitudes may be measured with the dysfunctional attitudes scale (DAS), a measure that is sensitive for detecting negativistic thinking during major depressive episodes (Simons et al. 1986; Weissman 1979), and also demonstrates very good internal consistency (Cronbach alpha = 0.85–0.87) (Cane et al. 1986; Oliver and Baumgart 1985) and has high test–retest reliability (Oliver and Baumgart 1985; Weissman 1979).

Two findings initially suggested a relationship between manipulations of extracellular serotonin and dysfunctional attitudes. The first is that dysfunctional attitudes normalise during the response to selective serotonin reuptake inhibitor (SSRI) treatment (Fava et al. 1994; Simons et al. 1986). The second is that raising extracellular serotonin after administration of intravenous d-fenfluramine rapidly shifts dysfunctional attitudes toward optimism in healthy individuals (Meyer et al. 2003). These results argue that one of the roles of serotonin is to modulate dysfunctional attitudes in humans. More recently, the rostral anterior cingulate cortex and subregions of prefrontal cortex (dorsolateral, and medial prefrontal cortex) have been demonstrated to participate in functions related to optimism and pessimism (Elliott et al. 2002; Sharot et al. 2007).

Direct evidence that serotonin is low in brain is difficult to obtain: Brain serotonin cannot be directly measured in vivo and it is likely, based upon animal simulations of postmortem delay, that serotonin levels are very unstable, even within 24 h of death (Kontur et al. 1994). Moreover, postmortem investigations reviewed by Mann et al. (1996), Stockmeier (2003) have difficultly sampling medication free subjects.

While one cannot measure extracellular serotonin directly during major depressive episodes with neuroimaging either, one may measure an index of regional 5-HT_{2A} receptor density such as 5-HT_{2A} BP or 5-HT_{2A} BP_{ND} (an index of specific binding relative to free and non-specific binding). 5-HT_{2A} density has an inverse relationship to extracellular serotonin such that binding increases when extracellular serotonin is chronically lowered (O'Regan et al. 1987; Roth et al. 1987; Stockmeier and Kellar 1986; Todd et al. 1995). Therefore, if extracellular serotonin loss occurred in the prefrontal and anterior cingulate cortex during MDE, increased 5-HT_{2A} BP_{ND} would occur in these regions. A review of the initial set of 5-HT_{2A} imaging studies of MDE find a reduction in those with recent antidepressant use, and no change in those with no recent antidepressant use (see Table 1) (Meyer 2008). One could interpret the reductions in 5-HT_{2A} binding in recently treated subjects as being consequent to recent antidepressant use. The other set of findings, reflecting the medication free state of MDE, which found no change in 5-HT_{2A} BP_{ND} would suggest either abandoning the notion of reduced extracellular serotonin in the prefrontal and anterior cingulate cortex during MDE or creating an alternative hypothesis.

One alternative perspective is that monoamine loss during MDD is heterogenous and that the loss is greatest in those with the most severe symptoms. The first investigations of this revision began with prefrontal cortex 5-HT_{2A} BP_{ND} measurement and its relationship to dysfunctional attitudes. A strong correlation was observed between severity of dysfunctional attitudes (pessimism) and elevation in prefrontal and anterior cortex 5-HT_{2A} BP_{ND}. Furthermore, cortex 5-HT_{2A} BP_{ND} was significantly elevated in subjects with MDE and severe pessimism (Meyer et al. 2003). For example, in the prefrontal cortex region of interest centred on Brodman's area 9, 5-HT_{2A} BP_{ND} was elevated 29 % in depressed subjects with dysfunctional attitude scores greater (more pessimistic) than the median for the group. There was also a strong, significant correlation between severity of pessimism and prefrontal cortex 5-HT_{2A} BP_{ND} (see Fig. 4). A study by Bhagwagar et al. replicated this relationship between dysfunctional attitudes severity and prefrontal cortex 5-HT_{2A} BP_{ND} in recovered depressed subjects(Bhagwagar et al. 2006). In another study of a large sample of healthy subjects, two personality factors related to pessimism, vulnerability and anxiety, also positively correlated with prefrontal cortex, temporal cortex and left insula 5-HT_{2A} BP_{ND} (Frokjaer et al. 2008).

These results provide an explanation to interpret the investigations of suicide victims which had been a key focus of the mood disorders field between the mid 1984 and 2000. At that time, the most consistent postmortem biological abnormality in suicide victims was increased serotonin$_{2A}$ receptor density in the prefrontal cortex, most commonly in Brodmann's area 9 (Arango et al. 1990, 1992; Arora and Meltzer 1989; Hrdina and Vu 1993; Mann et al. 1986; Pandey et al. 2002; Stanley

Table 1 Imaging studies of 5-HT_{2A} receptors in major depressive disorder [updated from Meyer (2008)]

Study	Method	Number of subjects	Medication free status	Result
D'Haenen et al. (1992)	[^{123}I] ketanserin SPECT	19 Depressed 10 Healthy	7 days	Greater in parietal cortex
Biver et al. (1997)	[^{18}F]altanserin PET	8 Depressed 22 Healthy	10 days	Lower in orbitofrontal cortex
Attar Levy et al. (1999)	[^{18}F] setoperone PET	7 Depressed 7 Healthy	Taking benzodiazepines	Lower in prefrontal cortex
Meyer et al. (1999)	[^{18}F] setoperone PET	14 Depressed 14 Healthy	3 months plus 5 half lives	No difference
Meltzer et al. (1999)	[^{18}F] altanserin PET	11 Depressed 11 Healthy	"untreated"	No difference
Yatham et al. (2000)	[^{18}F] setoperone PET	20 Depressed 20 Healthy	2 weeks	Decrease in all cortex
Messa et al. (2003)	[^{18}F] setoperone PET	19 Depressed 19 Healthy	Taking benzodiazepines	Decrease in all cortex
Meyer et al. (2003)[a]	[^{18}F] setoperone PET	22 Depressed 22 Healthy	6 months	Positive association with dysfunctional attitude severity in cortex
Mintun et al. (2004)[b]	[^{18}F]altanserin PET	46 Depressed 29 Healthy	4 weeks	Decrease in hippocampus
Bhagwagar et al. (2006)	[^{11}C] MDL100907	20 Recovered Depressed 20 Healthy	6 months	Positive association with dysfunctional attitude severity in prefrontal cortex; elevation in most cortex regions

[a] Subjects enroled in the study by Meyer et al. (1999) were also included in the expanded study by Meyer et al. (2003) of 5-HT_{2A} receptors and dysfunctional attitudes in subjects with depression as well as subjects with borderline personality disorder

[b] Findings appear largely driven by a single healthy subject with very high 5-HT_{2A} BP_{ND}

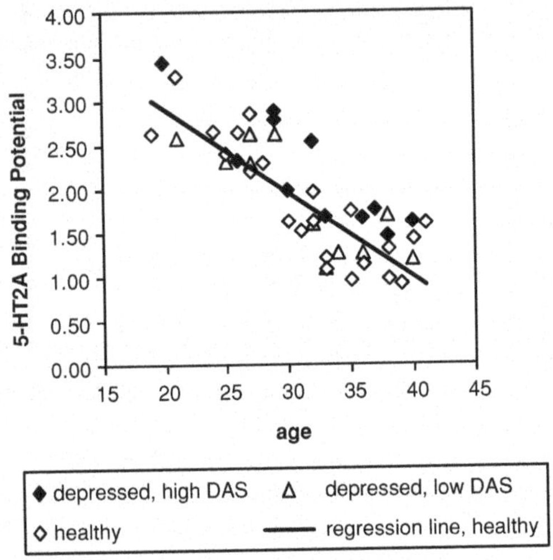

Fig. 4 5-HT_{2A} Binding Potential is Greater During Major Depressive Episodes (MDE) With Highly Abnormal Dysfunctional Attitudes. (5-HT_{2A} receptor binding potential in averaged bilateral middle frontal gyrus (Brodmann's area 9) is plotted against age to show the relationship between depressed and healthy subjects. The 22 depressed patients were divided into high and low dysfunctional attitudes scale (DAS) groups depending upon whether their DAS scores were above or below the median DAS score for the MDE group. This median score was 166. MDE subjects with high DAS scores had significantly higher 5-HT_{2A} receptor binding potential as compared to healthy subjects (ANCOVA (age covariate), diagnosis, $F_{1,19}=11$, $p=0.003$). (Age was an expected covariate in the model which was designed to identify a disease effect influencing 5-HT_{2A} receptor availability.))

and Mann 1983; Stockmeier 2003; Stockmeier et al. 1997; Turecki et al. 1999; Yates et al. 1990). At the time the studies reported the abnormality as alterations in serotonin$_2$ receptor binding, but it is generally accepted that these studies investigated serotonin$_{2A}$ receptors given that ligand binding to 5-HT_{2C} receptors in cortex is extremely low (Hoyer et al. 1986; Marazziti et al. 1999) and mRNA of 5-HT_{2B} receptors is extremely low in cortex (Schmuck et al. 1994). Although these findings occurred in studies in which the diagnosis of the suicide victim was unrestricted, these findings were more consistent in the subsample of studies of suicide victims who had major depressive disorder and were medication free (Hrdina and Vu 1993; Yates et al. 1990).

The investigations correlating severity of dysfunctional attitudes with greater 5-HT_{2A} BP_{ND} (Meyer et al. 2003; Bhagwagar et al. 2006) explains at a diagnostic and symptom specific level what clinical phenomenon was studied in these postmortem studies: Fifty percent of suicide victims have major depressive disorder (Barraclough et al. 1974; Robins et al. 1959). The dysfunctional attitudes scale is highly correlated with hopelessness measured with the Beck Hopelessness Scale (Bouvard et al. 1992; Cannon et al. 1999; DeRubeis et al. 1990; Norman et al. 1988). Given that hopelessness is a risk factor for suicide (Beck et al. 1985; Beck et al. 1989), it is likely that investigations of suicide victims reporting

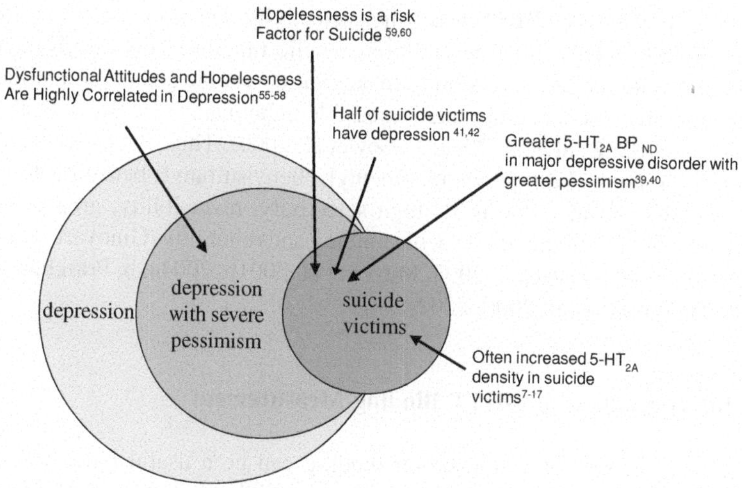

Fig. 5 Elevated prefrontal cortex 5-HT_{2A} density in suicide mainly reflects sampling from major depressive disorder with severe pessimism

increased 5-HT_{2A} BP_{ND} sampled depressed subjects with greater severity of pessimism. See Fig. 5 which represents the relationship of sampling studies for major depressive disorder, dysfunctional attitudes and suicide. Interestingly, while these findings are consistent with postmortem study in suicide victims, they are also consistent with a model of heterogeneous extracellular serotonin loss in prefrontal cortex in MDD such that extracellular serotonin is lowest in major depressive episodes with more severe pessimism.

3.2 *5-HTT Imaging and Pessimism*

3.2.1 5-HTT Radioligands

In 2000, the first high quality radiotracer for measuring indices of serotonin transporter binding occurred with the development of [^{11}C] DASB. The two previous radiotracers applied for in vivo imaging had significant limitations: 2-beta-carbomethoxy-3-beta-(4-iodophenyl)-tropane (β-CIT) single photon emission tomography (SPECT), the first technique developed had a specific binding signal that could be detected in the midbrain (Brucke et al. 1993; Innis et al. 1993; Kuikka et al. 1993), but it has almost equal affinity for the dopamine transporter as compared to the serotonin transporter (Carroll et al. 1995; Laruelle et al. 1994). Since there is high dopamine transporter density in the substantia nigra (Ciliax et al. 1999), the relative contributions of specific binding to dopamine and serotonin transporters cannot be differentiated in midbrain and it was the midbrain for which this radiotracer technique was applied as an index of serotonin transporter binding. The second applied radiotracer, $[^{11}C]$(+)McN5652, had better selectivity, but also had a low ratio of specific binding relative to free and non-specific binding. This disadvantage, in combination with modest reversibility makes valid, reliable quantitation difficult in regions other than the thalamus, and impossible in human cortex (Buck et al. 2000; Ikoma et al. 2002; Kent et al. 2002; Parsey et al. 2000). Thus, the radiotracer $[^{11}C]$ (DASB, 3-amino-4-(2-dimethylaminomethyl phenylsulfanyl)-benzonitrile) represented a major advance due to its high selectivity, reversibility, greater specific binding relative to free and non-specific binding and reliability (Ginovart et al. 2001; Houle et al. 2000; Ichise et al. 2003; Meyer et al. 2001b, 2004a, b; Praschak-Rieder et al. 2005; Wilson et al. 2000, 2002).

3.2.2 Interpretations of 5-HTT Binding Measurement

Measurement of serotonin transporter binding can be a useful index for several different models of disease that may affect extracellular serotonin levels. There are at least four such models may be considered in relationship to how serotonin transporter binding may be abnormal in a disease that lowers extracellular brain serotonin (Meyer 2007). These are referred to subsequently as models one through four. Abnormalities in serotonin transporter binding during major depressive episodes may be discussed in the context of these models.

Model one is a lesion model that reduces monoamine releasing neurons. In a lesion model, lowered 5-HTT binding measures occur. Model two is a secondary change in serotonin transporter binding as a sequelae to serotonin lowering via a different process. Model two is unlikely to be relevant for serotonin transporter binding. Acute reductions in serotonin have repeatedly shown reductions in 5-HTT mRNA (Linnet et al. 1995; Xiao et al. 1999; Yu et al. 1995). However, long-term reductions or elevations in serotonin typically show no effect upon regional 5-HTT density (Benmansour et al. 1999; Dewar et al. 1992; Graham et al. 1987). Model three is increased clearance of extracellular monoamine via greater monoamine transporter density. In model three, greater available serotonin transporter binding leads to greater clearance of monoamines from extracellular locations.

Model four is endogenous displacement and endogenous displacement is the property of a few radioligands to express different binding after short-term manipulations of their endogenous neurotransmitter, such that greater binding occurs during depletion of endogenous neurotransmitter and reduced binding occurs during elevations of endogenous neurotransmitter. For $[^{11}C]$ DASB, endogenous displacement may occur with large magnitude changes in extracellular 5-HT, but this is unlikely to occur with extracellular 5-HT changes that are physiologically tolerable for humans (Meyer 2007). For other PET radiotracers such as $[^{11}C]$HOMADAM, $[^{11}C]$ MADAM, $[^{11}C]$ (+) McN5652, or SPECT radiotracers $[^{123}I]$-B-CIT SPECT or $[^{123}I]$ ADAM SPECT, it is unknown whether endogenous levels of serotonin influence binding levels. This fourth model is unlikely to apply to PET imaging studies with $[^{11}C]$ DASB in humans, but it is unclear as to whether this model applies to other serotonin transporter radiotracers since the question has not been tested.

3.2.3 Dysfunctional Attitudes During Major Depressive Episodes and 5-HTT Binding

Dysfunctional attitudes, an index of pessimism, are elevated during major depressive disorder, and, as noted above, represent important symptoms because they generates sad mood and are strongly related to suicide. Hopelessness (Beck et al. 1985, 1989) and difficulty seeing positive reasons for living (Malone et al. 2000) are also significant risk factors for suicide and it has been demonstrated in four separate samples of subjects with major depressive episodes that greater hopelessness is highly positively correlated with greater severity of dysfunctional attitudes as measured with the dysfunctional attitudes scale (Bouvard et al. 1992; Cannon et al. 1999; DeRubeis et al. 1990; Norman et al. 1988).

In the two postmortem investigations of 5-HTT density in subjects with recent symptoms of depressive episodes no changes in 5-HTT density in the dorsal raphe or the locus coeruleus were found (Bligh-Glover et al. 2000; Klimek et al. 2003). Other postmortem investigations of 5-HTT density sampled subjects with a history of a depressive episode and these investigations usually studied the prefrontal cortex and/or dorsal raphe nucleus. Findings ranged from decreased 5-HTT density (Arango et al. 2001, Austin et al. 2002; Crow et al. 1984; Mann et al. 2000; Perry et al. 1983) to no difference in 5-HTT density (Hendricksen et al. 2004; Hrdina et al. 1990; Lawrence et al. 1990; Leake et al. 1991; Little et al. 1997). In several of these studies, subjects were medication free (Hrdina et al. 1990; Lawrence et al. 1990; Mann et al. 2000) and for many of these investigations, average postmortem delay was less than 1 day (Austin et al. 2002; Bligh-Glover et al. 2000; Klimek et al. 2003; Little et al. 1997; Mann et al. 2000; Perry et al. 1983). Further detail may be found in the review of Stockmeier (2003). Other sampling issues that influence postmortem investigations are inclusion of subjects with bipolar disorder and lack of differentiation between early versus late onset MDD. None of the postmortem studies investigated the relationship between 5-HTT binding and hopelessness or pessimism.

The first investigation of [^{11}C]DASB PET imaging of major depressive disorder examined the relationship of 5-HTT BP_{ND} to severity of dysfunctional attitudes and presence of a major depressive episode. Meyer et al., sampled 20 subjects with major depressive episodes (from early onset major depressive disorder) and 20 healthy controls (Meyer et al. 2004a). Subjects were medication free for at least 3 months, had no other comorbid axis I illnesses, were nonsmoking, and had early onset major depressive disorder. There was no difference in 5-HTT BP_{ND} in either cortical or subcortical regions (including medial prefrontal cortex, dorsolateral prefrontal cortex and anterior cingulate cortex) between the group with major depressive episodes and healthy controls. However, subjects with severely pessimistic dysfunctional attitudes who were in the midst of major depressive episodes had significantly higher 5-HTT BP_{ND}, compared to healthy in brain regions sampling serotonin nerve terminals (dorsolateral and medial prefrontal cortex, anterior cingulated cortex, thalamus, bilateral caudate and bilateral putamen). On average, 5-HTT BP_{ND} was 21 % greater in these regions in subjects who were in

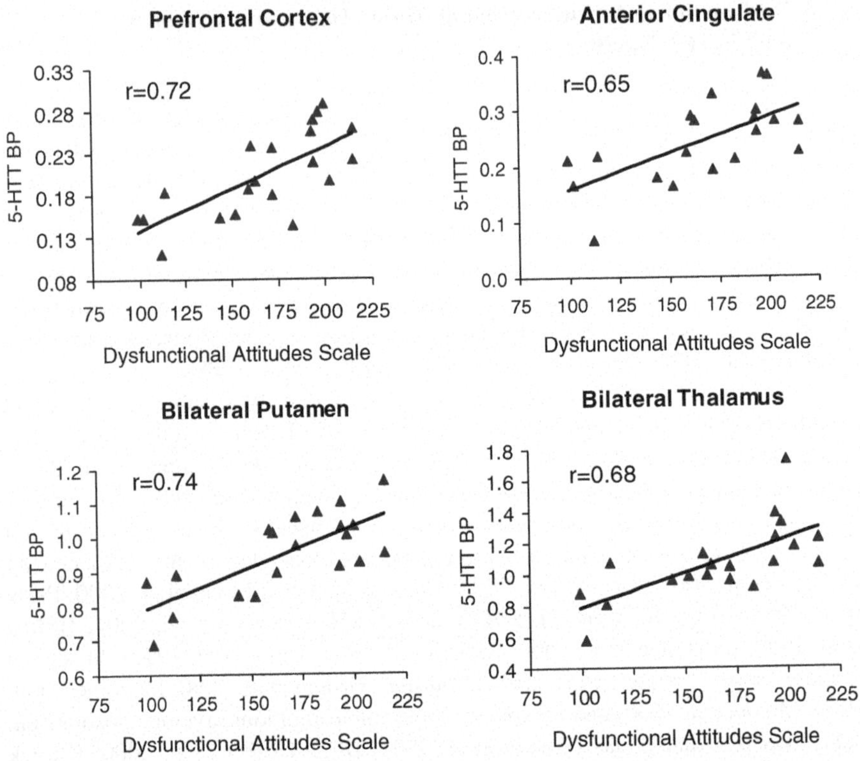

Fig. 6 Correlations between dysfunctional attitudes (DAS) and serotonin transporter binding potential (5-HTT BP) in some of the larger regions in depressed subjects. Highly significant correlations were found: prefrontal cortex (P = 0.0004), anterior cingulate (P = 0.002), bilateral putamen (P = 0.0002), bilateral thalamus (P = 0.001). Reprinted from the Archives of General Psychiatry

the midst of major depressive episodes with severely pessimistic dysfunctional attitudes. Moreover, within the major depressive episode group, greater 5-HTT BP_{ND} was highly correlated with more negativistic dysfunctional attitudes in the same brain regions (see Fig. 6). The interpretation was that given serotonin transporters have an important role in influencing extracellular serotonin, greater regional 5-HTT levels provide greater vulnerability to low extracellular 5-HT and symptoms of extremely negativistic dysfunctional attitudes. This interpretation, in subjects with high levels of pessimism during MDE, corresponds to the third model discussed earlier under "Interpretations of 5-HTT Binding Measurement".

Neuroimaging investigations, sampling subjects with early onset MDD who are medication free for greater than 2 months, are non-smoking and do not have comorbid axis I disorders, and also apply better quality radiotracer technology, tend to find either no change in regional 5-HTT binding or an increase in regional 5-HTT binding (Cannon et al. 2007; Herold et al. 2006; Ichimiya et al. 2002; Meyer et al. 2004a). Investigations which include sampling of subjects with late onset MDD,

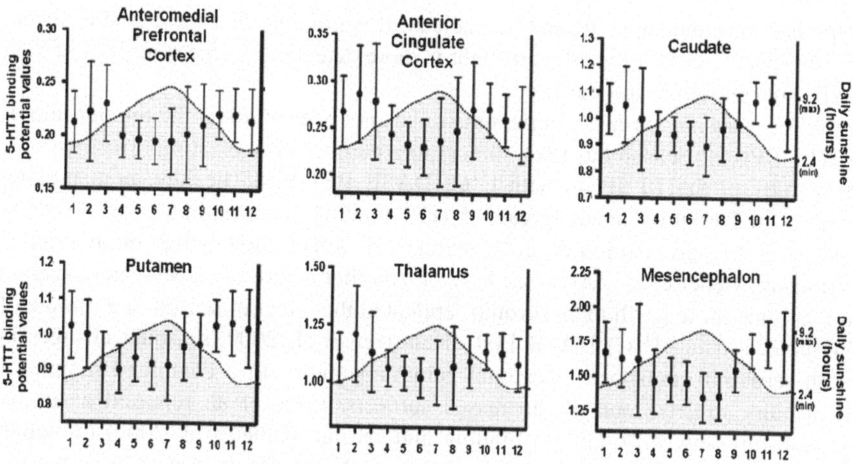

Fig. 7 Regional 5-HTT BP_{ND} versus month (n=88 healthy subjects). Reciprocal peaks and troughs of brain serotonin transporter binding and duration of sunshine in 88 healthy study participants. Serotonin transporter binding potential values were measured in six brain regions. Circles represent bimonthly moving average means. Error bars represent 95 per cent confidence intervals of the mean. The shaded areas represent the average duration of sunshine in Toronto, Ontario, Canada. Regional 5-HTT BP_{ND} was significantly greater in spring/summer than fall/winter by 10 to 16%, p<0.02 for each region. Differences in peak to trough ranged from 22 to 42%. Reprinted from Meyer et al., Archives of General Psychiatry 65(9):1072–1078

comorbid axis I psychiatric disorders, recent antidepressant use, current cigarette smoking and do not apply a selective radiotracer are more likely to report a reduction in regional 5-HTT binding (Joensuu et al. 2007; Malison et al. 1998; Newberg et al. 2005; Parsey et al. 2006a; Selvaraj et al. 2009). Only the first study of [^{11}C] DASB PET concurrently investigated a measure of pessimism or hopelessness.

3.2.4 Serotonin Transporter Binding and Seasonal Behaviour

Seasonal affective disorder is an important problem for countries with regions located at greater extremes of latitude. Rates of SAD typically range from 1 to 6 % in regions of 40′ latitude or greater (Magnusson 2000). Furthermore, at these latitudes, 25 % of healthy individuals experience lower mood, less energy, greater appetite and increased sleep in the winter (Kasper et al. 1989; Rosen et al. 1990).

There is significant evidence for seasonal variations in markers of serotonin physiology. In postmortem study of serotonin concentrations, Carlsson et al. reported a seasonal variation of serotonin levels in human hypothalamus with lower levels in late winter and higher levels in late summer (Carlsson et al. 1980). More recently, Lambert et al. found seasonal fluctuations in whole brain serotonin turnover in humans (Lambert et al. 2002). In rodents, reduced light exposure is associated with greater 5-HTT density (Rovescalli et al. 1989), lower 5-HT release (Blier et al. 1989) and greater 5-HT clearance (Rovescalli et al. 1989) in the hypothalamus and suprachiasmatic nucleus. It is possible that these findings were region specific or that the effect of light is more detectable in these regions of high 5-HT concentration and density.

While seasonal variation in pessimism has not been studied, the relationship of 5-HTT BP_{ND} to season has been subject to a number of interesting investigations. In a study of 5-HTT BP_{ND} with [^{11}C] DASB PET in 88 healthy, non-smoking humans in Toronto, Canada, greater 5-HTT binding occurred in the fall/winter as compared to spring/summer in a number of affect modulating brain regions (Praschak-Rieder et al. 2008) (see Fig. 7). Another centre in Copenhagen, located at a latitude more north than Toronto replicated the effect of season in a sample of 54 subjects using [^{11}C] DASB PET (Kalbitzer et al. 2009, seasonal changes in brain serotonin transporter, personal communication). In a combined sample of 49 healthy subjects and 49 depressed subjects, Ruhe et al. reported a similar relationship between midbrain binding and season (Ruhe et al. 2009) (although [^{123}I]B-CIT SPECT is not very selective for 5-HTT). Buchert et al. reported the same seasonal finding in the midbrain in a sample of 39 subjects using [^{11}C] McN5652 PET but not in the thalamus (Buchert et al. 2006). Although the Buchert study reported a positive result in only one of the two regions assayed, the sensitivity of this study to detect seasonal change was more modest: the sample is smaller than the other studies, and [^{11}C] McN5652 is a less sensitive technique than [^{11}C] DASB (Meyer 2007, 2008; Kent et al. 2002; Meyer et al. 2001b, 2004b). There are two studies in the literature that do not report greater 5-HTT binding in winter and both have a small sample size ($n = 12$ or less) (Koskela et al. 2008; Neumeister et al. 2000). To date, most studies of large sample size and reasonably northern latitude report greater 5-HTT BP_{ND} in most brain regions in the fall/winter as compared to spring/summer.

3.3 5-HT$_{1A}$ Receptor Binding and Anxiety

Most [^{11}C] WAY-100635 PET studies report lower 5-HT$_{1A}$ BP_{ND} in subregions of prefrontal cortex (dorsolateral, ventrolateral, orbitofrontal), anterior cingulate cortex, temporal cortex and raphe during major depressive episodes and continuance of this reduction during remission (Bhagwagar et al. 2004; Drevets et al. 1999; Sargent et al. 2000). There is one study with a different result and it may be that the selection of white matter as a reference region, and/or sampling characteristics may account for the difference (Parsey et al. 2006b). A potential problem with applying white matter as a reference tissue is that in contrast to using grey matter in cerebellar cortex, the properties of white matter are more likely to be different from grey matter, and the modelling method requires the assumption that the free and non-specific binding in the reference tissue is similar to grey matter tissue. Hence, a grey matter region with low specific binding is preferable as a reference tissue such as the cerebellar cortex excluding the vermis.

There is reason to consider that finding in MDD may actually be more strongly associated with anxiety, which often occurs during major depressive episodes or presence of comorbid anxiety disorders, which also frequently occur with MDD since reduced 5-HT_{1A} binding also occurs in affect modulating regions also occurs in anxiety disorders. In a [^{18}F]trans-4-fluoro-N-2-[4-(2-methoxyphenyl)piperazin-1-yl]ethyl]-N-(2-pyridyl) cyclohexanecarboxamide ([^{18}F]-FCWAY) study of panic disorder (with comorbid MDD in almost half the subject sample), significant reductions in 5-HT_{1A} binding in anterior cingulate, posterior cingulate and raphe regions were reported (Neumeister et al. 2004a). (Quantitation with this radiotracer is valid for subcortical regions but not cortical regions due to bone uptake of radiotracer metabolite.) A second study applying [^{11}C] WAY100635 reported similar findings with the greatest magnitude being in orbitofrontal cortex, temporal cortex and midbrain (Nash et al. 2008). In social anxiety disorder, reduced 5-HT_{1A} binding occurs in most brain regions (insula, anterior cingulate cortex, medial orbitofrontal cortex, amygdala and midbrain) (Lanzenberger et al. 2007). Since the neuroimaging studies of 5-HT_{1A} binding in MDD did not exclude comorbid anxiety disorders, one interpretation of the reduced 5-HT_{1A} binding in the brain regions assayed in MDD, particularly in orbitofrontal cortex, anterior cingulate cortex and midbrain, is that it reflects comorbid anxiety disorders. MDD with anxiety is often treatment resistant, and this interpretation is appealing because there could be an opportunity to target pathologies related to this abnormality in this treatment resistant subgroup (Trivedi et al. 2006).

A limitation of this interpretation is that a postmortem study with subjects who mainly did not have comorbid anxiety disorders reported decreased 5-HT_{1A} antagonist binding in orbitofrontal cortex (Stockmeier et al. 2009) and another explanation could be that abnormally low 5-HT_{1A} density in prefrontal and cingulate cortex is associated with both MDD and anxiety disorders. Future work should investigate the cellular mechanism, and identify the clinical subgroup of MDD with this finding so as to optimise therapeutics. One explanation is that lower 5-HT_{1A} receptor density reflects decreased 5-HT_{1A} receptors on GABA interneurons or the segments of pyramidal cell axons proximal to extensions from GABA containing chandelier interneurons (Stockmeier et al. 2009).

4 D_2, DAT and Motor Slowing

4.1 *Dorsal Striatal [^{11}C]Raclopride Binding and Motor Retardation in Major Depressive Disorder*

Among the symptoms of MDEs, motor retardation and anhedonia both reflect functions modulated by dopamine release in the dorsal putamen and nucleus accumbens respectively. The second symptom, anhedonia, is more difficult to investigate with neuroimaging markers of dopamine: The nucleus accumbens is a

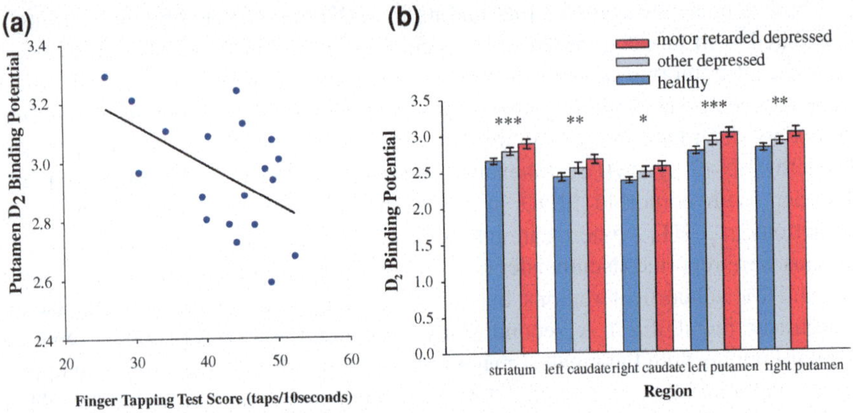

Fig. 8 (**a**) Correlation between bilateral putamen D_2 receptor binding potential and motor speed in depressed (n=19) subjects (The correlation was significant (r = -0.53, p = 0.02). In order to reduce variance related to age and gender, individual putamen D_2 BP values were normalized to age 30 using the slope of the linear decline in D_2 BP with age. Motor speed was measured with the finger tapping test. Individual finger tapping scores were normalized similarly to a 30 year old male (Meyer et al. 2006a, b). (**b**) Striatal D_2 Receptor Binding Potential in Motor Retarded Depressed (n=10), Other Depressed (n=11) and Healthy Subjects (n=21) (Striatal D_2 Receptor Binding Potential in Motor Retarded Depressed (n=10), Other Depressed (n=11) and Healthy Subjects (n=21) (D2 binding potential values were normalized to a 30-year-old subject using the slope of the age-related decline. Differences between the healthy group and the depressed subjects with motor retardation assessed by means of independent sample t tests. $*p \leq 0.05$ $**p \leq 0.01$ $***p \leq 0.005$)

smaller structure. Anhedonia is difficult to measure during major depressive episodes since it can be biased by negativistic perspective if the measure involves self report or motor speed if the measure involves movement. However, given recent advances in the development of neuroimaging markers for D_3 receptors for the ventral striatum, and ongoing development of quantitation of hedonic measures in depression, it is likely that future investigations will focus upon this direction (Willeit et al. 2008). The first symptom, motor retardation, is straightforward to investigate with quantitative measures because dorsal putamen is large, motor retardation can be measured with well validated motor tests such as the finger tapping test, and [^{11}C] raclopride has a strong specific binding signal in dorsal putamen. [^{11}C] raclopride is a widely available PET radiotracer selective for D_2 type receptors whose specific binding changes inversely relative to changes in neurotransmitter levels (Laruelle 2000).

In addition to the straightforward ability to quantify D_2 BP_{ND} in dorsal striatum, there were other reasons to investigate this biomarker in relation to motor retardation. One is that greater D_2 BP_{ND} and reduced dopamine levels occur in dorsal striatum in Parkinson's disease and other similar diseases that involve reduced movement speed (Kim et al. 2002; Kish et al. 1988). The second is that D_2 BP_{ND} found with [^{11}C]raclopride PET, is inversely proportional to extracellular dopamine levels in animal and human paradigms that manipulate dopamine levels (Laruelle 2000). The third reason is that decreased cerebrospinal fluid levels of the dopamine metabolite homovanillic acid is often reported during MDE with motor retardation, suggesting that some regions of the brain have lower levels of dopamine when motor retardation is present during major depressive episodes (Korf and Praag 1971; Post et al. 1973; Praag et al. 1975).

Unfortunately, early imaging studies of dopamine receptors did not sample depressed subjects who were medication free and non-smoking or addressing the confounding effect of age (Meyer 2008). Since D_2 BP_{ND} declines with age and motor retardation increases with age, to assess this relationship in a manner that addressed the age bias, one must correct for the effect of age upon each variable prior to investigating the relationship between the two. There is only one study that meets all of these criteria, and there is no other study meeting two of these three criteria (Meyer 2008). The main findings of the one, unbiased study, of striatal D_2 BP_{ND} and motor retardation were that the caudate and putamen D_2 BP_{ND} were elevated in the depressed group as compared to the healthy group, and that greater putamen D_2 BP_{ND} was significantly correlated with more severe motor retardation in the depressed group (Meyer et al. 2006b) (see Fig. 8). The findings support a specific, neuromodulatory role for striatal dopamine loss during MDE, particularly when motor retardation is present.

4.2 Striatal DAT Binding and Motor Retardation During Major Depressive Episodes

With regard to dopamine transporter (DAT) imaging in MDD, studies sampling medication free subjects typically report lower striatal DAT BP_{ND} (Meyer et al. 2001a; Neumeister et al. 2001; Sarchiapone et al. 2006), whereas studies sampling subjects with recent antidepressant treatment sometimes report higher striatal DAT binding (Brunswick et al. 2003). Most studies address the age-related decline in DAT and some address the confound of cigarette smoking. Mechanisms to explain reduced striatal DAT binding include lesions to dopamine releasing neurons or a downregulation model (in response to another monoamine lowering process). In contrast to the serotonin transporter, dopamine transporters in the striatum downregulate when subacute to chronic dopamine depletion occurs (Gordon et al. 1996; Han et al. 1999; Ikawa et al. 1993; Kilbourn et al. 1992).

The downregulation model may explain reduced striatal DAT binding during MDD since greater MAO-A levels occur in striatum in major depressive episodes and MAO-A metabolises dopamine (Meyer et al. 2006a, 2009). In addition, greater striatal D_2 BP_{ND} was found during major depressive episodes with $[^{11}C]$raclopride PET and this index is increased when extracellular dopamine is reduced (Laruelle 2000).

Also, the correlation between lower dorsal putamen DAT BP_{ND} values and less impaired performance on the finger tapping test has a particular interpretation (Meyer et al. 2001a). The finger tapping test is a measure of motor slowing in MDD (Meyer et al. 2001a, 2006b). Slower performance on the finger tapping test is correlated with greater putamen D_2 BP_{ND} during MDE and greater D_2 BP_{ND} from [^{11}C] raclopride occurs when extracellular dopamine is lower (Laruelle 2000). The data can be interpreted as follows: Major depressive episodes without motor retardation are associated with lower DAT BP_{ND} and demonstrate a compensatory protective mechanism (Meyer et al. 2001a). Downregulation of DAT occurs when dopamine is chronically low in striatum (Gordon et al. 1996; Han et al. 1999; Ikawa et al. 1993; Kilbourn et al. 1992), but reduced DAT levels decrease clearance of extracellular dopamine. Compared with the healthy state, the compensated state has only mildly reduced extracellular striatal dopamine concentrations with downregulated DAT. This is the process whereby dorsal striatal DAT BP_{ND} is decreased protects against motor slowing.

5 Behavioural Correlates of With Abnormalities in Glutatmate Regulation

Magnetic resonance spectroscopy (MRS) can measure ‘Glx’ in brain tissue, an index mainly composed of glutamate (intracellular and extracellular), and glutamine. Exposure to elevated glutamate levels has been proposed as a mechanism that leads to sad mood because treatment resistant MDD subjects often experience a rapid, short term, mood elevation after single dose ketamine, a NMDA receptor antagonist and some subtypes of MDE have a response to lamotrigine, a medication that can reduce glutamate release (Barbee and Jamhour 2002; Barbee et al. 2011; Zarate et al. 2006). Also, elevations in extracellular glutamate have been proposed as being relevant to mood symptoms because reductions in glia are reported in orbitofrontal, dorsolateral prefrontal and anterior cingulate cortex in MDD and glia clear glutamate via excitatory amino acid transporters (Rajkowska and Miguel-Hidalgo 2007).

There are several studies applying proton MRS to measure Glx, in reasonably large samples of MDD with region specific results. For example, greater Glx levels were reported in the occipital cortex, a reduction was reported in the medial prefrontal cortex and no change was reported in the pregenual cingulate cortex (Hasler et al. 2007; Sanacora et al. 2004; Walter et al. 2009). A reasonable explanation for the anatomical variation is that glutamate levels are generally elevated in the brain during MDD, and that regions with reduced glial cell density (Rajkowska and Miguel-Hidalgo 2007) have reduced Glx because the intracellular contribution of Glx signal are lower. Some level of homogenous abnormality in brain glutamate regulation during MDD is suggested by a recent [^{11}C]ABP688 study reporting reduced prefrontal cortex, cingulate cortex, insula, thalamus and hippocampus mGlu5 receptor binding in conjunction with a postmortem study of reduced mGlu5 density in prefrontal cortex (Deschwanden et al. 2011). The investigation of pregenual cingulate cortex did detect, in highly anhedonic depressed subjects, decreased glutamine, a metabolic product of glutamate (Walter et al. 2009) and the investigation reporting greater glutamate in the occipital cortex found the greatest levels in melancholia (Hasler et al. 2007; Sanacora et al. 2004; Walter et al. 2009). Ideally, future studies should aim to sample the regions previously investigated concurrently so as to optimally address the regional specificity of findings, and preferentially sample melancholic and anhedonic subjects so as to replicate the findings by Sanacora et al. (2004) and Walter et al.(2009). While there is evidence for disturbances in glutamate levels during major depressive episodes, the mechanism of the relationship to specific mood symptoms requires further study.

6 Conclusions

While neuroimaging is limited by the range in biomarkers available, its ability for in vivo measurement is bridging markers of neurochemistry and neuroplasticity to pathophysiological mechanism of symptom onset for major depressive disorder. For example, indices of greater monoamine oxidase A (MAO-A) level, particularly in the prefrontal and anterior cingulate cortex, are associated with depressed mood states, and high-risk states for onset of major depressive episodes. MAO-A metabolises monoamines, and greater metabolism of monoamines occurs when MAO-A is elevated in brain. Hence, greater levels of MAO-A may be viewed as a key monoamine lowering process during major depressive episodes.

Evidence to date suggests that ongoing deficiency of specific monoamines in specific regions is implicated in symptoms of disease. Lower extracellular serotonin is associated with greater pessimism in humans and chronic serotonin deficiency is associated with upregulation of 5-HT_{2A} (serotonin$_{2A}$) receptors in cortex. During major depressive episodes, when pessimism is more severe, greater 5-HT_{2A} binding occurs in prefrontal and anterior cingulate cortex. These results argue for a mechanism of lowering extracellular serotonin in the prefrontal and anterior cingulate cortex, consequent to elevated MAO-A level. The relationship between elevated 5-HTT binding greater pessimism during major depressive episodes suggests that greater 5-HTT density in the context of elevated MAO-A level further contributes to serotonin deficiency in these brain regions (see Fig. 9). A similar mechanism may explain the association between neuroimaging indices of greater dorsal striatal D_2 density, DAT density and symptoms of motor retardation: Greater MAO-A level and relatively greater DAT density lower extracellular dopamine in the dorsal striatum, leading to motor retardation (also see Fig. 9).

Specific mechanisms underlying other neuroimaging abnormalities continue to be investigated. Indices of greater 5-HT_{1A} density, particularly in the cingulate cortex, have been associated with major depressive disorder, and well as anxiety disorders, suggesting that this abnormality is mechanistically related to presence of anxiety symptoms. At this point, abnormalities in glutamate level have been most strongly associated with presence of major depressive episodes, with greater levels in occipital cortex, and reduced levels in prefrontal cortex. Future neurochemical imaging investigations will ultimately focus upon detecting the mechanisms that

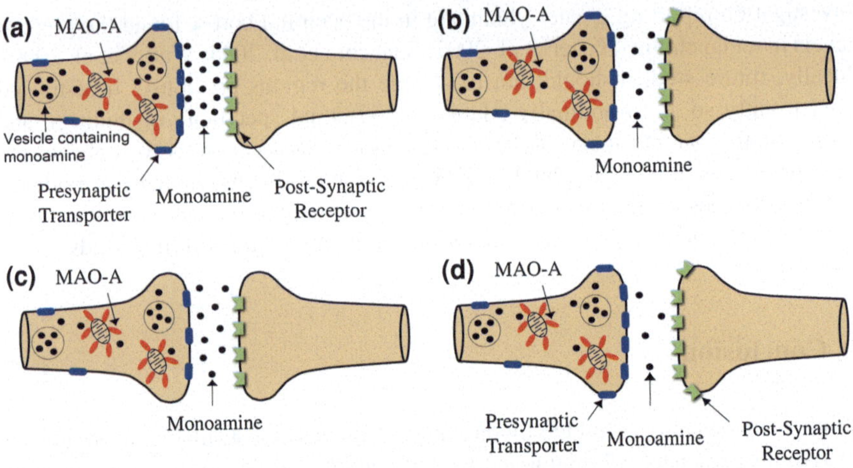

Fig. 9 Modern Model of Excessive Extracellular Monoamine Loss During Major Depressive Disorder (**a**) Monoamine release in a synapse in health. (**b**) During a major depressive episode, monoamine oxidase A (MAO-A) density is elevated resulting in greater metabolism of monoamines. Outcomes range from (**c**) to (**d**). (**c**) If the monoaminge transporter density is low during a major depressive episode, the effect of elevated MAO-A upon reducing extracellular monoamine is attenuated resulting in a moderate loss of monoamine, eventually resulting in a moderate rise in symptoms. (**d**) If the monoamine transporter density is not low during a major depressive episode, then there is no protection against the effect of elevated MAO-A. The extracellular concentration of the monoamine is severely reduced and rise in symptom s is severe. This model applies to regions whose functions are affected by changes in monoamine levels. Studies of transporter binding related binding to function (such as motor retardation to striatal DAT binding, and pessmism to prefrontal/anterior cingulate cortex 5-HTT binding. MAO-A in cell types not proximal to the release of the individual monoamine may play a role in its metabolism (for example MAO-A in norepinephrine releasing neurons, glia, and astrocytes may also play a role in the metabolism of serotonin)

predispose towards onset of MDE so as to create new biologically informed methods of prevention, and superior, more symptom targeted treatments based upon symptom specific underlying pathology.

References

American Psychiatric Association (1994) Diagnostic and statistical manual of mental disorders. IV edn. Washington, DC. Am Psychiatr Assoc 387

Anda RF, Williamson DF, Escobedo LG, Mast EE, Giovino GA, Remington PL (1990) Depression and the dynamics of smoking a national perspective. JAMA 264:1541–1545

Arango V, Ernsberger P, Marzuk PM, Chen JS, Tierney H, Stanley M, Reis DJ, Mann JJ (1990) Autoradiographic demonstration of increased serotonin 5-HT2 and beta- adrenergic receptor binding sites in the brain of suicide victims. Arch Gen Psychiatry 47:1038–1047

Arango V, Underwood M, Mann J (1992) Alterations in monoamine receptors in the brain of suicide victims. J Clin Psychopharm 12:8S–12S

Arango V, Underwood MD, Boldrini M, Tamir H, Kassir SA, Hsiung S, Chen JJ, Mann JJ (2001) Serotonin 1A receptors, serotonin transporter binding and serotonin transporter mRNA expression in the brainstem of depressed suicide victims. Neuro psychopharmacol 25:892–903

Arora RC, Meltzer HY (1989) Serotonergic measures in the brains of suicide victims: 5-HT2 binding sites in the frontal cortex of suicide victims and control subjects. Am J Psychiatry 146:730–736

Attar-Levy D, Martinot J-L, Blin J, Dao-Castellana M-H, Crouzel C, Mazoyer B, Poirer M-F, Bourdel M-C, Aymard N, Syrota A, Feline A (1999) The cortical serotonin2 receptors studied with positron emission tomography and [18F]-setoperone during depressive illness and antidepressant treatment with clomipramine. Biol Psychiatry 45:180–186

Austin MC, Whitehead RE, Edgar CL, Janosky JE, Lewis DA (2002) Localized decrease in serotonin transporter-immunoreactive axons in the prefrontal cortex of depressed subjects committing suicide. Neuroscience 114:807–815

Bacher I, Houle S, Xu X, Zawertailo L, Soliman A, Wilson AA, Selby P, George TP, Sacher J, Miler L, Kish SJ, Rusjan P, Meyer JH (2011) Monoamine oxidase a binding in the prefrontal and anterior cingulate cortices during acute withdrawal from heavy cigarette smoking. Arch Gen Psychiatry 68:817–826

Barbee JG, Jamhour NJ (2002) Lamotrigine as an augmentation agent in treatment-resistant depression. J Clin Psychiatry 63:737–741

Barbee JG, Thompson TR, Jamhour NJ, Stewart JW, Conrad EJ, Reimherr FW, Thompson PM, Shelton RC (2011) A double-blind placebo-controlled trial of lamotrigine as an antidepressant augmentation agent in treatment-refractory unipolar depression. J Clin Psychiatry 72:1405–1412

Barraclough B, Bunch J, Nelson B, Sainsbury P (1974) A hundred cases of suicide: clinical aspects. Br J Psychiatry 125:355–373

Barton DA, Esler MD, Dawood T, Lambert EA, Haikerwal D, Brenchley C, Socratous F, Hastings J, Guo L, Wiesner G, Kaye DM, Bayles R, Schlaich MP, Lambert GW (2008) Elevated brain serotonin turnover in patients with depression: effect of genotype and therapy. Arch Gen Psychiatry 65:38–46

Beck A, Steer R, Kovacs M, Garrison B (1985) Hopelessness and eventual suicide: a 10-year prospective study of patients hospitalized with suicidal ideation. Am J Psychiatry 142(5):559–563

Beck AT, Brown G, Steer RA (1989) Prediction of eventual suicide in psychiatric inpatients by clinical ratings of hopelessness. J Consult Clin Psychol 57:309–310

Benmansour S, Cecchi M, Morilak D, Gerhardt G, Javors M, Gould G, Frazer A (1999) Effects of chronic antidepressant treatments on serotonin transporter function. Density mRNA Level J Neurosci 19(23):10494–10501

Bhagwagar Z, Rabiner EA, Sargent PA, Grasby PM, Cowen PJ (2004) Persistent reduction in brain serotonin1A receptor binding in recovered depressed men measured by positron emission tomography with [11C]WAY-100635. Mol Psychiatry 9:386–392

Bhagwagar Z, Hinz R, Taylor M, Fancy S, Cowen P, Grasby P (2006) Increased 5-HT(2A) receptor binding in euthymic, medication-free patients recovered from depression: a positron emission study with [(11)C]MDL 100,907. Am J Psychiatry 163:1580–1587

Biver F, Wikler D, Lotstra F, Damhaut P, Goldman S, Mendlewicz J (1997) Serotonin 5-HT2 receptor imaging in major depression: focal changes in orbito-insular cortex. Br J Psychiatry 171:444–448

Blier P, Galzin AM, Langer SZ (1989) Diurnal variation in the function of serotonin terminals in the rat hypothalamus. J Neurochem 52:453–459

Bligh-Glover W, Kolli TN, Shapiro-Kulnane L, Dilley GE, Friedman L, Balraj E, Rajkowska G, Stockmeier CA (2000) The serotonin transporter in the midbrain of suicide victims with major depression. Biol Psychiatry 47:1015–1024

Bouvard M, Charles S, Guerin J, Aimard G, Cottraux J (1992) Study of beck's hopelessness scale. validation and factor analysis. Encephale 18:237–240

Breslau N, Peterson EL, Schultz LR, Chilcoat HD, Andreski P (1998) Major depression and stages of smoking a longitudinal investigation. Arch Gen Psychiatry 55:161–166

Brockington IF, Cernik KF, Schofield EM, Downing AR, Francis AF, Keelan C (1981) Puerperal psychosis phenomena and diagnosis. Arch Gen Psychiatry 38:829–833
Brucke T, Kornhuber J, Angelberger P, Asenbaum S, Frassine H, Podreka I (1993) SPECT imaging of dopamine and serotonin transporters with [123I]beta-CIT. binding kinetics in the human brain. J Neural Transm Gen Sect 94:137–146
Brunswick DJ, Amsterdam JD, Mozley PD, Newberg A (2003) Greater availability of brain dopamine transporters in major depression shown by [99m Tc]TRODAT-1 SPECT imaging. Am J Psychiatry 160:1836–1841
Buchert R, Schulze O, Wilke F, Berding G, Thomasius R, Petersen K, Brenner W, Clausen M (2006) Is correction for age necessary in SPECT or PET of the central serotonin transporter in young, healthy adults? J Nucl Med 47:38–42
Buck A, Gucker PM, Schonbachler RD, Arigoni M, Kneifel S, Vollenweider FX, Ametamey SM, Burger C (2000) Evaluation of serotonergic transporters using PET and [11C](+)McN-5652: assessment of methods. J Cereb Blood Flow Metab 20:253–262
Cane D, Olinger L, Gotlib I, Kuiper N (1986) Factor structure of the dysfunctional attitude scale in a student population. J Clin Psychol 42:307–309
Cannon B, Mulroy R, Otto MW, Rosenbaum JF, Fava M, Nierenberg AA (1999) Dysfunctional attitudes and poor problem solving skills predict hopelessness in major depression. J Affect Disord 55:45–49
Cannon DM, Ichise M, Rollis D, Klaver JM, Gandhi SK, Charney DS, Manji HK, Drevets WC (2007) Elevated serotonin transporter binding in major depressive disorder assessed using positron emission tomography and [(11)C]DASB; comparison with bipolar disorder. Biol Psychiatry 62:870–877
Carey MP, Kalra DL, Carey KB, Halperin S, Richards CS (1993) Stress and unaided smoking cessation: a prospective investigation. J Consult Clin Psychol 61:831–838
Carlsson A, Svennerholm L, Winblad B (1980) Seasonal and circadian monoamine variations in human brains examined post mortem. Acta Psychiatr Scand Suppl 280:75–85
Carroll FI, Kotian P, Dehghani A, Gray JL, Kuzemko MA, Parham KA, Abraham P, Lewin AH, Boja JW, Kuhar MJ (1995) Cocaine and 3 beta-(4′-substituted phenyl)tropane-2 beta-carboxylic acid ester and amide analogues. New high-affinity and selective compounds for the dopamine transporter. J Med Chem 38:379–388
Ciliax BJ, Drash GW, Staley JK, Haber S, Mobley CJ, Miller GW, Mufson EJ, Mash DC, Levey AI (1999) Immunocytochemical localization of the dopamine transporter in human brain. J Comp Neurol 409:38–56
Crow TJ, Cross AJ, Cooper SJ, Deakin JF, Ferrier IN, Johnson JA, Joseph MH, Owen F, Poulter M, Lofthouse R et al (1984) Neurotransmitter receptors and monoamine metabolites in the brains of patients with Alzheimer-type dementia and depression, and suicides. Neuropharmacology 23:1561–1569
DeRubeis RJ, Evans MD, Hollon SD, Garvey MJ, Grove WM, Tuason VB (1990) How does cognitive therapy work? Cognitive change and symptom change in cognitive therapy and pharmacotherapy for depression. J Consult Clin Psychol 58:862–869
Deschwanden A, Karolewicz B, Feyissa AM, Treyer V, Ametamey SM, Johayem A, Burger C, Auberson YP, Sovago J, Stockmeier CA, Buck A, Hasler G (2011) Reduced metabotropic glutamate receptor 5 density in major depression determined by [11C]ABP688 PET and postmortem study. Am J Psychiatry 168:727–734
Dewar KM, Grondin L, Carli M, Lima L, Reader TA (1992) [3H]paroxetine binding and serotonin content of rat cortical areas, hippocampus, neostriatum, ventral mesencephalic tegmentum, and midbrain raphe nuclei region following p-chlorophenylalanine and p-chloroamphetamine treatment. J Neurochem 58:250–257
D-haenen H, Bossuyt A, Mertens J, Bossuyt-Piron C, Gijesmans M, Kaufman L (1992) SPECT imaging of serotonin2 receptors in depression. Psychiatry Res Neuroimaging 45:227–237
Drevets WC, Frank E, Price JC, Kupfer DJ, Holt D, Greer PJ, Huang Y, Gautier C, Mathis C (1999) PET imaging of serotonin 1A receptor binding in depression. Biol Psychiatry 46:1375–1387

Elliott R, Rubinsztein JS, Sahakian BJ, Dolan RJ (2002) The neural basis of mood-congruent processing biases in depression. Arch Gen Psychiatry 59:597–604
Fava M, Bless E, Otto M, Pava J, Rosenbaum J (1994) Dysfunctional attitudes in major depression changes with pharmacotherapy. J Nerv Ment Dis 182(1):45–49
Fowler JS, Volkow ND, Wang GJ, Pappas N, Logan J, Shea C, Alexoff D, MacGregor RR, Schlyer DJ, Zezulkova I, Wolf AP (1996) Brain monoamine oxidase A inhibition in cigarette smokers. Proc Natl Acad Sci U S A 93:14065–14069
Freis ED (1954) Mental depression in hypertensive patients treated for long periods with large doses of reserpine. N Engl J Med 251:1006–1008
Frokjaer VG, Mortensen EL, Nielsen FA, Haugbol S, Pinborg LH, Adams KH, Svarer C, Hasselbalch SG, Holm S, Paulson OB, Knudsen GM (2008) Frontolimbic serotonin 2A receptor binding in healthy subjects is associated with personality risk factors for affective disorder. Biol Psychiatry 63:569–576
Garvey AJ, Bliss RE, Hitchcock JL, Heinold JW, Rosner B (1992) Predictors of smoking relapse among self-quitters: a report from the normative aging study. Addict Behav 17:367–377
Ginovart N, Wilson AA, Meyer JH, Hussey D, Houle S (2001) Positron emission tomography quantification of [(11)C]-DASB binding to the human serotonin transporter: modeling strategies. J Cereb Blood Flow Metab 21:1342–1353
Gordon I, Weizman R, Rehavi M (1996) Modulatory effect of agents active in the presynaptic dopaminergic system on the striatal dopamine transporter. Eur J Pharmacol 298:27–30
Graham D, Tahraoui L, Langer SZ (1987) Effect of chronic treatment with selective monoamine oxidase inhibitors and specific 5-hydroxytryptamine uptake inhibitors on [3H]paroxetine binding to cerebral cortical membranes of the rat. Neuropharmacology 26:1087–1092
Han S, Rowell PP, Carr LA (1999) D2 autoreceptors are not involved in the down-regulation of the striatal dopamine transporter caused by alpha-methyl-p-tyrosine. Res Commun Mol Pathol Pharmacol 104:331–338
Hasler G, van der Veen JW, Tumonis T, Meyers N, Shen J, Drevets WC (2007) Reduced prefrontal glutamate/glutamine and gamma-aminobutyric acid levels in major depression determined using proton magnetic resonance spectroscopy. Arch Gen Psychiatry 64:193–200
Hasler G, Fromm S, Carlson PJ, Luckenbaugh DA, Waldeck T, Geraci M, Roiser JP, Neumeister A, Meyers N, Charney DS, Drevets WC (2008) Neural response to catecholamine depletion in unmedicated subjects with major depressive disorder in remission and healthy subjects. Arch Gen Psychiatry 65:521–531
Hendricksen M, Thomas AJ, Ferrier IN, Ince P, O'Brien JT (2004) Neuropathological study of the dorsal raphe nuclei in late-life depression and Alzheimer's disease with and without depression. Am J Psychiatry 161:1096–1102
Herold N, Uebelhack K, Franke L, Amthauer H, Luedemann L, Bruhn H, Felix R, Uebelhack R, Plotkin M (2006) Imaging of serotonin transporters and its blockade by citalopram in patients with major depression using a novel SPECT ligand [(123)I]-ADAM. J Neural Transm 113:659–670
Houle S, Ginovart N, Hussey D, Meyer J, Wilson A (2000) Imaging the serotonin transporter with positron emission tomography: initial human studies with [11C]DAPP and [11C]DASB. Eur J Nucl Med 27:1719–1722
Hoyer D, Pazos A, Probst A, Palacios JM (1986) Serotonin receptors in the human brain. II. characterization and autoradiographic localization of 5-HT1C and 5-HT2 recognition sites. Brain Res 376:97–107
Hrdina P, Vu T (1993) Chronic fluoxetine treatment upregulates 5-HT uptake sites and 5-HT2 receptors in rat brain: an autoradiographic study. Synapse 14:324–331
Hrdina P, Foy B, Hepner A, Summers R (1990) Antidepressant binding sites in brain: autoradiographic comparison of [3H]paroxetine and [3H]imipramine localization and relationship to serotonin transporter. J Pharm Exp Ther 252:410–418
Ichimiya T, Suhara T, Sudo Y, Okubo Y, Nakayama K, Nankai M, Inoue M, Yasuno F, Takano A, Maeda J, Shibuya H (2002) Serotonin transporter binding in patients with mood disorders: a PET study with [11C](+)McN5652. Biol Psychiatry 51:715–722

Ichise M, Liow JS, Lu JQ, Takano A, Model K, Toyama H, Suhara T, Suzuki K, Innis RB, Carson RE (2003) Linearized reference tissue parametric imaging methods: application to [11C]DASB positron emission tomography studies of the serotonin transporter in human brain. J Cereb Blood Flow Metab 23:1096–1112

Ikawa K, Watanabe A, Kaneno S, Toru M (1993) Modulation of [3H]mazindol binding sites in rat striatum by dopaminergic agents. Eur J Pharmacol 250:261–266

Ikoma Y, Suhara T, Toyama H, Ichimiya T, Takano A, Sudo Y, Inoue M, Yasuno F, Suzuki K (2002) Quantitative analysis for estimating binding potential of the brain serotonin transporter with [11 C]McN5652. J Cereb Blood Flow Metab 22:490–501

Innis RB, Seibyl JP, Scanley BE, Laruelle M, Abi-Dargham A, Wallace E, Baldwin RM, Zea-Ponce Y, Zoghbi S, Wang S et al (1993) Single photon emission computed tomographic imaging demonstrates loss of striatal dopamine transporters in Parkinson disease. Proc Natl Acad Sci U S A 90:11965–11969

Joensuu M, Tolmunen T, Saarinen PI, Tiihonen J, Kuikka J, Ahola P, Vanninen R, Lehtonen J (2007) Reduced midbrain serotonin transporter availability in drug-naive patients with depression measured by SERT-specific [(123)I] nor-beta-CIT SPECT imaging. Psychiatry Res 154:125–131

Johnson S, Stockmeier CA, Meyer JH, Austin MC, Albert PR, Wang J, May WL, Rajkowska G, Overholser JC, Jurjus G, Dieter L, Johnson C, Sittman DB, Ou XM (2011) The reduction of R1, a novel repressor protein for monoamine oxidase A, in major depressive disorder. Neuropsychopharmacology 36(10):2139–2148

Kasper S, Wehr TA, Bartko JJ, Gaist PA, Rosenthal NE (1989) Epidemiological findings of seasonal changes in mood and behavior. A telephone survey of Montgomery county. Md Arch Gen Psychiatry 46:823–833

Kendell RE, Chalmers JC, Platz C (1987) Epidemiology of puerperal psychoses. Br J Psychiatry 150:662–673

Kenford SL, Smith SS, Wetter DW, Jorenby DE, Fiore MC, Baker TB (2002) Predicting relapse back to smoking: contrasting affective and physical models of dependence. J Consult Clin Psychol 70:216–227

Kent JM, Coplan JD, Lombardo I, Hwang DR, Huang Y, Mawlawi O, Van Heertum RL, Slifstein M, Abi-Dargham A, Gorman JM, Laruelle M (2002) Occupancy of brain serotonin transporters during treatment with paroxetine in patients with social phobia: a positron emission tomography study with 11C McN 5652. Psychopharmacology (Berl) 164:341–348

Kilbourn MR, Sherman PS, Pisani T (1992) Repeated reserpine administration reduces in vivo [18F]GBR 13119 binding to the dopamine uptake site. Eur J Pharmacol 216:109–112

Kim YJ, Ichise M, Ballinger JR, Vines D, Erami SS, Tatschida T, Lang AE (2002) Combination of dopamine transporter and D2 receptor SPECT in the diagnostic evaluation of PD, MSA, and PSP. Mov Disord 17:303–312

Kish SJ, Shannak K, Hornykiewicz O (1988) Uneven pattern of dopamine loss in the striatum of patients with idiopathic Parkinson's disease. Pathophysiologic and clinical implications. N Engl J Med 318:876–880

Klimek V, Roberson G, Stockmeier CA, Ordway GA (2003) Serotonin transporter and MAO-B levels in monoamine nuclei of the human brainstem are normal in major depression. J Psychiatr Res 37:387–397

Konradi C, Svoma E, Jellinger K, Riederer P, Denney R, Thibault J (1988) Topographic immunocytochemical mapping of monoamine oxidase-A, monoamine oxidase-B and tyrosine hydroxylase in human post mortem brain stem. Neuroscience 26:791–802

Konradi C, Kornhuber J, Froelich L, Fritze J, Heinsen H, Beckmann H, Schulz E, Riederer P (1989) Demonstration of monoamine oxidase-A and -B in the human brainstem by a histochemical technique. Neuroscience 33:383–400

Kontur PJ, al-Tikriti M, Innis RB, Roth RH (1994) Postmortem stability of monoamines, their metabolites, and receptor binding in rat brain regions. J Neurochem 62:282–290

Korf J, Praag HMV (1971) Retarded depression and the dopamine metabolism. Psychopharmacologia 19:199–203

Koskela A, Kauppinen T, Keski-Rahkonen A, Sihvola E, Kaprio J, Rissanen A, Ahonen A (2008) Brain serotonin transporter binding of [123I]ADAM: within-subject variation between summer and winter data. Chronobiol Int 25:657–665

Kuikka JT, Bergstrom KA, Vanninen E, Laulumaa V, Hartikainen P, Lansimies E (1993) Initial experience with single-photon emission tomography using iodine-123-labelled 2 beta-carbomethoxy-3 beta-(4-iodophenyl) tropane in human brain. Eur J Nucl Med 20:783–786

Lambert GW, Reid C, Kaye DM, Jennings GL, Esler MD (2002) Effect of sunlight and season on serotonin turnover in the brain. Lancet 360:1840–1842

Lanzenberger RR, Mitterhauser M, Spindelegger C, Wadsak W, Klein N, Mien LK, Holik A, Attarbaschi T, Mossaheb N, Sacher J, Geiss-Granadia T, Kletter K, Kasper S, Tauscher J (2007) Reduced serotonin-1A receptor binding in social anxiety disorder. Biol Psychiatry 61:1081–1089

Laruelle M (2000) Imaging synaptic neurotransmission with in vivo binding competition techniques: a critical review. J Cereb Blood Flow Metab 20:423–451

Laruelle M, Giddings SS, Zea-Ponce Y, Charney DS, Neumeyer JL, Baldwin RM, Innis RB (1994) Methyl 3 beta-(4-[125I]iodophenyl)tropane-2 beta-carboxylate in vitro binding to dopamine and serotonin transporters under "physiological" conditions. J Neurochem 62:978–986

Law M, Tang JL (1995) An analysis of the effectiveness of interventions intended to help people stop smoking. Arch Intern Med 155:1933–1941

Lawrence KM, De Paermentier F, Cheetham SC, Crompton MR, Katona CL, Horton RW (1990) Brain 5-HT uptake sites, labelled with [3H]paroxetine, in antidepressant-free depressed suicides. Brain Res 526:17–22

Leake A, Fairbairn AF, McKeith IG, Ferrier IN (1991) Studies on the serotonin uptake binding site in major depressive disorder and control post-mortem brain: neurochemical and clinical correlates. Psychiatry Res 39:155–165

Linnet K, Koed K, Wiborg O, Gregersen N (1995) Serotonin depletion decreases serotonin transporter mRNA levels in rat brain. Brain Res 697:251–253

Liotti M, Mayberg HS, McGinnis S, Brannan SL, Jerabek P (2002) Unmasking disease-specific cerebral blood flow abnormalities: mood challenge in patients with remitted unipolar depression. Am J Psychiatry 159:1830–1840

Little KY, McLauglin DP, Ranc J, Gilmore J, Lopez JF, Watson SJ, Carroll FI, Butts JD (1997) Serotonin transporter binding sites and mRNA levels in depressed persons committing suicide. Biol Psychiatry 41:1156–1164

Luque JM, Kwan SW, Abell CW, Da Prada M, Richards JG (1995) Cellular expression of mRNAs encoding monoamine oxidases A and B in the rat central nervous system. J Comp Neurol 363:665–680

Magnusson A (2000) An overview of epidemiological studies on seasonal affective disorder. Acta Psychiatr Scand 101:176–184

Malison RT, Price LH, Berman R, van Dyck CH, Pelton GH, Carpenter L, Sanacora G, Owens MJ, Nemeroff CB, Rajeevan N, Baldwin RM, Seibyl JP, Innis RB, Charney DS (1998) Reduced brain serotonin transporter availability in major depression as measured by [123I]-2 beta-carbomethoxy-3 beta-(4-iodophenyl)tropane and single photon emission computed tomography [see comments]. Biol Psychiatry 44:1090–1098

Malone KM, Oquendo MA, Haas GL, Ellis SP, Li S, Mann JJ (2000) Protective factors against suicidal acts in major depression: reasons for living. Am J Psychiatry 157:1084–1088

Mann JJ, Stanley M, McBride PA, McEwen BS (1986) Increased serotonin2 and beta-adrenergic receptor binding in the frontal cortices of suicide victims. Arch Gen Psychiatry 43:954–959

Mann J, Underwood M, Arango V (1996) Postmortem studies of suicide victims biology of schizophrenia and affective disorders. American Psychiatric Press, Washington, pp 197–221

Mann JJ, Huang YY, Underwood MD, Kassir SA, Oppenheim S, Kelly TM, Dwork AJ, Arango V (2000) A serotonin transporter gene promoter polymorphism (5-HTTLPR) and prefrontal cortical binding in major depression and suicide [see comments]. Arch Gen Psychiatry 57:729–738

Marazziti D, Rossi A, Giannaccini G, Zavaglia KM, Dell'Osso L, Lucacchini A, Cassano GB (1999) Distribution and characterization of [3H]mesulergine binding in human brain postmortem. Eur Neuropsychopharmacol 10:21–26

Meltzer C, Price J, Mathis C, Greer P, Cantwell M, Houck P, Mulsant B, Ben-Eliezer D, Lopresti B, DeKosky S, Reynolds C (1999) PET imaging of serotonin type 2A receptors in late-life neuropsychiatric disorders. Am J Psych 156(12):1871–1878

Messa C, Colombo C, Moresco RM, Gobbo C, Galli L, Lucignani G, Gilardi MC, Rizzo G, Smeraldi E, Zanardi R, Artigas F, Fazio F (2003) 5-HT(2A) receptor binding is reduced in drug-naive and unchanged in SSRI-responder depressed patients compared to healthy controls: a PET study. Psychopharmacology (Berl) 167:72–78

Meyer JH (2007) Imaging the serotonin transporter during major depressive disorder and antidepressant treatment. J Psychiatry Neurosci 32:86–102

Meyer JH (2008) Applying neuroimaging ligands to study major depressive disorder. Semin Nucl Med 38:287–304

Meyer J, Kapur S, Houle S, DaSilva J, Owczarek B, Brown G, Wilson A, Kennedy S (1999) Prefrontal cortex 5-HT2 receptors in depression: a [18F] setoperone PET imaging study. Am J Psychiatry 156:1029–1034

Meyer JH, Kruger S, Wilson AA, Christensen BK, Goulding VS, Schaffer A, Minifie C, Houle S, Hussey D, Kennedy SH (2001a) Lower dopamine transporter binding potential in striatum during depression. Neuroreport 12:4121–4125

Meyer JH, Wilson AA, Ginovart N, Goulding V, Hussey D, Hood K, Houle S (2001b) Occupancy of serotonin transporters by paroxetine and citalopram during treatment of depression: a [(11)C]DASB PET imaging study. Am J Psychiatry 158:1843–1849

Meyer JH, McMain S, Kennedy SH, Korman L, Brown GM, DaSilva JN, Wilson AA, Blak T, Eynan-Harvey R, Goulding VS, Houle S, Links P (2003) Dysfunctional attitudes and 5-HT(2) receptors during depression and self-harm. Am J Psychiatry 160:90–99

Meyer JH, Houle S, Sagrati S, Carella A, Hussey DF, Ginovart N, Goulding V, Kennedy J, Wilson AA (2004a) Brain serotonin transporter binding potential measured with carbon 11-labeled DASB positron emission tomography: effects of major depressive episodes and severity of dysfunctional attitudes. Arch Gen Psychiatry 61:1271–1279

Meyer JH, Wilson AA, Sagrati S, Hussey D, Carella A, Potter WZ, Ginovart N, Spencer EP, Cheok A, Houle S (2004b) Serotonin transporter occupancy of five selective serotonin reuptake inhibitors at different doses: an [11C]DASB positron emission tomography study. Am J Psychiatry 161:826–835

Meyer JH, Ginovart N, Boovariwala A, Sagrati S, Hussey D, Garcia A, Young T, Praschak-Rieder N, Wilson AA, Houle S (2006a) Elevated monoamine oxidase a levels in the brain: an explanation for the monoamine imbalance of major depression. Arch Gen Psychiatry 63:1209–1216

Meyer JH, McNeely HE, Sagrati S, Boovariwala A, Martin K, Verhoeff NP, Wilson AA, Houle S (2006b) Elevated putamen D(2) receptor binding potential in major depression with motor retardation: an [11C]raclopride positron emission tomography study. Am J Psychiatry 163:1594–1602

Meyer JH, Wilson AA, Sagrati S, Miler L, Rusjan P, Bloomfield PM, Clark M, Sacher J, Voineskos AN, Houle S (2009) Brain monoamine oxidase A binding in major depressive disorder: relationship to selective serotonin reuptake inhibitor treatment, recovery, and recurrence. Arch Gen Psychiatry 66:1304–1312

Mintun MA, Sheline YI, Moerlein SM, Vlassenko AG, Huang Y, Snyder AZ (2004) Decreased hippocampal 5-HT2A receptor binding in major depressive disorder: in vivo measurement with [18F]altanserin positron emission tomography. Biol Psychiatry 55:217–224

Moll G, Moll R, Riederer P, Gsell W, Heinsen H, Denney RM (1990) Immunofluorescence cytochemistry on thin frozen sections of human substantia nigra for staining of monoamine oxidase A and monoamine oxidase B: a pilot study. J Neural Transm Suppl 32:67–77

Nash JR, Sargent PA, Rabiner EA, Hood SD, Argyropoulos SV, Potokar JP, Grasby PM, Nutt DJ (2008) Serotonin 5-HT1A receptor binding in people with panic disorder: positron emission tomography study. Br J Psychiatry 193:229–234

Neumeister A, Pirker W, Willeit M, Praschak-Rieder N, Asenbaum S, Brucke T, Kasper S (2000) Seasonal variation of availability of serotonin transporter binding sites in healthy female subjects as measured by [123I]-2 beta-carbomethoxy-3 beta-(4-iodophenyl)tropane and single photon emission computed tomography. Biol Psychiatry 47:158–160

Neumeister A, Willeit M, Praschak-Rieder N, Asenbaum S, Stastny J, Hilger E, Pirker W, Konstantinidis A, Kasper S (2001) Dopamine transporter availability in symptomatic depressed patients with seasonal affective disorder and healthy controls. Psychol Med 31:1467–1473

Neumeister A, Bain E, Nugent AC, Carson RE, Bonne O, Luckenbaugh DA, Eckelman W, Herscovitch P, Charney DS, Drevets WC (2004a) Reduced serotonin type 1A receptor binding in panic disorder. J Neurosci 24:589–591

Neumeister A, Nugent AC, Waldeck T, Geraci M, Schwarz M, Bonne O, Bain EE, Luckenbaugh DA, Herscovitch P, Charney DS, Drevets WC (2004b) Neural and behavioral responses to tryptophan depletion in unmedicated patients with remitted major depressive disorder and controls. Arch Gen Psychiatry 61:765–773

Newberg AB, Amsterdam JD, Wintering N, Ploessl K, Swanson RL, Shults J, Alavi A (2005) 123I-ADAM binding to serotonin transporters in patients with major depression and healthy controls: a preliminary study. J Nucl Med 46:973–977

Norman W, Miller I, Dow M (1988) Characteristics of depressed patients with elevated levels of dysfunctional cognitions. Cog Ther Res 12:39–51

O'Hara MW, Swain A (1996) Rates and risk of postpartum depression-a meta analysis. Int Rev Psychiatry 8:37–54

O'Hara MW, Schlechte JA, Lewis DA, Wright EJ (1991) Prospective study of postpartum blues. Biologic and psychosocial factors. Arch Gen Psychiatry 48:801–806

Oliver J, Baumgart E (1985) The dysfunctional attitude scale: psychometric properties and relation to depression in an unselected adult population. Cog Ther Res 9:161–167

O'Regan D, Kwok RP, Yu PH, Bailey BA, Greenshaw AJ, Boulton AA (1987) A behavioural and neurochemical analysis of chronic and selective monoamine oxidase inhibition. Psychopharmacology 92:42–47

Ou XM, Chen K, Shih JC (2006) Monoamine oxidase A and repressor R1 are involved in apoptotic signaling pathway. Proc Natl Acad Sci U S A 103:10923–10928

Pandey GN, Dwivedi Y, Rizavi HS, Ren X, Pandey SC, Pesold C, Roberts RC, Conley RR, Tamminga CA (2002) Higher expression of serotonin 5-HT(2A) receptors in the postmortem brains of teenage suicide victims. Am J Psychiatry 159:419–429

Parsey RV, Kegeles LS, Hwang DR, Simpson N, Abi-Dargham A, Mawlawi O, Slifstein M, Van Heertum RL, Mann JJ, Laruelle M (2000) In vivo quantification of brain serotonin transporters in humans using [11C]McN 5652. J Nucl Med 41:1465–1477

Parsey RV, Hastings RS, Oquendo MA, Huang YY, Simpson N, Arcement J, Huang Y, Ogden RT, Van Heertum RL, Arango V, Mann JJ (2006a) Lower serotonin transporter binding potential in the human brain during major depressive episodes. Am J Psychiatry 163:52–58

Parsey RV, Oquendo MA, Ogden RT, Olvet DM, Simpson N, Huang YY, Van Heertum RL, Arango V, Mann JJ (2006b) Altered serotonin 1A binding in major depression: a [carbonyl-C-11]WAY100635 positron emission tomography study. Biol Psychiatry 59:106–113

Perry EK, Marshall EF, Blessed G, Tomlinson BE, Perry RH (1983) Decreased imipramine binding in the brains of patients with depressive illness. Br J Psychiatry 142:188–192

Post RM, Kotin J, Goodwin FK, Gordon EK (1973) Psychomotor activity and cerebrospinal fluid amine metabolites in affective illness. Am J Psychiatry 130:67–72

Praag HM, Korf J, Lakke J, Schut T (1975) Dopamine metabolism in depressions, psychoses, and Parkinson's disease: the problem of the specificity of biological variables in behaviour disorders. Psychol Med 5:138–146

Praschak-Rieder N, Wilson AA, Hussey D, Carella A, Wei C, Ginovart N, Schwarz MJ, Zach J, Houle S, Meyer JH (2005) Effects of tryptophan depletion on the serotonin transporter in healthy humans. Biol Psychiatry 58:825–830

Praschak-Rieder N, Willeit M, Wilson AA, Houle S, Meyer JH (2008) Seasonal variation in human brain serotonin transporter binding. Arch Gen Psychiatry 65:1072–1078

Pratt L, Brody D (2010) Depression and smoking in the U.S. household population aged 20 and over, 2005–2008. NCHS Data Brief 34:1–8

Rajkowska G, Miguel-Hidalgo JJ (2007) Gliogenesis and glial pathology in depression. CNS Neurol Disord Drug Targets 6:219–233

Robins E, Murphy G, Wilkinson R, Gassner S, Kayes J (1959) Some clinical considerations in the prevention of suicide based on a study of 134 successful suicides. Am J Publ Health 49(7):888–899

Rommelspacher H, Meier-Henco M, Smolka M, Kloft C (2002) The levels of norharman are high enough after smoking to affect monoamineoxidase B in platelets. Eur J Pharmacol 441:115–125

Rosen LN, Targum SD, Terman M, Bryant MJ, Hoffman H, Kasper SF, Hamovit JR, Docherty JP, Welch B, Rosenthal NE (1990) Prevalence of seasonal affective disorder at four latitudes. Psychiatry Res 31:131–144

Roth B, McLean S, Zhu X, Chuang D (1987) Characterization of two [3H]ketanserin recognition sites in rat striatum. J Neurochem 49:1833–1838

Rovescalli AC, Brunello N, Riva M, Galimberti R, Racagni G (1989) Effect of different photoperiod exposure on [3H]imipramine binding and serotonin uptake in the rat brain. J Neurochem 52:507–514

Ruhe HG, Mason NS, Schene AH (2007) Mood is indirectly related to serotonin, norepinephrine and dopamine levels in humans: a meta-analysis of monoamine depletion studies. Mol Psychiatry 12:331–359

Ruhe HG, Booij J, Reitsma JB, Schene AH (2009) Serotonin transporter binding with [123I]beta-CIT SPECT in major depressive disorder versus controls: effect of season and gender. Eur J Nucl Med Mol Imaging 36:841–849

Sacher J, Wilson A, Houle S, Hassan S, Rusjan P, Bloomfield P, Stewart D, Meyer J (2010) Elevated brain monoamine oxidase A binding in early postpartum. Arch Gen Psychiatry 67(5):468–474

Sanacora G, Gueorguieva R, Epperson CN, Wu YT, Appel M, Rothman DL, Krystal JH, Mason GF (2004) Subtype-specific alterations of gamma-aminobutyric acid and glutamate in patients with major depression. Arch Gen Psychiatry 61:705–713

Sarchiapone M, Carli V, Camardese G, Cuomo C, Di Giuda D, Calcagni ML, Focacci C, De Risio S (2006) Dopamine transporter binding in depressed patients with anhedonia. Psychiatry Res 147:243–248

Sargent PA, Kjaer KH, Bench CJ, Rabiner EA, Messa C, Meyer J, Gunn RN, Grasby PM, Cowen PJ (2000) Brain serotonin1A receptor binding measured by positron emission tomography with [11C]WAY-100635: effects of depression and antidepressant treatment. Arch Gen Psychiatry 57:174–180

Saura J, Kettler R, Da Prada M, Richards JG (1992) Quantitative enzyme radioautography with 3H-Ro 41-1049 and 3H-Ro 19-6327 in vitro: localization and abundance of MAO-A and MAO-B in rat CNS, peripheral organs, and human brain. J Neurosci 12:1977–1999

Saura J, Bleuel Z, Ulrich J, Mendelowitsch A, Chen K, Shih JC, Malherbe P, Da Prada M, Richards JG (1996) Molecular neuroanatomy of human monoamine oxidases A and B revealed by quantitative enzyme radioautography and in situ hybridization histochemistry. Neuroscience 70:755–774

Schmuck K, Ullmer C, Engels P, Lubbert H (1994) Cloning and functional characterization of the human 5-HT2B serotonin receptor. FEBS Lett 342:85–90

Selvaraj S, Venkatesha Murthy N, Bhagwagar Z, Bose SK, Hinz R, Grasby PM, Cowen PJ (2009) Diminished brain 5-HT transporter binding in major depression: a positron emission tomography study with [(11)C]DASB. Psychopharmacology 213(2-3):555–562

Sharot T, Riccardi AM, Raio CM, Phelps EA (2007) Neural mechanisms mediating optimism bias. Nature 450:102–105

Simons AD, Murphy GE, Levine JL, Wetzel RD (1986) Cognitive therapy and pharmacotherapy for depression. Sustained improvement over one year. Arch Gen Psychiatry 43:43–48

Stanley M, Mann JJ (1983) Increased serotonin-2 binding sites in frontal cortex of suicide victims. Lancet 1:214–216

Stockmeier CA (2003) Involvement of serotonin in depression: evidence from postmortem and imaging studies of serotonin receptors and the serotonin transporter. J Psychiatr Res 37:357–373

Stockmeier CA, Kellar KJ (1986) In vivo regulation of the serotonin-2 receptor in rat brain. Life Sci 38:117–127

Stockmeier CA, Dilley GE, Shapiro LA, Overholser JC, Thompson PA, Meltzer HY (1997) Serotonin receptors in suicide victims with major depression. Neuropsychopharmacology 16:162–173

Stockmeier CA, Howley E, Shi X, Sobanska A, Clarke G, Friedman L, Rajkowska G (2009) Antagonist but not agonist labeling of serotonin-1A receptors is decreased in major depressive disorder. J Psychiatr Res 43:887–894

Todd KG, McManus DJ, Baker GB (1995) Chronic administration of the antidepressants phenelzine, desipramine, clomipramine, or maprotiline decreases binding to 5-hydroxytryptamine2A receptors without affecting benzodiazepine binding sites in rat brain. Cell Mol Neurobiol 15:361–370

Tom SM, Fox CR, Trepel C, Poldrack RA (2007) The neural basis of loss aversion in decision-making under risk. Science 315:515–518

Trivedi MH, Rush AJ, Wisniewski SR, Nierenberg AA, Warden D, Ritz L, Norquist G, Howland RH, Lebowitz B, McGrath PJ, Shores-Wilson K, Biggs MM, Balasubramani GK, Fava M (2006) Evaluation of outcomes with citalopram for depression using measurement-based care in STAR*D: implications for clinical practice. Am J Psychiatry 163:28–40

Turecki G, Briere R, Dewar K, Antonetti T, Lesage AD, Seguin M, Chawky N, Vanier C, Alda M, Joober R, Benkelfat C, Rouleau GA (1999) Prediction of level of serotonin 2A receptor binding by serotonin receptor 2A genetic variation in postmortem brain samples from subjects who did or did not commit suicide. Am J Psychiatry 156:1456–1458

Verhoeff NP, Kapur S, Hussey D, Lee M, Christensen B, Psych C, Papatheodorou G, Zipursky RB (2001) A simple method to measure baseline occupancy of neostriatal dopamine d(2) receptors by dopamine in vivo in healthy subjects. Neuropsychopharmacology 25:213–223

Walter M, Henning A, Grimm S, Schulte RF, Beck J, Dydak U, Schnepf B, Boeker H, Boesiger P, Northoff G (2009) The relationship between aberrant neuronal activation in the pregenual anterior cingulate, altered glutamatergic metabolism, and anhedonia in major depression. Arch Gen Psychiatry 66:478–486

Weissman A (1979) The dysfunctional attitude scale: a validation study. Dissertation Abstr Int 40:1389B–1390B

Willeit M, Ginovart N, Graff A, Rusjan P, Vitcu I, Houle S, Seeman P, Wilson AA, Kapur S (2008) First human evidence of d-amphetamine induced displacement of a D2/3 agonist radioligand: a [11C]-(+)-PHNO positron emission tomography study. Neuropsychopharmacology 33:279–289

Wilson A, Schmidt M, Ginovart N, Meyer J, Houle S (2000) Novel radiotracers for imaging the serotonin transporter by positron emission tomography: synthesis, radiosynthesis, in vitro and ex vivo evaluation of [11C]-labelled 2-(Phenylthio) araalkylamines. J Med Chem 43:3103–3110

Wilson AA, Ginovart N, Hussey D, Meyer J, Houle S (2002) In vitro and in vivo characterisation of [11C]-DASB: a probe for in vivo measurements of the serotonin transporter by positron emission tomography. Nucl Med Biol 29:509–515

World Health Organization (2008) The global burden of disease: 2004 update. Department of Health Statistics and Informatics, Information Evidence and Research Cluster, WHO, Geneva

Xiao Q, Pawlyk A, Tejani-Butt SM (1999) Reserpine modulates serotonin transporter mRNA levels in the rat brain. Life Sci 64:63–68

Yates M, Leake A, Candy JM, Fairbairn AF, McKeith IG, Ferrier IN (1990) 5HT2 receptor changes in major depression. Biol Psychiatry 27:489–496

Yatham LN, Liddle PF, Shiah IS, Scarrow G, Lam RW, Adam MJ, Zis AP, Ruth TJ (2000) Brain serotonin2 receptors in major depression: a positron emission tomography study. Arch Gen Psychiatry 57:850–858

Youdim MB, Edmondson D, Tipton KF (2006) The therapeutic potential of monoamine oxidase inhibitors. Nat Rev Neurosci 7:295–309

Young SN, Smith SE, Pihl RO, Ervin FR (1985) Tryptophan depletion causes a rapid lowering of mood in normal males. Psychopharmacology 87:173–177

Yu A, Yang J, Pawlyk AC, Tejani-Butt SM (1995) Acute depletion of serotonin down-regulates serotonin transporter mRNA in raphe neurons. Brain Res 688:209–212

Zarate CA Jr, Singh JB, Carlson PJ, Brutsche NE, Ameli R, Luckenbaugh DA, Charney DS, Manji HK (2006) A randomized trial of an N-methyl-D-aspartate antagonist in treatment-resistant major depression. Arch Gen Psychiatry 63:856–864

Inflammation and Depression

Patricia A. Zunszain, Nilay Hepgul and Carmine M. Pariante

Abstract Major depressive disorder (MDD) is a complex illness and it is likely that alterations in several interacting systems underlie its pathogenesis. Numerous hypotheses have been proposed to elucidate its origins. The inflammatory hypothesis emphasises the role of psycho-neuroimmunological dysfunctions. This is based on several observations: subsets of MDD patients have an altered peripheral immune system, with impaired cellular immunity and increased levels of proinflammatory cytokines; cytokines can influence neurotransmitter metabolism, neuroendocrine function and regional brain activity, all of which are relevant to depression; acute administration of cytokines causes sickness behaviour which shares features with depression, and patients undergoing cytokine treatment develop depressive symptoms. In this chapter, we discuss the evidence linking inflammation and MDD, looking at data from clinical and animal studies, the role of stress, possible mechanisms and the involvement of genetic polymorphisms. Further understanding of pathways involved is still needed. This will be vital for the identification of new drug targets and preventative strategies.

Keywords Cytokines · Inflammatory · Neuroendocrine · Neurotransmitter · Stress

P. A. Zunszain · N. Hepgul · C. M. Pariante (✉)
Section of Perinatal Psychiatry and Stress, Psychiatry and Immunology (SPI-lab),
Department of Psychological Medicine, Institute of Psychiatry,
King's College London, London, UK
e-mail: carmine.pariante@kcl.ac.uk

P. A. Zunszain (✉)
e-mail: patricia.zunszain@kcl.ac.uk

Curr Topics Behav Neurosci (2013) 14: 135–151
DOI: 10.1007/7854_2012_211

Published Online: 3 May 2012

Contents

1 Introduction

Major depressive disorder (MDD) is a highly prevalent, heterogeneous and recurrent neuropsychiatric condition characterised by a broad range of symptoms, including altered mood and cognitive functions. It has a multifactorial aetiology originating from the interaction between environmental and genetic factors and presents frequent comorbidity. Since MDD is a complex disorder, it is likely that alterations in several interacting systems underlie its pathogenesis. Therefore, numerous hypotheses have been proposed to elucidate its origins. One of these is the inflammatory hypothesis, initially suggested as the macrophage theory of depression (Smith 1991), and now also known as the malaise or cytokine theory of depression (Maes et al. 2009; Miller et al. 2009; Ur et al. 1992). This particular hypothesis emphasises the role of psycho-neuroimmunological dysfunctions where there is activation of the immune system.

Indeed, subsets of MDD patients have an altered peripheral immune system, which includes both impaired cellular immunity, and an increase in the levels of proinflammatory cytokines. Furthermore, studies indicate that innate immune cytokines can influence pathophysiological domains, such as neurotransmitter metabolism, neuroendocrine function and regional brain activity, all of which are relevant to depression (Dantzer et al. 2008; Schiepers et al. 2005). Moreover, administration of high levels of proinflammatory cytokines has been shown to cause changes in behaviour, such as low mood, fatigue, anxiety, sleep disturbances, anhedonia and cognitive dysfunction, all of which closely resemble symptoms observed in MDD (Capuron and Miller 2004; Pollak and Yirmiya 2002). We will discuss each of these observations in further detail, looking at evidence from clinical and animal studies, the role of stress, possible mechanisms and the involvement of genetic polymorphisms.

2 Cytokine Abnormalities in Depression

Several biomarkers of inflammation have been measured in MDD. Some of the most frequently reported are cytokines produced by innate immune cells, including interleukin-1 (IL-1), IL-6 and tumour necrosis factor-alpha (TNF-α). Also of importance is the acute-phase C-reactive protein (CRP) that is produced by hepatocytes in response to IL-6. Recent meta-analyses have reported positive associations of all these immune molecules with depression, in both serum and plasma (Dowlati et al. 2010; Howren et al. 2009). It is important to mention, though, that these increased levels are much lower than those found in autoimmune or infectious diseases. Some of these increases have also been measured in the brain. For example, levels of IL-1β have been described as elevated in cerebrospinal fluid (CSF) of MDD patients (Levine et al. 1999). Associations with changes in cellular immunity, including reduced natural killer cell activity and lymphocyte proliferation have also been reported (Blume et al. 2011; Zorrilla et al. 2001). Additionally, postmortem gene analyses in the prefrontal cortex of MDD patients have suggested an upregulation of a variety of pro and antiinflammatory cytokines (Shelton et al. 2011). Depressed patients have also shown higher levels of other acute phase proteins, chemokines and cellular adhesion molecules including α-1-acid glycoprotein, α-1-antichymotrypsin, haptoglobin, human macrophage chemoattractant protein-1 (MCP-1), soluble intracellular adhesion molecule-1 (sICAM-1) and E-selectin (Raison et al. 2006).

But do some or any of these changes occur before (therefore leading to), or after (therefore as a consequence of) depression? That is, is an altered immune system an "at-risk" or a "state" condition? The issue is still unresolved as the evidence is conflicting. For a long time the brain was considered, from an immunological point of view, to be a privileged organ. We now know that proinflammatory cytokines are produced in response to insults not only by immune cells but also by neurons and neural stem cells (Tsakiri et al. 2008; Zunszain et al. 2012). Cytokines can act directly on the brain causing changes known as "sickness behaviour", a coordinated set of subjective, behavioural and physiological changes that develop in sick individuals during the course of an infection (Dantzer 2004). If the activation of the peripheral immune system is prolonged, the immune signalling to the brain can lead to an exacerbation of sickness behaviour and to the development of depressive symptoms in vulnerable individuals (Dantzer et al. 2008). The therapeutic administration of cytokines and several animal models, discussed below, support this view. Further support for an aetiological role of inflammation in depression comes from longitudinal studies of humans. For example, a decade-long study found that serum levels of high sensitivity CRP in women is an independent risk marker for de novo MDD (Pasco et al. 2010). Levels of CRP and IL-6 have been shown to predict the subsequent development of depression in another 12-year follow-up study (Gimeno et al. 2008). On the other hand, some observations support an "after" or "state" condition. A recent longitudinal study (Copeland et al. 2012) recruited young children relatively free of immune and psychiatric disorders, and followed them up until age 21. CRP levels were not associated with later depression status, while depression was associated with later CRP levels. This relationship was more obvious after cumulative episodes of depression. These results suggest that depression precedes inflammation. Clearly, more studies are needed.

3 Evidence for a Cytokine Involvement

3.1 Therapeutic Cytokine Administration

Chronic therapeutic administration of interferon-alpha (IFN-α) has shown to lead to depressive symptomatology in up to 50 % of patients, who display similar biological alterations as those found in MDD (Asnis and De La Garza 2006; Capuron and Miller 2004; Lotrich et al. 2007; Maddock et al. 2005; Raison et al. 2005). Therefore, IFN-α-induced depression has been used as a model to identify the specific changes in the immune system that may be involved in instigating the behavioural changes that lead to depression. IFN-α is a cytokine released by the innate immune system in response to viral infections, and has been shown to acutely induce the production and release of other innate immune cytokines such as IL-6 and TNF-α (Prather et al. 2009; Raison et al. 2010a). The high rate of depression during antiviral treatment with IFN-α is consistent with the evidence that increased inflammatory processes, activated by chronic psychosocial stress, participate in the pathogenesis of MDD (Miller et al. 2009). Specifically, treatment with IFN-α is associated with increased levels of a number of proinflammatory cytokines in the serum and CSF, such as ICAM-1, IL-1, IL-1 receptor antagonist, IL-6, IL-8, MCP-1 and TNF-α, as well as their soluble receptors. It is of note that IFN-α induces changes in inflammatory biomarkers in all patients, with patients developing depression showing even larger changes in these biomarkers when compared with patients who do not develop depression (Capuron and Miller 2004; Raison et al. 2008; Wichers et al. 2007).

3.2 Immune Stimulation

In comparison to the chronic exposure to IFN-α, acute immune stimulation has also been shown to lead to emotional symptoms indicative of depression. For example, a single exposure to *Salmonella typhi* vaccine led to more negative mood which correlated with increases in IL-6 (Wright et al. 2005), and to slower reaction times in a cognitive task (Brydon et al. 2008). In addition, subjects receiving *Salmonella abortus equi* endotoxin showed a significant increase in their levels of anxiety and depressed mood, together with a decrease of verbal and nonverbal memory functions (Reichenberg et al. 2001). These changes correlated with increased circulating levels of TNF-α, soluble TNF receptors, IL-6, IL-1 receptor antagonist and cortisol. Furthermore, a single *Escherichia coli* endotoxin exposure led to feelings of social disconnection and depressed mood, together with significant increases in IL-6 and TNF-α (Eisenberger et al. 2010b). Moreover, short-term exposure to low-dose *Escherichia coli* endotoxin produced increases in depressed mood over time in healthy individuals, by altering reward-related neural responses (Eisenberger et al. 2010a). Interestingly, subchronic pretreatment with an antidepressant blunted the mood changes induced by the endotoxin without affecting the immune activation (Hannestad et al. 2011).

3.3 Animal Models

Animal studies have provided useful evidence for a putative role of cytokines in the context of depression, as administration of different proinflammatory cytokines in rodents results in sickness behaviour. Animals display a range of sickness symptoms which include decreased interest in exploration, depressed motor activity, withdrawal from social activities, reduced food and water intake, deficient cognition, hyperalgesia and hypersomnia (Dantzer 2001; Yirmiya et al. 2002). In a more indirect way, mice subjected to chronic mild stress (CMS) show increased IL-1β levels in the hippocampus and in parallel, changes in behaviour resembling depressive symptoms, such as decreased sucrose preference and reduced social exploration. In contrast, mice with deletion of the IL-1 receptor do not display such behavioural changes (Goshen et al. 2008). In another indirect study, constant darkness produced depression-like behaviour in mice with concomitant elevated levels of IL-6 in plasma together with increased IL-6 protein levels in the hippocampus (Monje et al. 2011). Furthermore, exposure to yet another cytokine, TNF-α, also produced depressive-like states in mice, which were then blocked by coadministration of an anti-TNF-α antibody (Kaster et al. 2012). Interestingly, treatment of animals with antidepressants can relieve some of the symptoms of cytokine-induced sickness behaviour (Pollak and Yirmiya 2002).

4 The Role of Stress

It is now accepted that psychosocial stress can precipitate episodes of depression (Kendler et al. 1999). Concomitantly, episodes of stress can induce inflammatory processes (Black 2003). For example, chronic stress, including difficult caregiving and hostile marital relationships, has been associated with increased levels of CRP (Miller 2008). In particular, a stressful childhood seems to produce neuroendocrine and immunological abnormalities that are thought to mediate the development of a proinflammatory phenotype in adulthood (Chida et al. 2007; Elenkov 2008).

In a study using the Trier Social Stress Test, an acute stressor, men with MDD and a history of early life stress displayed an exaggerated IL-6 response, together with increased DNA binding of the key proinflammatory transcription nuclear factor NF-κB in peripheral blood mononuclear cells (Pace et al. 2006). Additionally, greater acute IL-6 release and higher IL-6 concentrations over time have been observed in individuals with moderate to severe childhood maltreatment (Carpenter et al. 2010). Furthermore, a history of childhood trauma correlates with higher levels of CRP, as well as an increased reactivity of the hypothalamic pituitary adrenal (HPA) axis in adulthood (Danese et al. 2008).

A disturbance in HPA axis functionality (see mechanisms below) is one of the characteristic features of MDD (Miller et al. 2009; Pariante 2003, 2006; Stetler and Miller 2011) and appears to be involved also in IFN-α-induced depression. Hyperactivity of the HPA axis in MDD can be demonstrated by increased circulating cortisol and nonsuppression of cortisol following administration of dexamethasone (Holsboer et al. 1982). Interestingly, while dexamethasone administration significantly suppressed mitogen-induced IL-1β production in healthy controls, MDD patients exhibited dexamethasone nonsuppression (Maes et al. 1991). Overall, these changes are said to confer an immune resistance to glucocorticoids. An alteration of glucocorticoid receptor (GR) function might play a central role in this glucocorticoid resistance, or impairment of the glucocorticoid-mediated negative feedback of the HPA axis (Carvalho and Pariante 2008). IFN-α has also been shown to induce a progressive increase in cortisol output during the day, accompanied by a reduction in the cortisol awakening response (Wichers et al. 2007). Additionally, depression following IFN-α treatment has been associated with an increase in evening cortisol levels and a consequent flattening of the cortisol rhythm (that is, a smaller difference between the morning peak and the evening trough) (Raison et al. 2010a).

5 Cytokines Effects in the Brain

5.1 Alterations in Neurotransmitter Systems

Cytokines affect the synthesis, release and reuptake of all neurotransmitters associated with the development and maintenance of affective disorders (Anisman et al. 2008; Dunn 2006; Miller et al. 2009). The most studied has been serotonin, and in particular, the function of its transporter, which regulates serotonin uptake into presynaptic neurons. For example, IL-1β has been shown to increase serotonin uptake in the brain (Zhu et al. 2010), while IFN-α has shown attenuation of the expression of 5HT1A, one of the serotonin receptors (Cai et al. 2005). IFN-α treated patients have shown low levels of serum serotonin, with this effect being more pronounced in those who developed depression (Bonaccorso et al. 2002; Wichers et al. 2005). Moreover, the severity of the depression correlated with higher CSF levels of IL-6 and lower levels of the serotonin metabolite 5-hydroxyindoleacetic acid, indicating reduced brain serotonergic activity (Raison et al. 2009). Also of importance is dopamine. Reduced CSF concentrations of homovanillic acid, one of the metabolites of dopamine, have been associated with depressive-like behaviour in primates, secondary to IFN-α administration (Felger et al. 2007). In rats, administration of IFN-α has also led to decreased concentrations of dopamine in the CSF, in association with the production of nitric oxide (Kitagami et al. 2003). Finally, effects on norepinephrine have also been observed, with peripheral and central administration of IL-1 or high doses of TNF-α inducing norepinephrine release in the brain, most markedly in the hypothalamus (Dunn 2006).

5.2 *Alterations of Neuroendocrine Pathways*

Numerous cytokines are able to activate the HPA axis (for a comprehensive review, see Turnbull and Rivier (1999)). Regulation of the HPA axis starts by release of corticotropin-releasing hormone (CRH) from the paraventricular nucleus of the hypothalamus. CRH leads to the release of adrenocorticotropic hormone (ACTH) in the pituitary, inducing discharge of cortisol from the adrenal cortex. In normal individuals, cortisol exerts a negative feedback mechanism that controls its production, by regulating the synthesis of CRH. This feedback mechanism can be altered by cytokines. Acute cytokine administration in humans has shown to cause both elevated plasma levels of ACTH and cortisol, and also increases in CRH release (Besedovsky and del Rey 1996; Pace et al. 2007; Pariante and Miller 2001). Interestingly, the acute ACTH and cortisol response that follows a first injection of IFN-α in cancer patients correlated with a later development of depression (Capuron et al. 2003).

When administered chronically, though, cytokines are associated with a flattening of the cortisol response and also increased evening cortisol levels (Raison et al. 2010a). Inflammatory cytokines can disrupt GR function, while decreasing GR expression (Pace et al. 2007). For example, cytokines have been found to inhibit GR function in cytokine-releasing immune cells by acting on GR translocation or on GR-mediated gene transcription, resulting in a deregulated production of other proinflammatory cytokines. Mechanisms described include activation of signalling pathways, such as p38 mitogen-activated protein kinase (MAPK), NF-κB, signal transducers and activators of transcription and cyclooxygenase (Dantzer et al. 2008; Miller et al. 1999; Pace et al. 2007; Pariante et al. 1999). Interestingly, glucocorticoids such as cortisol have initially been found to inhibit the inflammatory response and the production and release of proinflammatory cytokines (Rhen and Cidlowski 2005).

Since communication occurs between the endocrine, immune, and central nervous system, an activation of the inflammatory responses can affect neuroendocrine processes, and vice versa (Zunszain et al. 2011). Several mechanisms may

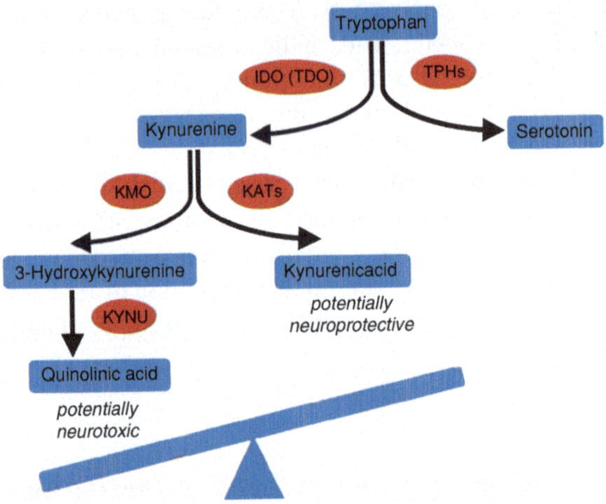

Fig. 1 Simplified kynurenine pathway of tryptophan metabolism, *IDO* indolamine-2,3-dioxygenase, *KAT* kynurenine aminotransferase, *KMO* kynurenine 3-monooxygenase or kynurenine 3-hydroxylase, *KYNU* kynureninase, *TDO* tryptophan-2,3-dioxygenase, *TPH* tryptophan hydroxylase

be involved. A study of peripheral blood monocytes of caregivers of patients with malignant brain cancer supports the idea that chronic stress is accompanied by GR resistance and activation of proinflammatory pathways (Miller et al. 2008). These monocytes showed diminished expression of transcripts bearing response elements for glucocorticoids, and heightened expression of transcripts with response elements for NF-κB (Hayden et al. 2006). These findings suggest that stress produces functional resistance to glucocorticoids, which enables activation of proinflammatory transcription control pathways, even in the absence of decreased GR mRNA expression or excess cortisol production. This evidence suggests that HPA axis hyperactivity and inflammation might be part of the same pathophysiological process: HPA axis hyperactivity is a marker of glucocorticoid resistance, implying ineffective action of glucocorticoid hormones on target tissues, which could lead to immune activation; and, equally, inflammation could stimulate HPA axis activity via both a direct action of cytokines on the brain and by inducing glucocorticoid resistance.

5.3 *Imbalances in the Kynurenine Pathway*

Several proinflammatory cytokines can induce the enzyme indoleamine 2,3-dioxygenase (IDO) (Wirleitner et al. 2003), which converts the essential amino acid tryptophan into kynurenine, as seen in Fig. 1. Kynurenine is a precursor of the bioactive metabolites quinolinic acid (QUIN) and kynurenic acid (KYNA). QUIN is an N-methyl-D-aspartate (NMDA) receptor agonist, potentially neurotoxic and thus potentially contributing to depression (Hashimoto 2009; Muller and Schwarz 2007; Wichers et al. 2005), whereas KYNA is an NMDA receptor antagonist generally considered neuroprotective (Moroni 1999; Nozaki and Beal 1992).

Following IDO activation, both the relative balance between QUIN and KYNA and the reduced peripheral availability of tryptophan (putatively leading to reduced serotonin synthesis in the brain) have been proposed to be of significance in depression and neurodegeneration (Capuron et al. 2002; Maes et al. 2009; McNally et al. 2008; Muller and Schwarz 2007; Myint and Kim 2003; Myint et al. 2007; Raison et al. 2010b; Wichers et al. 2005). Animal data supports the involvement of IDO. For example, peripheral administration of lipopolysaccharide activates IDO leading to a depressive-like syndrome in mice, which is prevented by blockade of IDO activation (O'Connor et al. 2009). In patients with IFN-α-induced depression, both reduced levels of serum tryptophan and serotonin, together with elevated levels of serum tryptophan metabolites have been reported (Bonaccorso et al. 2002; Wichers et al. 2005). Furthermore, increased levels of kynurenine and QUIN which correlate with depressive symptoms have been observed in both blood and CSF of IFN-α treated patients (Capuron et al. 2002; Raison et al. 2010b).

5.4 *Neurogenesis, Neuroplasticity and Neurodegeneration*

Proinflammatory cytokines influence neuronal functioning through changes in apoptosis, oxidative stress and metabolic derangement, as well as by impairing processes of synaptic plasticity and neurogenesis (Hayley et al. 2005; Leonard and Maes 2012; Maes et al. 2009; McAfoose and Baune 2009; Song and Wang 2011). For example, IL-1β and TNF-α have been shown to inhibit long-term potentiation, a form of neuronal plasticity believed to underlie learning and memory (Cunningham et al. 1996), both of which are frequently affected in MDD. IL-1β also decreased hippocampal neurogenesis, both in animal (Goshen et al. 2008; Koo and Duman 2008; Kuzumaki et al. 2010) and human in vitro models (Zunszain et al. 2012). TNF-α and IFN-γ have shown inhibition of neural progenitor cell proliferation (Ben-Hur et al. 2003). Furthermore, TNF-α induced neuronal cell death through blockade of the glutamate transporter activity, thereby potentiates glutamate neurotoxicity (Zou and Crews 2005). Indirectly, the detrimental effects of stress on neurogenesis have been reversed by hippocampal transplantation of neural progenitor cells that overexpress IL-1 receptor antagonist (Ben Menachem-Zidon et al. 2008), or by using IL-1 receptor knockout mice (Koo and Duman 2009). Concomitantly, this blockade of cytokine action has led to a decrease of depressive symptoms in these models.

6 Role of Polyunsaturated Fatty Acids

Polyunsaturated Fatty Acids (PUFAs) play an important role in MDD (Freeman et al. 2006). Lower levels of ω-3 PUFA levels may contribute to depression. Administration of ω-3 PUFAs significantly improved depressive symptoms in patients with major depression (Lesperance et al. 2011) and in elderly depressed females (Rondanelli et al. 2011). A meta-analysis of ω-3 PUFAs in depression showed that eicosapentaenoic acid (EPA) has significant antidepressant activity (Lin and Su 2007). In animal models, ω-3 PUFAs have also been shown to have antidepressant effects as demonstrated by an improvement in the forced swimming test in rats (Huang et al. 2008). One possible explanation for the association between depleted ω-3 PUFAs and depression, and the clinical efficacy of ω-3 PUFAs administration, is the fact that ω-3 and ω-6 PUFAs modulate immune functions (Leonard and Maes 2012). ω-3 PUFAs, like EPA, attenuate prostaglandin E2 (PGE2) synthesis and the production of monocytic and T cell cytokines, including IL-1, IL-6, TNF-α and IFN-α (Su 2009). The depletion of ω-3 PUFAs and the relatively higher ω-6 contents in depression may cause an increased production of proinflammatory cytokines and T cell cytokines, and therefore take part in the immune pathophysiology of depression (Leonard and Maes 2012).

7 Genetics

Genetic factors have an impact on several aspects regulated by cytokines including their own regulation, the response to antidepressants and to PUFAs. For example, healthy controls with the short allele of the serotonin transporter gene promoter region have a higher IL-6/IL-10 ratio than healthy controls with the long allele. As a higher IL-6/IL-10 ratio indicates a proinflammatory state, patients with the short allele may be at higher risk of developing depression (Caspi et al. 2003). Polymorphisms of the serotonin transporter gene have also been found to be associated with an increased number of depressive symptoms and elevated plasma IL-6 levels (Bull et al. 2009). MDD patients who have specific polymorphisms of the IL-1β gene are also less likely to achieve remission under adequate antidepressant treatment (Baune et al. 2010). Polymorphisms in two key enzymes of PUFA metabolism, phospholipase A2 (PLA2) and cyclooxygenase 2 (COX-2), are also relevant. Together with the erythrocyte levels of the three main PUFAs, docosahexaenoic acid (DHA), EPA and arachidonic acid (AA), patients who develop IFN-α-induced depression have a higher frequency of particular PLA2 or COX-2 genotypes (Su et al. 2010). Interestingly, these "at risk" genotypes are also associated with lower levels of the antiinflammatory PUFAs, DHA and EPA, at baseline or during IFN-α treatment. This once again suggests that increased reactivity of inflammatory processes (or, in this case, lack of "restrain" on the inflammatory processes by the lower PUFAs levels) is fundamental in the development of the depressive symptoms. Results may be population and/or race-dependent. While in Caucasians a polymorphism in the promoter region of the gene encoding IDO was found to be associated with depressive symptoms after IFN-α treatment (Smith et al. 2011), no link was found in a Brazilian population (Galvao-de Almeida et al. 2011).

8 Final Remarks and Future Research

Although the evidence linking inflammation and MDD is increasing, the role of immune response dysfunctions in the pathogenesis of this disorder is still not totally clear. The use of high throughput technologies will allow collection of genetic, gene expression and protein data from the blood and brain of depressed patients, as well as well-validated animal models of depression, which will hopefully result in a better understanding of the link. A clear description of pathways involved will be vital for the identification of new drug targets. Interestingly, antiinflammatory drugs such as aspirin (Mendlewicz et al. 2006), celecoxib (Muller et al. 2006) and etanercept (Tyring et al. 2006) have already been shown to have antidepressant effects or improve antidepressant response. Therefore, normalisation of activated inflammation pathways may be one novel mechanism of antidepressant action. Furthermore, given the relative contribution of psychosocial stress to both inflammation and depression, psychotherapeutic strategies aimed at reducing stress may indirectly affect inflammation, and could therefore be considered as alternative treatment and preventative strategies. Additionally, it is also possible that changes in inflammation are critical in only a subset of patients, where an altered response may be sufficient to cause the full symptoms of MDD, so a better classification of subtypes of MDD will shed light on this. Moreover, if chronic inflammatory changes represent a common feature of depression, this could predispose depressed patients to neurodegenerative changes later in life. Indeed, there is clinical evidence that depression is a common antecedent of Alzheimer's disease and may be an early manifestation of dementia before the cognitive declines become apparent (Byers and Yaffe 2011; Leonard 2007). Finally, considering the impact of cytokines on all neurotransmitter systems, a better understanding of the consequences of altered inflammatory responses may help not only MDD but also other neuropsychiatric illnesses.

Acknowledgments CMP has been funded by the UK MRC since 1999, first as a Clinical Training Fellow, and currently as an MRC Clinician Scientist Fellow. His research is also funded by the NIHR South London and Maudsley NHS Foundation Trust and Institute of Psychiatry Specialist Biomedical Research Centre for Mental Health, the NARSAD, the APIRE, the British Academy and the European Union Framework 7. PZ is funded by a NARSAD Young Investigator Award. The authors have no relevant financial interest to disclose.

References

Anisman H, Merali Z, Hayley S (2008) Neurotransmitter, peptide and cytokine processes in relation to depressive disorder: comorbidity between depression and neurodegenerative disorders. Prog Neurobiol 85:1–74

Asnis GM, De La Garza R 2nd (2006) Interferon-induced depression in chronic hepatitis C: a review of its prevalence, risk factors, biology, and treatment approaches. J Clin Gastroenterol 40:322–335

Baune BT, Dannlowski U, Domschke K, Janssen DG, Jordan MA, Ohrmann P, Bauer J, Biros E, Arolt V, Kugel H, Baxter AG, Suslow T (2010) The interleukin 1 beta (IL1B) gene is associated with failure to achieve remission and impaired emotion processing in major depression. Biol Psychiatry 67:543–549

Ben-Hur T, Ben-Menachem O, Furer V, Einstein O, Mizrachi-Kol R, Grigoriadis N (2003) Effects of proinflammatory cytokines on the growth, fate, and motility of multipotential neural precursor cells. Mol Cell Neurosci 24:623–631

Ben Menachem-Zidon O, Goshen I, Kreisel T, Ben Menahem Y, Reinhartz E, Ben Hur T, Yirmiya R (2008) Intrahippocampal transplantation of transgenic neural precursor cells overexpressing interleukin-1 receptor antagonist blocks chronic isolation-induced impairment in memory and neurogenesis. Neuropsychopharmacology 33:2251–2262

Besedovsky HO, del Rey A (1996) Immune-neuro-endocrine interactions: facts and hypotheses. Endocr Rev 17:64–102

Black PH (2003) The inflammatory response is an integral part of the stress response: implications for atherosclerosis, insulin resistance, type II diabetes and metabolic syndrome X. Brain Behav Immun 17:350–364

Blume J, Douglas SD, Evans DL (2011) Immune suppression and immune activation in depression. Brain Behav Immun 25:221–229

Bonaccorso S, Marino V, Puzella A, Pasquini M, Biondi M, Artini M, Almerighi C, Verkerk R, Meltzer H, Maes M (2002) Increased depressive ratings in patients with hepatitis C receiving interferon-alpha-based immunotherapy are related to interferon-alpha-induced changes in the serotonergic system. J Clin Psychopharmacol 22:86–90

Brydon L, Harrison NA, Walker C, Steptoe A, Critchley HD (2008) Peripheral inflammation is associated with altered substantia nigra activity and psychomotor slowing in humans. Biol Psychiatry 63:1022–1029

Bull SJ, Huezo-Diaz P, Binder EB, Cubells JF, Ranjith G, Maddock C, Miyazaki C, Alexander N, Hotopf M, Cleare AJ, Norris S, Cassidy E, Aitchison KJ, Miller AH, Pariante CM (2009) Functional polymorphisms in the interleukin-6 and serotonin transporter genes, and depression and fatigue induced by interferon-alpha and ribavirin treatment. Mol Psychiatry 14:1095–1104

Byers AL, Yaffe K (2011) Depression and risk of developing dementia. Nat Rev Neurol 7:323–331

Cai W, Khaoustov VI, Xie Q, Pan T, Le W, Yoffe B (2005) Interferon-alpha-induced modulation of glucocorticoid and serotonin receptors as a mechanism of depression. J Hepatol 42: 880–887

Capuron L, Raison CL, Musselman DL, Lawson DH, Nemeroff CB, Miller AH (2003) Association of exaggerated HPA axis response to the initial injection of interferon-alpha with development of depression during interferon-alpha therapy. Am J Psychiatry 160:1342–1345

Capuron L, Miller AH (2004) Cytokines and psychopathology: lessons from interferon-alpha. Biol Psychiatry 56:819–824

Capuron L, Ravaud A, Neveu PJ, Miller AH, Maes M, Dantzer R (2002) Association between decreased serum tryptophan concentrations and depressive symptoms in cancer patients undergoing cytokine therapy. Mol Psychiatry 7:468–473

Carpenter LL, Gawuga CE, Tyrka AR, Lee JK, Anderson GM, Price LH (2010) Association between plasma IL-6 response to acute stress and early-life adversity in healthy adults. Neuropsychopharmacology 35:2617–2623

Carvalho LA, Pariante CM (2008) In vitro modulation of the glucocorticoid receptor by antidepressants. Stress 11:411–424

Caspi A, Sugden K, Moffitt TE, Taylor A, Craig IW, Harrington H, McClay J, Mill J, Martin J, Braithwaite A, Poulton R (2003) Influence of life stress on depression: moderation by a polymorphism in the 5-HTT gene. Science 301:386–389

Chida Y, Sudo N, Sonoda J, Hiramoto T, Kubo C (2007) Early-life psychological stress exacerbates adult mouse asthma via the hypothalamus-pituitary-adrenal axis. Am J Respir Crit Care Med 175:316–322

Copeland WE, Shanahan L, Worthman C, Angold A, Costello EJ (2012) Cumulative depression episodes predict later C-reactive protein levels: a prospective analysis. Biol Psychiatry 71: 15–21

Cunningham AJ, Murray CA, O'Neill LA, Lynch MA, O'Connor JJ (1996) Interleukin-1 beta (IL-1 beta) and tumour necrosis factor (TNF) inhibit long-term potentiation in the rat dentate gyrus in vitro. Neurosci Lett 203:17–20

Danese A, Moffitt TE, Pariante CM, Ambler A, Poulton R, Caspi A (2008) Elevated inflammation levels in depressed adults with a history of childhood maltreatment. Arch Gen Psychiatry 65:409–415

Dantzer R (2001) Cytokine-induced sickness behavior: mechanisms and implications. Ann N Y Acad Sci 933:222–234

Dantzer R (2004) Cytokine-induced sickness behaviour: a neuroimmune response to activation of innate immunity. Eur J Pharmacol 500:399–411

Dantzer R, O'Connor JC, Freund GG, Johnson RW, Kelley KW (2008) From inflammation to sickness and depression: when the immune system subjugates the brain. Nat Rev Neurosci 9:46–56

Dowlati Y, Herrmann N, Swardfager W, Liu H, Sham L, Reim EK, Lanctot KL (2010) A meta-analysis of cytokines in major depression. Biol Psychiatry 67:446–457

Dunn AJ (2006) Effects of cytokines and infections on brain neurochemistry. Clin Neurosci Res 6:52–68

Eisenberger NI, Berkman ET, Inagaki TK, Rameson LT, Mashal NM, Irwin MR (2010a) Inflammation-induced anhedonia: endotoxin reduces ventral striatum responses to reward. Biol Psychiatry 68:748–754

Eisenberger NI, Inagaki TK, Mashal NM, Irwin MR (2010b) Inflammation and social experience: an inflammatory challenge induces feelings of social disconnection in addition to depressed mood. Brain Behav Immun 24:558–563

Elenkov IJ (2008) Neurohormonal-cytokine interactions: implications for inflammation, common human diseases and well-being. Neurochem Int 52:40–51

Felger JC, Alagbe O, Hu F, Mook D, Freeman AA, Sanchez MM, Kalin NH, Ratti E, Nemeroff CB, Miller AH (2007) Effects of interferon-alpha on rhesus monkeys: a nonhuman primate model of cytokine-induced depression. Biol Psychiatry 62:1324–1333

Freeman MP, Hibbeln JR, Wisner KL, Davis JM, Mischoulon D, Peet M, Keck PE Jr, Marangell LB, Richardson AJ, Lake J, Stoll AL (2006) Omega-3 fatty acids: evidence basis for treatment and future research in psychiatry. J Clin Psychiatry 67:1954–1967

Galvao-de Almeida A, Quarantini LC, Sampaio AS, Lyra AC, Parise CL, Parana R, de Oliveira IR, Koenen KC, Miranda-Scippa A, Guindalini C (2011) Lack of association of indoleamine 2,3-dioxygenase polymorphisms with interferon-alpha-related depression in hepatitis C. Brain Behav Immun 25:1491–1497

Gimeno D, Marmot MG, Singh-Manoux A (2008) Inflammatory markers and cognitive function in middle-aged adults: the Whitehall II study. Psychoneuroendocrinology 33:1322–1334

Goshen I, Kreisel T, Ben-Menachem-Zidon O, Licht T, Weidenfeld J, Ben-Hur T, Yirmiya R (2008) Brain interleukin-1 mediates chronic stress-induced depression in mice via adrenocortical activation and hippocampal neurogenesis suppression. Mol Psychiatry 13:717–728

Hannestad J, DellaGioia N, Ortiz N, Pittman B, Bhagwagar Z (2011) Citalopram reduces endotoxin-induced fatigue. Brain Behav Immun 25:256–259

Hashimoto K (2009) Emerging role of glutamate in the pathophysiology of major depressive disorder. Brain Res Rev 61:105–123

Hayden MS, West AP, Ghosh S (2006) NF-kappaB and the immune response. Oncogene 25:6758–6780

Hayley S, Poulter MO, Merali Z, Anisman H (2005) The pathogenesis of clinical depression: stressor- and cytokine-induced alterations of neuroplasticity. Neuroscience 135:659–678

Holsboer F, Liebl R, Hofschuster E (1982) Repeated dexamethasone suppression test during depressive illness. Normalisation of test result compared with clinical improvement. J Affect Disord 4:93–101

Howren MB, Lamkin DM, Suls J (2009) Associations of depression with C-reactive protein, IL-1, and IL-6: a meta-analysis. Psychosom Med 71:171–186
Huang SY, Yang HT, Chiu CC, Pariante CM, Su KP (2008) Omega-3 fatty acids on the forced-swimming test. J Psychiatr Res 42:58–63
Kaster MP, Gadotti VM, Calixto JB, Santos AR, Rodrigues AL (2012) Depressive-like behavior induced by tumor necrosis factor-alpha in mice. Neuropharmacology 62:419–426
Kendler KS, Karkowski LM, Prescott CA (1999) Causal relationship between stressful life events and the onset of major depression. Am J Psychiatry 156:837–841
Kitagami T, Yamada K, Miura H, Hashimoto R, Nabeshima T, Ohta T (2003) Mechanism of systemically injected interferon-alpha impeding monoamine biosynthesis in rats: role of nitric oxide as a signal crossing the blood-brain barrier. Brain Res 978:104–114
Koo JW, Duman RS (2008) IL-1beta is an essential mediator of the antineurogenic and anhedonic effects of stress. Proc Natl Acad Sci U S A 105:751–756
Koo JW, Duman RS (2009) Interleukin-1 receptor null mutant mice show decreased anxiety-like behavior and enhanced fear memory. Neurosci Lett 456:39–43
Kuzumaki N, Ikegami D, Imai S, Narita M, Tamura R, Yajima M, Suzuki A, Miyashita K, Niikura K, Takeshima H, Ando T, Ushijima T, Suzuki T (2010) Enhanced IL-1beta production in response to the activation of hippocampal glial cells impairs neurogenesis in aged mice. Synapse 64:721–728
Leonard BE (2007) Inflammation, depression and dementia: are they connected? Neurochem Res 32:1749–1756
Leonard B, Maes M (2012) Mechanistic explanations how cell-mediated immune activation, inflammation and oxidative and nitrosative stress pathways and their sequels and concomitants play a role in the pathophysiology of unipolar depression. Neurosci Biobehav Rev 36:764–785
Lesperance F, Frasure-Smith N, St-Andre E, Turecki G, Lesperance P, Wisniewski SR (2011) The efficacy of omega-3 supplementation for major depression: a randomized controlled trial. J Clin Psychiatry 72:1054–1062
Levine J, Barak Y, Chengappa KN, Rapoport A, Rebey M, Barak V (1999) Cerebrospinal cytokine levels in patients with acute depression. Neuropsychobiology 40:171–176
Lin PY, Su KP (2007) A meta-analytic review of double-blind, placebo-controlled trials of antidepressant efficacy of omega-3 fatty acids. J Clin Psychiatry 68:1056–1061
Lotrich FE, Rabinovitz M, Gironda P, Pollock BG (2007) Depression following pegylated interferon-alpha: characteristics and vulnerability. J Psychosom Res 63:131–135
Maddock C, Landau S, Barry K, Maulayah P, Hotopf M, Cleare AJ, Norris S, Pariante CM (2005) Psychopathological symptoms during interferon-alpha and ribavirin treatment: effects on virologic response. Mol Psychiatry 10:332–333
Maes M, Minner B, Suy E, D'Hondt P, Jacobs MP, Raus J (1991) Cortisol escape from suppression by dexamethasone during depression is strongly predicted by basal cortisol hypersecretion and increasing age combined. Psychoneuroendocrinology 16:295–310
Maes M, Yirmyia R, Noraberg J, Brene S, Hibbeln J, Perini G, Kubera M, Bob P, Lerer B, Maj M (2009) The inflammatory and neurodegenerative (I&ND) hypothesis of depression: leads for future research and new drug developments in depression. Metab Brain Dis 24:27–53
McAfoose J, Baune BT (2009) Evidence for a cytokine model of cognitive function. Neurosci Biobehav Rev 33:355–366
McNally L, Bhagwagar Z, Hannestad J (2008) Inflammation, glutamate, and glia in depression: a literature review. CNS Spectr 13:501–510
Mendlewicz J, Kriwin P, Oswald P, Souery D, Alboni S, Brunello N (2006) Shortened onset of action of antidepressants in major depression using acetylsalicylic acid augmentation: a pilot open-label study. Int Clin Psychopharmacol 21:227–231
Miller AH (2008) Inflammation versus glucocorticoids as purveyors of pathology during stress: have we reached the tipping point? Biol Psychiatry 64:263–265
Miller AH, Pariante CM, Pearce BD (1999) Effects of cytokines on glucocorticoid receptor expression and function. Glucocorticoid resistance and relevance to depression. Adv Exp Med Biol 461:107–116

Miller GE, Chen E, Sze J, Marin T, Arevalo JM, Doll R, Ma R, Cole SW (2008) A functional genomic fingerprint of chronic stress in humans: blunted glucocorticoid and increased NF-kappaB signaling. Biol Psychiatry 64:266–272

Miller AH, Maletic V, Raison CL (2009) Inflammation and its discontents: the role of cytokines in the pathophysiology of major depression. Biol Psychiatry 65:732–741

Monje FJ, Cabatic M, Divisch I, Kim EJ, Herkner KR, Binder BR, Pollak DD (2011) Constant darkness induces IL-6-dependent depression-like behavior through the NF-kappaB signaling pathway. J Neurosci 31:9075–9083

Moroni F (1999) Tryptophan metabolism and brain function: focus on kynurenine and other indole metabolites. Eur J Pharmacol 375:87–100

Muller N, Schwarz MJ (2007) The immune-mediated alteration of serotonin and glutamate: towards an integrated view of depression. Mol Psychiatry 12:988–1000

Muller N, Schwarz MJ, Dehning S, Douhe A, Cerovecki A, Goldstein-Muller B, Spellmann I, Hetzel G, Maino K, Kleindienst N, Moller HJ, Arolt V, Riedel M (2006) The cyclooxygenase-2 inhibitor celecoxib has therapeutic effects in major depression: results of a double-blind, randomized, placebo controlled, add-on pilot study to reboxetine. Mol Psychiatry 11:680–684

Myint AM, Kim YK (2003) Cytokine-serotonin interaction through IDO: a neurodegeneration hypothesis of depression. Med Hypotheses 61:519–525

Myint AM, Kim YK, Verkerk R, Scharpe S, Steinbusch H, Leonard B (2007) Kynurenine pathway in major depression: evidence of impaired neuroprotection. J Affect Disord 98:143–151

Nozaki K, Beal MF (1992) Neuroprotective effects of L-kynurenine on hypoxia-ischemia and NMDA lesions in neonatal rats. J Cereb Blood Flow Metab 12:400–407

O'Connor JC, Lawson MA, Andre C, Moreau M, Lestage J, Castanon N, Kelley KW, Dantzer R (2009) Lipopolysaccharide-induced depressive-like behavior is mediated by indoleamine 2,3-dioxygenase activation in mice. Mol Psychiatry 14:511–522

Pace TW, Mletzko TC, Alagbe O, Musselman DL, Nemeroff CB, Miller AH, Heim CM (2006) Increased stress-induced inflammatory responses in male patients with major depression and increased early life stress. Am J Psychiatry 163:1630–1633

Pace TW, Hu F, Miller AH (2007) Cytokine-effects on glucocorticoid receptor function: relevance to glucocorticoid resistance and the pathophysiology and treatment of major depression. Brain Behav Immun 21:9–19

Pariante CM (2003) Depression, stress and the adrenal axis. J Neuroendocrinol 15:811–812

Pariante CM (2006) The glucocorticoid receptor: part of the solution or part of the problem? J Psychopharmacol 20:79–84

Pariante CM, Miller AH (2001) Glucocorticoid receptors in major depression: relevance to pathophysiology and treatment. Biol Psychiatry 49:391–404

Pariante CM, Pearce BD, Pisell TL, Sanchez CI, Po C, Su C, Miller AH (1999) The proinflammatory cytokine, interleukin-1alpha, reduces glucocorticoid receptor translocation and function. Endocrinology 140:4359–4366

Pasco JA, Nicholson GC, Williams LJ, Jacka FN, Henry MJ, Kotowicz MA, Schneider HG, Leonard BE, Berk M (2010) Association of high-sensitivity C-reactive protein with de novo major depression. Br J Psychiatry 197:372–377

Pollak Y, Yirmiya R (2002) Cytokine-induced changes in mood and behaviour: implications for 'depression due to a general medical condition', immunotherapy and antidepressive treatment. Int J Neuropsychopharmacol 5:389–399

Prather AA, Rabinovitz M, Pollock BG, Lotrich FE (2009) Cytokine-induced depression during IFN-alpha treatment: the role of IL-6 and sleep quality. Brain Behav Immun 23:1109–1116

Raison CL, Demetrashvili M, Capuron L, Miller AH (2005) Neuropsychiatric adverse effects of interferon-alpha: recognition and management. CNS Drugs 19:105–123

Raison CL, Capuron L, Miller AH (2006) Cytokines sing the blues: inflammation and the pathogenesis of depression. Trends Immunol 27:24–31

Raison CL, Borisov AS, Woolwine BJ, Massung B, Vogt G, Miller AH (2008) Interferon-alpha effects on diurnal hypothalamic-pituitary-adrenal axis activity: relationship with proinflammatory cytokines and behavior. Mol Psychiatry 15:535–547

Raison CL, Borisov AS, Majer M, Drake DF, Pagnoni G, Woolwine BJ, Vogt GJ, Massung B, Miller AH (2009) Activation of central nervous system inflammatory pathways by interferon-alpha: relationship to monoamines and depression. Biol Psychiatry 65:296–303

Raison CL, Borisov AS, Woolwine BJ, Massung B, Vogt G, Miller AH (2010a) Interferon-alpha effects on diurnal hypothalamic-pituitary-adrenal axis activity: relationship with proinflammatory cytokines and behavior. Mol Psychiatry 15:535–547

Raison CL, Dantzer R, Kelley KW, Lawson MA, Woolwine BJ, Vogt G, Spivey JR, Saito K, Miller AH (2010b) CSF concentrations of brain tryptophan and kynurenines during immune stimulation with IFN-alpha: relationship to CNS immune responses and depression. Mol Psychiatry 15:393–403

Reichenberg A, Yirmiya R, Schuld A, Kraus T, Haack M, Morag A, Pollmacher T (2001) Cytokine-associated emotional and cognitive disturbances in humans. Arch Gen Psychiatry 58:445–452

Rhen T, Cidlowski JA (2005) Antiinflammatory action of glucocorticoids–new mechanisms for old drugs. N Engl J Med 353:1711–1723

Rondanelli M, Giacosa A, Opizzi A, Pelucchi C, La Vecchia C, Montorfano G, Negroni M, Berra B, Politi P, Rizzo AM (2011) Long chain omega 3 polyunsaturated fatty acids supplementation in the treatment of elderly depression: effects on depressive symptoms, on phospholipids fatty acids profile and on health-related quality of life. J Nutr Health Aging 15:37–44

Schiepers OJ, Wichers MC, Maes M (2005) Cytokines and major depression. Prog Neuropsychopharmacol Biol Psychiatry 29:201–217

Shelton RC, Claiborne J, Sidoryk-Wegrzynowicz M, Reddy R, Aschner M, Lewis DA, Mirnics K (2011) Altered expression of genes involved in inflammation and apoptosis in frontal cortex in major depression. Mol Psychiatry 16:751–762

Smith RS (1991) The macrophage theory of depression. Med Hypotheses 35:298–306

Smith AK, Simon JS, Gustafson EL, Noviello S, Cubells JF, Epstein MP, Devlin DJ, Qiu P, Albrecht JK, Brass CA, Sulkowski MS, McHutchinson JG, Miller AH (2011) Association of a polymorphism in the indoleamine-2,3-dioxygenase gene and interferon-alpha-induced depression in patients with chronic hepatitis C. Mol Psychiatry. [Epub ahead of print]

Song C, Wang H (2011) Cytokines mediated inflammation and decreased neurogenesis in animal models of depression. Prog Neuropsychopharmacol Biol Psychiatry 35:760–768

Stetler C, Miller GE (2011) Depression and hypothalamic-pituitary-adrenal activation: a quantitative summary of four decades of research. Psychosom Med 73:114–126

Su KP (2009) Biological mechanism of antidepressant effect of omega-3 fatty acids: how does fish oil act as a 'mind-body interface'? Neurosignals 17:144–152

Su KP, Huang SY, Peng CY, Lai HC, Huang CL, Chen YC, Aitchison KJ, Pariante CM (2010) Phospholipase A2 and cyclooxygenase 2 genes influence the risk of interferon-alpha-induced depression by regulating polyunsaturated fatty acids levels. Biol Psychiatry 67:550–557

Tsakiri N, Kimber I, Rothwell NJ, Pinteaux E (2008) Differential effects of interleukin-1 alpha and beta on interleukin-6 and chemokine synthesis in neurones. Mol Cell Neurosci 38:259–265

Turnbull AV, Rivier CL (1999) Regulation of the hypothalamic-pituitary-adrenal axis by cytokines: actions and mechanisms of action. Physiol Rev 79:1–71

Tyring S, Gottlieb A, Papp K, Gordon K, Leonardi C, Wang A, Lalla D, Woolley M, Jahreis A, Zitnik R, Cella D, Krishnan R (2006) Etanercept and clinical outcomes, fatigue, and depression in psoriasis: double-blind placebo-controlled randomised phase III trial. Lancet 367:29–35

Ur E, White PD, Grossman A (1992) Hypothesis: cytokines may be activated to cause depressive illness and chronic fatigue syndrome. Eur Arch Psychiatry Clin Neurosci 241:317–322

Wichers MC, Koek GH, Robaeys G, Verkerk R, Scharpe S, Maes M (2005) IDO and interferon-alpha-induced depressive symptoms: a shift in hypothesis from tryptophan depletion to neurotoxicity. Mol Psychiatry 10:538–544

Wichers MC, Kenis G, Koek GH, Robaeys G, Nicolson NA, Maes M (2007) Interferon-alpha-induced depressive symptoms are related to changes in the cytokine network but not to cortisol. J Psychosom Res 62:207–214

Wirleitner B, Neurauter G, Schrocksnadel K, Frick B, Fuchs D (2003) Interferon-gamma-induced conversion of tryptophan: immunologic and neuropsychiatric aspects. Curr Med Chem 10: 1581–1591

Wright CE, Strike PC, Brydon L, Steptoe A (2005) Acute inflammation and negative mood: mediation by cytokine activation. Brain Behav Immun 19:345–350

Yirmiya R, Winocur G, Goshen I (2002) Brain interleukin-1 is involved in spatial memory and passive avoidance conditioning. Neurobiol Learn Mem 78:379–389

Zhu CB, Lindler KM, Owens AW, Daws LC, Blakely RD, Hewlett WA (2010) Interleukin-1 receptor activation by systemic lipopolysaccharide induces behavioral despair linked to MAPK regulation of CNS serotonin transporters. Neuropsychopharmacology 35:2510–2520

Zorrilla EP, Luborsky L, McKay JR, Rosenthal R, Houldin A, Tax A, McCorkle R, Seligman DA, Schmidt K (2001) The relationship of depression and stressors to immunological assays: a meta-analytic review. Brain Behav Immun 15:199–226

Zou JY, Crews FT (2005) TNF alpha potentiates glutamate neurotoxicity by inhibiting glutamate uptake in organotypic brain slice cultures: neuroprotection by NF kappa B inhibition. Brain Res 1034:11–24

Zunszain PA, Anacker C, Cattaneo A, Carvalho LA, Pariante CM (2011) Glucocorticoids, cytokines and brain abnormalities in depression. Prog Neuropsychopharmacol Biol Psychiatry 35:722–729

Zunszain PA, Anacker C, Cattaneo A, Choudhury S, Musaelyan K, Myint AM, Thuret S, Price J, Pariante CM (2012) Interleukin-1beta: a new regulator of the kynurenine pathway affecting human hippocampal neurogenesis. Neuropsychopharmacology 37:939–949

Major Depression: A Role for Hippocampal Neurogenesis?

Margherita M. Lee, Andreas Reif and Angelika G. Schmitt

Abstract Since its discovery in mammals, adult neurogenesis, the process of generating functional neurons from neural progenitor cells in the adult brain, has inspired numerous animal studies. These have revealed that adult neurogenesis is a highly regulated phenomenon. Enriched environment, exercise and learning for instance, are positive regulators while stress and age are major negative regulators. Stressful life events are not only shown to reduce adult neurogenesis levels but are also discussed to be a key element in the development of various neuropsychiatric disorders such as depression. Interestingly, altered monoaminergic brain levels resulting from antidepressant treatment are shown to have a strong reinforcing effect on adult neurogenesis. Additionally, disturbed adult neurogenesis, possibly resulting in a malfunctioning hippocampus, may contribute to the cognitive deficits and reduced hippocampal volumes observed in depressed patients. Hence, the question arises as to whether disturbed adult neurogenesis and the etiopathogenesis of depression are causally linked. In this chapter, we discuss the possible causal interrelation of disturbed adult neurogenesis and the etiopathogenesis of depression as well as the possibility that adult neurogenesis is not exclusively linked to depression but is also linked to other psychiatric disorders including schizophrenia and neurodegenerative diseases like Alzheimer's disease. Additionally, we look at the functional relevance of adult neurogenesis in different species, upon which we base our discussion as to whether adult neurogenesis could be causally linked to

M. M. Lee (✉) · A. Reif · A. G. Schmitt
Department of Psychiatry Psychosomatics and Psychotherapy,
Julius-Maximilians-University of Würzburg, Würzburg, Germany
e-mail: Margherita.lee@stud-mail.uni-wuerzburg.de

A. Reif
e-mail: Reif_A@klinik.uni-wuerzburg.de

A. G. Schmitt
e-mail: Schmitt_A3@klinik.uni-wuerzburg.de

Curr Topics Behav Neurosci (2013) 14: 153–179
DOI: 10.1007/7854_2012_226

Published Online: 18 August 2012

the development of certain brain disorders in humans, or whether it is only an epiphenomenon.

Keywords Adult neurogenesis · Hippocampus · Subgranular zone · Neural Progenitor Cells · Stress

Contents

1 Introduction

Neurogenesis constitutes the process in which neural stem or progenitor cells give rise to new nerve cells during ontogenesis. Adult neurogenesis, the birth of new neurons in the adult brain, was first discovered in the 1960s by Joseph Altman (Altman 1962; Altman and Das 1965, 1967) and forgotten soon after, until it was rediscovered in the 1980s and 1990s in various species from songbirds to primates including humans (Goldman and Nottebohm 1983; Barnea and Nottebohm 1996; Eriksson et al. 1998; Gould et al. 1999b). Substantial production of new neurons in the adult mammalian brain is mainly restricted to the olfactory system, namely the subventricular zone of the lateral ventricle, and the dentate gyrus (DG) of the hippocampal formation. Not all newly formed cells survive, many of them die within weeks after generation (Dayer et al. 2003). Hippocampal adult neurogenesis is restricted to a relatively limited area, the subgranular zone (SGZ) of the DG (red line in Fig. 1). Currently, adult neurogenesis is thought to consist of several developmental stages (Kempermann et al. 2004; Ming and Song 2005) that are characterized by morphologically distinct cells (inset in Fig. 1). Two different types of adult neural stem cells (NSCs) can be found in the SGZ of the DG (no. 1 and 2 in Fig. 1). Type 1 cells represent radial glia-like cells, and type 2 cells are non-radial neural progenitor cells (NPCs). Both are capable of self-replication but possibly have a reciprocal lineage relationship. They proliferate and their offspring (No. 3 in Fig. 1) differentiate and migrate into the granule cell layer of the DG, and show low proliferative potential. Later, these cells extend their dendrites toward the molecular layer of the DG, and their axons extend into the hilus and the Cornu

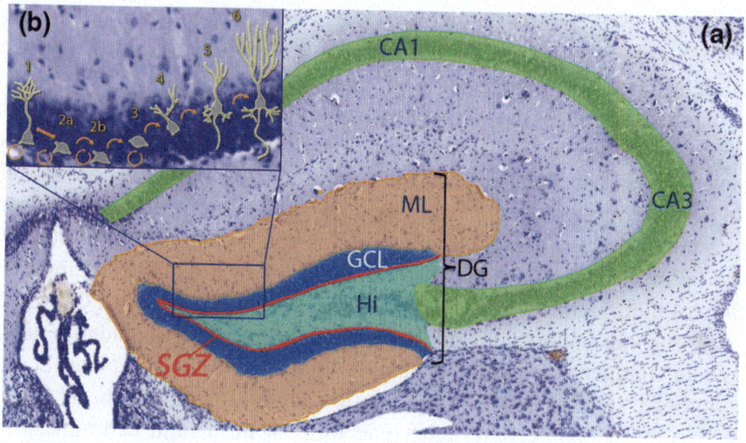

Fig. 1 *The hippocampus, one of the major sites of adult neurogenesis.* **a** Photomicrograph of a Nissl-stained coronal section displaying the mouse hippocampal formation with color accented subregions. The dentate gyrus (DG), which consists of the molecular layer (ML), the granule cell layer (GCL), and the hilus (Hi), as well as the hippocampus proper, which can be subdivided tangentially into the cornu ammonis (CA) sectors CA1, CA2 (not indicated), and CA3 are displayed. The subgranular zone (SGZ), a narrow layer of cells located between the GCL and hilus of the DG, contains adult progenitor cells and is accentuated with red. **b** Schematic representation of cells in different adult neurogenesis stages. The generation of new neurons in the DG can be divided into at least 6 stages. Two different types of adult neural stem cells reside at the SGZ of the DG. Cell No. 1 represents type 1 (radial) neural stem cells, and cells No. 2a/b represent non-radial neural stem cells, which possibly have a reciprocal lineage relationship and are capable of self-replication. Cell No. 3 represents migratory progenitor cells, which show little proliferative action. Cells No. 4–6 represent different stages of postmitotic newborn cells fated to finally become mature granule cells

ammonis area (CA) 3 (No. 4–6 in Fig. 1). Finally, the adult newborn neurons become functionally integrated into the hippocampal network, receiving inputs from the entorhinal cortex and sending outputs to hippocampal area CA3 and the hilus. Although, the physiological and behavioral role of adult neurogenesis is still under debate (Kempermann 2008; Deng et al. 2010; Burghardt et al. 2012; Glasper et al. 2012), nowadays, it is well accepted that it carries on from the late embryonic stage through to old age. The discovery of hippocampal adult neurogenesis has led to an explosion of research in neuroscience over the past two decades.

Typically adult neurogenesis is detected via immunohistochemistry. This can either be performed using 5-bromo-2-deoxyuridine (BrdU) antibodies following intraperitoneal (i.p.) BrdU administration (Gratzner 1982) or using antibodies detecting different endogenous adult neurogenesis markers without the need for stressful i.p. treatment of the respective animals (von Bohlen und Halbach 2011). BrdU can be given either once or a number of times, then incorporating into replicating DNA in place of thymidine (also called deoxythymidine). Neurons that have incorporated the injected BrdU during mitosis and which have subsequently differentiated into neurons, can be identified via double-labeling with antibodies against neuron markers such as NeuN. The use of various antibodies detecting endogenous adult neurogenesis allows the further division of adult neurogenesis into different developmental stages, and allows the monitoring of neurogenesis at these different stages in the same hippocampus using serial sections. This is possible since the various developmental stages correlate with the expression of different markers (von Bohlen und Halbach 2011). For the detection of cell proliferation at the initial phase of the birth of new neurons (adult neurogenesis) antibodies against the cell cycle marker Ki67 is commonly used (Kee et al. 2002). It should be mentioned, however, that antibodies against Ki67 detect mitotic cells regardless of their identity or fate (Gerdes et al. 1991; Scholzen and Gerdes 2000). Other very popular endogenous adult neurogenesis markers include (i) nestin, an intermediate filament expressed in many, if not all, NPCs, (ii) NeuroD, a basic helix-loop-helix protein and a differentiation factor for neurogenesis in diverse species, (iii) the polysialylated embryonic form of the neural cell adhesion molecule (NCAM), abbreviated to PSA-NCAM which has been found in cells that seem to be NPCs related to neural stem cells, iv) TUC (TOAD [turned on after division]/Ulip [UNC-33- like protein]/CRMP [collapsin response-mediator protein])-4 can be used as a marker for early postmitotic neurons but seems also to be expressed in mitotic cells during neurogenesis, and (v) doublecortin (DCX), a protein that promotes microtubule polymerization, which is present in migrating neuroblasts and young neurons (for review see: von Bohlen und Halbach 2011). As DNA labeling by BrdU is potentially toxic, and therefore a confounding factor in some experiments, the importance of these adult neurogenesis-related antibodies for further investigations of the adult neurogenesis phenomenon is often underemphasized.

In the late 1990s, studies relying on the BrdU integration method, brought about a number of discoveries regarding the regulation of adult neurogenesis. Among the positive regulators are enriched environment, voluntary exercise, as well as learning and memory tasks (Kempermann et al. 1997; Gould et al. 1999a; van Praag et al. 1999, 2005). In 1997, Kempermann and his colleagues discovered that mice exposed to an enriched environment, namely a homecage equipped with a running-wheel, toys and tunnels, displayed significantly more new neurons than mice housed in a conventional cage (Kempermann et al. 1997). In addition, Henriette van Praag and co-workers demonstrated that voluntary running resulted in an increase of BrdU-positive cells, indicating that physical activity can regulate hippocampal neurogenesis (van Praag et al. 1999). Other regulatory factors are neurotransmitters such as serotonin, which can both enhance the proliferation of NPCs and the production of new neurons (Gould 1999), gonadal hormones (Spritzer and Galea 2007; Galea 2008; Spritzer et al. 2011), and trophic factors such as brain-derived neurotrophic factor (BDNF) (Bekinschtein et al. 2011). Moreover, chronic antidepressant treatments markedly stimulate hippocampal neurogenesis (Malberg et al. 2000; Paizanis et al. 2007). Negative regulators of adult neurogenesis include binge alcohol exposure in adolescence (Morris et al. 2010) and oxidative stress (Taupin 2010), to give just a few examples. However, the most prominent negative regulators of adult neurogenesis are age (Seki and

Arai 1995; Kuhn et al. 1996; Kempermann et al. 1998; Amrein et al. 2011) and stress exposure during all ontogenetic phases (Gould et al. 1998; Gould and Tanapat 1999; Mirescu et al. 2004; Lucassen et al. 2010a; Hanson et al. 2011).

In addition to a genetic predisposition, stress can be looked upon as a key vulnerability factor in the development of various neuropsychiatric disorders. The intense regulation by stress (among other factors), strongly implicates adult neurogenesis in a variety of pathophysiological mechanisms. Therefore, one of the topics under intense research is the correlation between adult neurogenesis and the pathophysiology of psychiatric disorders like fear and anxiety (Santarelli et al. 2003; Saxe et al. 2006; Revest et al. 2009), major depression (Boldrini et al. 2009; Boldrini and Arango 2010; Lucassen et al. 2010b; Anacker et al. 2011; Snyder et al. 2011), and schizophrenia (Reif et al. 2007). The possible relevance of hippocampal adult neurogenesis for pathological mechanisms underlying major depression as well as other neuropsychiatric disorders is critically discussed in this review.

2 Adult Neurogenesis, Stress and Hippocampal Volume

Stress can be defined as a condition where an environmental demand exceeds the natural regulatory capacity of an organism, particularly in situations that include unpredictability and uncontrollability (Koolhaas et al. 2011). Moreover, stress can be looked upon as a significant causal agent, in the aetiology of a number of neuropsychiatric disorders such as schizophrenia (Wahlberg et al. 1997; van Os et al. 2010), but predominantly in the development of anxiety disorders and depression (Caspi et al. 2003; van Praag 2004; Pittenger and Duman 2008; Saveanu and Nemeroff 2012). The experience of a stressful situation results in the stimulation of the hypothalamic-pituitary-adrenal (HPA) axis. The HPA axis activity is governed by the secretion of corticotropin releasing hormone/factor (CRH/CRF) and vasopressin from the hypothalamus, which in turn activates the secretion of adrenocorticotrophic hormone (ACTH) from the pituitary to stimulate the secretion of the glucocorticoids (cortisol in humans and corticosterone in rodents) from the adrenal cortex. Released glucocorticoids then interact with their receptors in multiple body compartments including the brain (e.g. hippocampus, hypothalamus) where they serve a vital function in feedback inhibition of their own secretion. Additionally, glucocorticoids regulate neuronal survival, neurogenesis and hippocampal volume, as well as the acquisition of new memories and the emotional appraisal of events (Pariante and Lightman 2008).

HPA hyperactivity is regarded as causally linked to depression and to the modes of action of antidepressants (Holsboer 2001). Altered feedback inhibition in depressed patients resulting in HPA hyperactivity was demonstrated by use of the combined dexamethasone (DEX)/CRH test, a refined laboratory test for psychiatric disorders). For this test, patients are administered DEX the night prior to testing and then CRH the following afternoon (Heuser et al. 1994). Measurement of plasma cortisol and ACTH levels before, during, and after CRH administration revealed, that depressed patients release significantly more cortisol and ACTH after DEX and a CRH challenge in comparison with age-matched controls. This so-called DEX/CRH test phenomenon supports the assumption that psychiatric patients are prone to blunted glucocorticoid feedback regulation during the acute illness episode. Diminished glucocorticoid receptor (GR) expression or function has been postulated as causative factor (Heuser et al. 1994).

Based on these findings, Ridder et al. generated GR-heterozygous mutant mice (GR ±), which have a 50 % reduction of GR protein levels in the brain (Ridder et al. 2005). Similar to depressed patients, these mice display reduced feedback inhibition of the HPA axis and a pathological DEX/CRH test in which higher levels of circulating glucocorticoids could be measured in response to the CRH challenge compared to GR- wildtype littermates. Elevated levels of glucocorticoids, for instance as found in chronically stressed individuals, stimulate hippocampal glutamate release (Moghaddam et al. 1994) which in turn diminish stem cell proliferation in the adult DG (Gould and Tanapat 1999). Thus, it does not come as a surprise that GR +/− mice display significantly less BrdU-positive cells in the DG compared to wildtype mice (Kronenberg et al. 2009). Behaviorally, GR +/− mice show increased learned helplessness, a well-established analog of depression-like behavior in mice (Ridder et al. 2005). In their study published in 2007, Thomas and co-workers were able to show that acute psychosocial stress leads to a reduced number of new neurons in the DG of rats. Interestingly, stem cell proliferation itself remained unaltered, whereas short- and long-term survival of the newborn cells was decreased (Thomas et al. 2007). This implicates that acute social stress has a sustained impact on the integration of newborn cells into the hippocampal neural network by diminishing their chance of survival. On the other hand, a recent study showed that coping with intermittent social stress, which is an essential aspect of living in complex social environments, stimulates adult neurogenesis in adult monkeys (Lyons et al. 2010). On the basis of this finding, one could speculate that stress coping follows similar mechanisms as living in an enriched environment, which is known to prompt adult neurogenesis (Kempermann et al. 1997).

HPA axis dysregulation and the adverse effect of increased glucocorticoid concentrations on adult neurogenesis may be one out of several possible causes of the well-documented phenomenon of a decreased hippocampal volume in patients with mood disorders. Several forms of stress, and especially repeated periods of stress, are suggested to cause a reduction of hippocampal volume. Among these are traumatic life events and early life stress such as childhood maltreatment which are often found to be a key component in the life history of patients with mood disorders such as post-traumatic stress-disorder and major depression (Smith 2005; Pittenger and Duman 2008; Teicher et al. 2012). In these subjects, and especially in patients with a history of several depressive episodes, magnetic resonance imaging (MRI) studies revealed significantly smaller hippocampal volumes (especially on the left side) in comparison to controls but with no differences in total cerebral volumes (Sheline 1996; Bremner et al. 2000). This result was later confirmed in two meta-analyses of MRI studies of hippocampal volume in depression (Campbell et al. 2004; Videbech and Ravnkilde 2004). Moreover, it was found that significant neuromorphologic abnormalities in the hippocampus, as well as the dorso-lateral prefrontal cortex and anterior cingulum, of patients with depression were associated with a more severe outcome (Frodl et al. 2008a, b). On the other hand, patients treated repeatedly with antidepressants, such as the SSRIs sertraline, citalopram and paroxetine, the tricyclic antidepressants amitriptyline, amitriptylinoxide and doxepin, or other newer antidepressants like venlafaxine, reboxetine, and mirtazapine, showed a significant volume increase in the left hippocampus compared to the baseline volume (Frodl et al. 2008a). However, neuroimaging and *post mortem* studies also report an average hippocampal volume loss of 4–8 % in schizophrenic patients compared to controls (Geuze et al. 2005; Honea et al. 2005). Notably, reduced brain and/or hippocampal volumes could be already detected in patients at the first episode of schizophrenia, which points to biochemical changes inducing a morphological sequelae in the already predisposed brains or at least in brains of patients with a short disease story (Steen et al. 2006; Watson et al. 2012).

Much has been speculated about the possible mechanisms underpinning these structural changes. Sheline and co-workers suggested that the hippocampal atrophy reflects a neuronal loss due to a progressive process associated with glucocorticoid neurotoxicity (Sheline 1996). To understand how the hippocampus responds at the cellular and subcellular level to glucocorticoids, and how such changes are related to volume measures, Tata and co-workers carried out a study of the effects of long-term glucocorticoids on hippocampal CA3 volume, and identified altered dendritic and glial processes as well as altered numbers and sizes of synapses (Tata et al. 2006; Tata and Anderson 2010). These authors suggested that corticosterone caused astrocytes to take up a larger proportion of the tissue, with the result that the tissue volume was made up of a smaller proportion of neuronal processes and a pronounced loss of synapses. On the other hand, Stockmeier et al. reported in their *post mortem* study of hippocampus tissue that in major depression, the packing density of glia, pyramidal neurons, and granule cell neurons was significantly increased in all hippocampal subfields including the DG, and that pyramidal neuron soma size was significantly decreased as well (Stockmeier et al. 2004). These authors suggested that a significant reduction in neuropil in major depression may account for the decreased hippocampal volume detected by neuroimaging. In addition, they speculated that differential shrinkage of frozen sections of the hippocampus may be caused by an altered water content in hippocampus in major depression (Stockmeier et al. 2004). The findings on hippocampal volume loss due to stress and/or depression, combined with the numerous findings of a stress-induced decrease of adult neurogenesis, have inspired the idea that these two phenomena are causally linked (Jacobs 2002). Czeh and Lucassen, however, argued that either a massive neuronal loss or a suppression of adult neurogenesis in the DG can be excluded, as no major cell loss was apparent in human *post mortem* brain tissue of severely depressed patients (Czeh and Lucassen 2007). Moreover, they argued that in humans, adult neurogenesis adds too few neurons per day for the suppression of this process to provide a significant contribution to the 10–15 % reduction in hippocampal volume in depressed patients determined by the meta-analyses of Campbell et al. (2004) and Videbech and Ravnkilde (2004). However, they also considered that the slowing down of neuronal turnover in the DG by altering both cytogenesis and apoptosis would affect the composition of neurons in the DG, and may have an impact on the connectivity, input, and output properties of the hippocampal circuit, and thus hippocampal function.

3 Antidepressants and Adult Neurogenesis

The stimulatory effect of antidepressant treatment on adult neurogenesis is observed across different species, from rodents (Madsen et al. 2000; Malberg et al. 2000), tree shrews (Czeh et al. 2001), and monkeys (Perera et al. 2007) through to humans (Boldrini et al. 2009). Numerous studies have provided evidence that major classes of antidepressants, including selective serotonin reuptake inhibitors (SSRIs) like fluoxetine and selective norepinephrine reuptake inhibitors such as reboxetine (Malberg et al. 2000; Kodama et al. 2004; Namestkova et al. 2005; Encinas et al. 2006), increase adult neurogenesis. However, different studies have obtained different results, depending on certain parameters including the type of antidepressant and length of treatment. For instance, upregulated NPC proliferation rate and immature neuron survival rate was only achieved by chronic (14 or more days) but not acute (<10 days) treatment with SSRIs and the tricyclic antidepressant imipramine (as reviewed by Hanson et al. 2011), which is consistent with the time course for the therapeutic action of antidepressants. Administration of SSRIs results in elevated extracellular levels of 5-HT, which has been shown to have a positive effect on adult neurogenesis (Gould 1999; Brezun and Daszuta 2000; Banasr et al. 2004). Several studies including the work of Santarelli using 5-HT_{1A} receptor knockout mice (Santarelli et al. 2003), suggest this positive effect of 5-HT and SSRIs on neurogenesis is primarily mediated by 5-HT_{1A} receptor activation. These data are in accord with findings that during brain development, 5-HT is a key regulator of cell division and to modulate different processes including neurogenesis, apoptosis, axon branching, and dendritogenesis (Gaspar et al. 2003). Moreover, adverse experiences during early life and adulthood have been shown to alter the brain 5-HT system, and abnormalities in 5-HT are also strongly linked to the damaging effects of stress and, particularly, stress-induced neuropsychiatric disorders (Lowry et al. 2005; Holmes 2008; Lanfumey et al. 2008).

Among non-drug treatments that have behavioral antidepressant effects and also elevate adult neurogenesis are electroconvulsive shock (Madsen et al. 2000; Malberg et al. 2000; Hellsten et al. 2002), sleep deprivation (Grassi Zucconi et al. 2006), environmental enrichment (Veena et al. 2009), and exercise (van Praag et al. 1999; van Praag et al. 2005; Hanson et al. 2011). Conditioned inhibition of fear (or learned safety) which was shown to reduce depression-like behavior in mice, also promoted the survival of newborn cells in the DG (Pollak et al. 2008). Interestingly, this type of behavioral antidepressant increased the expression of BDNF in the hippocampus and lead to altered expression of genes involved in the dopamine and neuropeptide but not 5-HT system in the basolateral amygdala (Pollak et al. 2008). After several preclinical reports suggested that melatonin, the main product of the pineal gland, itself could be effective as an antidepressant (Overstreet et al. 1998, Papp et al. 2003), Ramirez-Rodriguez and co-workers looked for a possible influence of melatonin on adult neurogenesis. They revealed that NPC proliferation in the DG was not modified by acute or chronic melatonin treatment but the survival of new neurons was promoted by melatonin resulting in a net-increase of the production of new mature neurons (Ramirez-Rodriguez et al. 2009). However, repetitive transcranial magnetic stimulation which has a clinically observed antidepressant effect and attenuates the HPA axis in some studies, did not increase adult neurogenesis (Czeh et al. 2002). So far, the vast majority of substances with antidepressant effects but not all non-drug antidepressant treatments are reported to increase adult neurogenesis (Sahay and Hen 2007; Kempermann 2008; Hanson et al. 2011).

In contrast to antidepressants, studies of antipsychotic drugs, encompassing both acute and chronic treatment regimes, found no change in levels of stem cell proliferation in rodents (Wakade et al. 2002; Halim et al. 2004; Schmitt et al. 2004; Wang et al. 2004), although one such study demonstrated an increase in adult neurogenesis (Dawirs et al. 1998).

Whether adult neurogenesis may be a prerequisite for antidepressant action is not yet clear. The group of Santarelli for instance showed that disrupting antidepressant-induced adult neurogenesis using X-ray irradiation blocks neurogenic and behavioral responses of mice to antidepressants (Santarelli et al. 2003). Also the positive effect of environment enrichment on behavioral traits developed due to chronic social stress was found to be critically dependent on adult neurogenesis (Schloesser et al. 2010). Consequently, Meshi and co-workers investigated the causality of these effects by testing the effect of environmental enrichment on spatial learning and anxiety-like behavior in X-ray irradiated mice. In contrast to the study of Santarelli et al. (2003), Meshi and co-workers found that adult neurogenesis was not required for the behavioral effects of environmental enrichment on spatial learning, habituation to an unfamiliar environment, and conflict-based anxiety (Meshi et al. 2006).

A different explanation for the actions of antidepressant treatment on the hippocampus is given by Kobayashi et al. (2010). Specifically, these authors showed that chronic treatment of adult mice with fluoxetine-induced morphological and electrophysiological features of immature granule cells such as reduced expression of calbindin and reduced synaptic facilitation of dentate-to-CA3 signal transmission (Kobayashi et al. 2010). It was argued that these changes cannot be explained simply by an increase in newly generated immature neurons, but rather identify the phenomenon as "dematuration" of mature granule cells. Thus, it was concluded that 5-HT-targeted antidepressants reversed the established state of neuronal maturation in the adult hippocampus (Kobayashi et al. 2010). This effect of restoring neurons to a postnatal state by fluoxetine is not completely new. For instance, earlier Maya Vetencourt and colleagues showed that chronic administration of fluoxetine to adult rats restored plasticity in the visual cortex, a property which is usually restricted to a critical period during postnatal development, and is normally absent in the adult brain because of a decline of plasticity due to the maturation of intracortical inhibition (Maya Vetencourt et al. 2008).

4 Possible Function of Adult-Generated New Neurons in the Hippocampus

Although many studies show a clear link between environmental manipulations and changes in adult neurogenesis, it is unclear which aspect of the respective experience causes these changes, and which brain processes causing changes in behavior are modulated by altered adult neurogenesis. It is established that newborn granule cells finally integrate into the hippocampal networks, but there are still open questions concerning the functional integration of these neurons into behaviorally relevant neural networks. Thus, the real functional significance of adult neurogenesis, and specifically newborn neurons in the DG, is still under debate.

The involvement of the hippocampus, one of two brain sites where neurogenesis continues throughout life, in learning processes and memory formation has long been recognized (O'Keefe and Nadel 1978). Acquired memory initially depends on the hippocampus for the process of permanent memory formation. The hippocampus is an essential structure for specific types of memory, such as episodic memory and spatial memory (Squire et al. 1992, 1993). With regard to spatial learning, hippocampal neurons are thought to form a cognitive map of the environment based on multiple cues. Moreover, recent studies in animals and humans point to a role of the hippocampus in pattern separation, that is the disambiguation of related stimuli often during emotional learning (Snyder and Cameron 2012). Through its interactions with brain structures associated with emotion such as the amygdala, the hippocampus is also implicated in emotional behavior. Lesions of the amygdala, for instance, do not impair spatial maze learning but disrupts fear conditioning. Moreover, activation of the amygdala during strong emotional stimuli like those experienced in fear conditioning can enhance hippocampal-dependent learning (Cahill and McGaugh 1998; Bannerman et al. 2002). Thus, it is not surprising that the hippocampus is also an integral part of stress reactions; the region expresses high levels of glucocorticoid receptors and contributes to negative feedback mechanisms of the HPA axis (Jacobson and Sapolsky 1991).

Moser and Moser suggested that the hippocampus is functionally differentiated along its dorsoventral (septotemporal) axis (Moser and Moser 1998). Anatomical studies performed in the rat indicate that cortical and subcortical connections of the dorsal hippocampus (ventral hippocampus in primate) and ventral hippocampus (dorsal hippocampus in primates) are different (Swanson and Cowan 1977; van Groen and Wyss 1990). In rodents, spatial memory and episodic memory appears to depend on dorsal and not ventral hipocampus (Moser et al. 1995; Davachi and Wagner 2002; Broadbent et al. 2004). On the other hand, the ventral but not dorsal hippocampus is thought to mediate stress responses and therefore be involved in emotional behavior (Henke 1990; Herman et al. 1998; Fanselow and Dong 2010; Segal et al. 2010).

Increasing evidence suggests that the production of new neurons in the adult hippocampus participates in these different learning processes and memory formations. For example, the work of Kitamura and co-workers (2009) suggests that the level of adult neurogenesis in hippocampus plays a role determining the hippocampal-dependent period of memory. Using two different mouse models in which adult neurogenesis was physically (X-ray irradiation) or genetically (transgenic mice overexpressing follistatin in a forebrain-specific manner which depletes neurogenesis in hippocampus) suppressed, it was found that decreased adult neurogenesis was associated with a prolonged hippocampal-dependent period of associative fear memory (Kitamura et al. 2009). Conversely, increased adult neurogenesis resulting from voluntary exercise was found to speed up the decay rate of hippocampal-dependent memory (Kitamura et al. 2009).

Animal studies have shown a correlation between increased neurogenesis and improved performance on hippocampal-dependent learning tasks such as spatial learning (Gould et al. 1999a; Nilsson et al. 1999; van Praag et al. 1999; Drapeau et al. 2003; Ramirez-Amaya et al. 2006; Dupret et al. 2008; Imayoshi et al. 2008; Zhang et al. 2008; Deng et al. 2009) and contextual fear learning (Pham et al. 2005; Saxe et al. 2006). During memory formation the DG has recently been proposed to play a role in the process of transforming similar representations or memories into highly dissimilar, non-overlapping representations, or, in other words—the formation of distinct and orthogonal representations of mnemonic information. Experimental studies revealed that adult mice with increased hippocampal neurogenesis were more efficient at differentiating between overlapping contextual representations, suggestive of enhanced pattern separation (Sahay et al. 2011). On the other hand, mice with ablated hippocampal neurogenesis showed impaired spatial discrimination (Clelland et al. 2009). Thus, newborn neurons may be necessary for normal pattern separation function in the DG of adult mice (Clelland et al. 2009; Sahay et al. 2011). Burghardt and co-workers went one step further and suggested that adult neurogenesis contributes to cognitive flexibility (Burghardt et al. 2012). Specifically, this group used variants of the active place avoidance task in combination with two independent neurogenesis ablation techniques (X-ray irradiation and transgenic mice) to reveal that adult neurogenesis was necessary for changing a learned response to a stimulus-evoked memory (Burghardt et al. 2012).

Recent studies suggest a role for neurogenesis in the adult hippocampus in emotional memory formation. It is suggested that new neurons may influence learning and memory primarily in tasks that involve significant emotional arousal, or co-activation of the hippocampus and amygdala (Saxe et al. 2006). This is an important issue as impaired regulation of emotional memory is a feature of several affective disorders such as anxiety, depression and post-traumatic stress disorder. Both the amygdala and hippocampus are key elements in the neural circuitry generating and processing of emotions, and emotional memory. The interdependency of hippocampal adult neurogenesis and the activity of specific amygdala nuclei in emotional memory circuits was clearly shown by Kirby et al. (2012). In particular, lesions of the basolateral amygdala, but not of the central amygdala nucleus, resulted in reduced levels of hippocampal neurogenesis in the adult rat. Lesion of the basolateral amygdala nucleus shown to selectively prevent the activation of NPC proliferation in response to a fear conditioning task (Kirby et al. 2012). Snyder and co-workers suggested that new DG neurons play an important role in the buffering of stress responses, which is especially meaningful (in addition to their direct function in encoding of information) in the context of learning and memory processes under novel or aversive conditions (Snyder et al. 2011).

Although it has been clearly shown that adult neurogenesis is involved in different types of learning and memory processes, the mechanism, role and importance of newborn cells at different differentiation stages, and their regulation has to be further clarified. Moreover, the cascade of events linking the birth of new neurons to learning processes is unclear, as is the way in which new neurons are integrated into and/or removed from behaviorally relevant hippocampal neural networks. First insights into these issues were gained from a study of Dupret and co-workers on the effect of spatial learning in the Morris water maze on cell birth and death in the rodent hippocampus. This work revealed that spatial learning is associated with three different events: learning promoted survival of relatively mature neurons, apoptosis of more immature cells, and finally, proliferation of neural precursors (Dupret et al. 2007). The impact of the maturation stage of adult-born dentate granule cells on cognition is further underscored by the work of Deng et al. (2009). They found that mice with a reduced population of adult-born DG cells during an immature stage of development displayed deficits in both long-term memory retention in the Morris water maze and extinction of spatial preference and context-evoked fear (Deng et al. 2009). In contrast to the positive effect of spatial learning on NPC proliferation, contextual fear learning (at least) transiently suppressed cell proliferation with no change in cell survival, and differentiation in the DG of adult rats (Pham et al. 2005). This inhibition of NPC proliferation during the contextual fear conditioning process appears to be due specifically to the formation of an association between the context and shock during training, which is an amygdala-dependent function.

An important question to be answered in the future is whether adult neurogenesis is essential for appropriate learning processes and memory formation not only in rodents, but also in humans. Our knowledge of adult neurogenesis is largely derived from rodent or non-human primate studies (for review see: Kempermann 2002; Ming and Song 2005, 2011; Snyder and Cameron 2012), with there being only a handful of relevant studies in humans (Eriksson et al. 1998; Reif et al. 2006; Boldrini et al. 2009; Coras et al. 2010; Knoth et al. 2010; Lucassen et al. 2010b). There are only a few clues as to how far the situation in animal models really reflects that of the adult and aging human hippocampus. The ongoing birth of new neurons in the adult hippocampus is most prominent in rodents, but adult neurogenesis levels vary considerably among species, being lowest or missing in primates and bats, respectively (Amrein et al. 2011). As shown in Fig. 2 our own study revealed that the level of proliferating (Ki-67 positive) cells in DG of humans is reduced by a factor of 50 compared to mice (Lee et al. 2010). So far only a fraction of mammalian species has been investigated, therefore, further comparative studies are needed in order to recognize whether adult neurogenesis has a common function across species, or whether it mediates species-specific hippocampal functions as may be the case for spatial learning in rodents (Dupret et al. 2008; Imayoshi et al. 2008; Zhang et al. 2008; Deng et al. 2009). Across all species investigated so far, adult neurogenesis declines strongly from infancy to midlife (Amrein et al. 2011). As stated by Altman and colleagues in 1973, this infers a role for neurogenesis in transforming juvenile demeanor from an unpredictable to a predictable nature, typically characterizing the behavior of mammals once they have reached their reproductive phase in life. Additionally, any attempts to link differences in adult neurogenesis with hippocampus-dependent actions, encounter the problem that primates, and humans in particular, show comparatively low levels of adult neurogenesis (Eriksson et al. 1998; Gould 1999). Moreover, adult neurogenesis seems to become rudimentary in humans after 30 years of age (Fahrner et al. 2007). The question is how the activity-dependent incorporation of a low number of neurons into the hippocampal network can lead to a functional benefit during the lifespan of an individual. The so-called "neurogenic reserve" hypothesis suggested by Kempermann in 2008 addresses this problem by suggesting that adult neurogenesis in hippocampus might produce an activity- and experience-dependent potential for sustained cellular plasticity with increasing age, thereby providing such greater capacity and efficiency (Kempermann 2008). As shown in animal experiments, but not yet in humans, both physical and cognitive 'activity' reduces the age-dependent decline in NPC proliferation in the DG (Kempermann and Gage 1999; Kempermann et al. 2002; Kronenberg et al. 2006). Future studies are needed to test the "neurogenic reserve" hypothesis also in humans.

In addition to the decline of adult neurogenesis with aging, the distribution pattern of adult neurogenesis-related cell types (e.g. proliferating NPCs, adult-born immature and mature neurons) changes as the individuals grow older. Both, immunohistochemistry against Ki-67 as well as BrdU-labeling in the brains of young (postnatal day 5–19) Wistar rats, revealed that cell proliferation in early postnatal neurogenesis, a transition state between the embryonic and adult neurogenesis, occurred mainly in the hilus and only partly in the SGZ (Namba et al. 2005). In rats older than P19, most proliferating cells were found in the SGZ (Namba et al. 2005). This principle could be replicated for

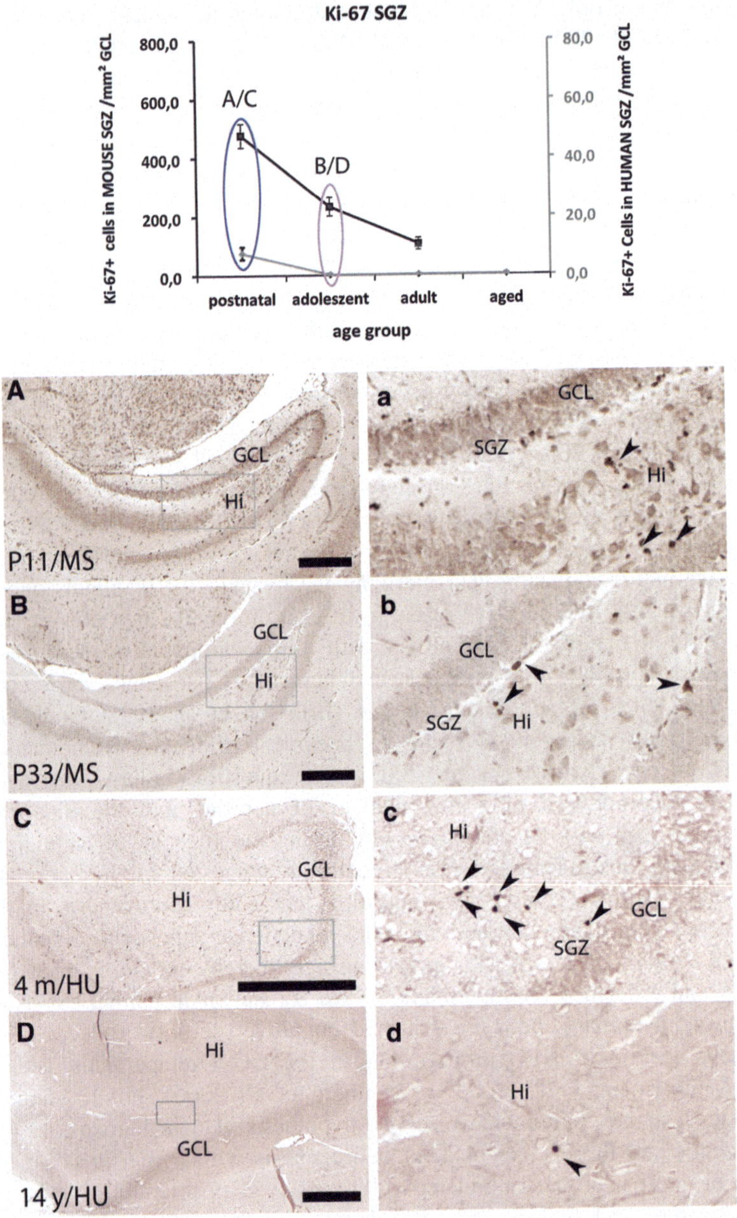

mice in our own study, however, not for humans (Lee et al. 2010). So far it seems that in humans most of the proliferating cells—detected with Ki-67—can be found in the hilus across all ages, from infants to seniors (Lee et al. 2010) (see Fig. 2).

◀**Fig. 2** *Hippocampal adult neurogenesis in mice and men during ontogeny. Top* Diagram displaying the density of Ki-67-positive proliferating cells (cells/mm^2 granule cell layer, GCL) in the subgranular zone (SGZ) of humans and mice of the age groups postnatal, adolescent, adult, and aged (only humans). Density of proliferating cells in the dentate gyrus (DG) of mice is up to 50-fold higher compared to humans in corresponding age groups (postnatal; adolescent; adult). *Bottom* Representative images of immunostained hippocampal sections using antibodies detecting Ki-67, a marker for proliferating cells. Ki-67-positive cells in the murine (A,a; B,b) and human (C,c; D,d) DG of roughly corresponding age groups. With regard to brain development postnatal day (P)11 in mice approximately corresponds to 4 months (m) of age in humans (HU) and both together represent the postnatal group. P33/MS roughly corresponds to 14 years (y) of age in Humans (14y/HU), which represent the adolescent group. Note, that the number of Ki-67-immunoreactive cells (*arrowheads* in a,b,c,d) decreases along aging, and that in the DG of the P33 old mouse these proliferating cells primarily are localized in the SGZ (B,b), and not in the Hilus (Hi), as it is shown in stained human sections of both ages (C,c; D,d) as well as in the DG of the P11 old mouse (A,a). GCL = granule cell layer; Scale bars in A/B = 250 μm and equals 100 μm in a/b; Scale bars in C/D = 1000 μm and equals 100 μm in c/d

5 Extended Adult Neurogenesis Hypothesis: Finding Links Between Adult Neurogenesis and Depression, as Well as Various Brain Disorders

So far only three studies have been published that use *post mortem* human brain tissue to investigate the possible link between adult neurogenesis and major depression. In 2006, we set out to reveal whether hippocampal stem cell proliferation was reduced in depressive patients, and also facilitated by antidepressant treatment. We also investigated whether adult neurogenesis is altered in schizophrenia (Reif et al. 2006). To study the proliferation of progenitor cells in the DG, one of the first steps of adult neurogenesis, we measured the proliferation marker Ki-67 on sections of the anterior hippocampus using quantitative immunocytochemistry. The findings suggested that the progenitor cell number was reduced in *post mortem* brains from schizophrenics, but not patients with depression. Moreover, DG progenitor proliferation did not seem to be modified by antidepressant drug treatment (Reif et al. 2006).

Boldrini and colleagues also hypothesized that antidepressant treatment increases NPC number and cell division in the human DG, and that dividing cells are fewer in subjects with depression (Boldrini et al. 2009). In order to test this hypothesis, whole frozen hippocampi from both antidepressant-treated and untreated subjects with major depression as well as control subjects were fixed, sectioned, and immunostained for NPCs and dividing cell markers (nestin and Ki-67, respectively), NeuN (neuronal nuclear antigen; mature neuron marker), and GFAP (glial fibrillary acidic protein; astrocyte marker), under single- or double-labeling conditions. Two findings of the study were gender and age differences. Thus, nestin-positive NPCs decreased with age, and females had a greater number of nestin-positive NPCs than males. These results accord with earlier animal studies which found that estrogen enhanced NPC proliferation during proestrus in female rats, resulting in more immature neurons in the hippocampal formation of females compared with males (Tanapat et al. 1999; Galea 2008). Also an age-related decrease in NPC proliferation is reported in rodents (Kuhn et al. 1996). The age-related changes were confirmed for humans in another study, which established that certain neurogenesis-associated features originally identified in rodents, also showed qualitative and quantitative age-related changes in the adult human hippocampus (Knoth et al. 2010). The number of Ki-67-positive dividing cells was not altered by either gender or age. An important finding of Boldrini et al. was that depressed subjects treated with an SSRI had a greater number of nestin-positive NPCs than untreated depressed subjects and controls, but the number of NPCs was not different between depressed patients treated with an SSRI or tricylic antidepressant. Dividing cell number was greater in depressed subjects treated with a tricylic antidepressant than untreated or SSRI-treated depressed subjects and controls. The increase of both nestin-positive NPCs and Ki-67-positive dividing cells in antidepressant treated subjects was primarily restricted to the rostral (anterior) DG. The study found no significant difference in NPCs or dividing cell number between untreated depressed subjects and healthy controls, which corresponds to findings of Reif and co-workers (Reif et al. 2006). Nevertheless, the data in the Boldrini study suggests that the number of NPCs in depressed subjects was higher than in controls, although this difference was not statistically significant. Taking into account that the majority of the subjects died by suicide, these results question whether the reported antidepressant-induced increase in adult neurogenesis in humans is a true effect.

To add to the studies of adult neurogenesis in *post mortem* tissue obtained form subjects with depression, Lucassen, and co-workers found decreased numbers of progenitor cells but no effect of antidepressant treatment in the hippocampus of elderly depressed patients compared to non-depressed elderly controls. This effect was found primarily in MCM2 (minichromosome maintenance protein 2)-immunoreactive progenitor cells but not PH3 (phosphorylated histone 3)-positive proliferating cells (Lucassen et al. 2010b). The authors explained the discrepancies between the different human *post mortem* studies (dealing with the same questions) on the basis of anatomical-, medication- and age-differences between the subjects in the various studies as well as the possibility of neurogenesis-independent mechanisms of antidepressant action. As far as the number of MCM2-positive proliferating cells is concerned, neither an effect of gender nor age was found (Lucassen et al. 2010b).

All the above-mentioned studies of neurogenesis in *post mortem* human tissue have their limitations. One is that some of the antigens measured in these studies such as Ki-67 (Reif et al. 2006; Boldrini et al. 2009) and MCM2 (Lucassen et al. 2010b), are not only expressed in proliferating NPC but in any proliferating cell, including astroglia, microglia, or endothelial cells. This makes it impossible to determine how many of the immunoreactive cells actually differentiate into new neurons. The Boldrini study tried to circumvent this problem by counting cells that were immunoreactive for the NPC-marker nestin (Boldrini et al. 2009). However, as it is unlikely that all of these cells survive, even this marker tells us little about the number of new functional neurons and the lack or increase thereof in either untreated or treated depressed subjects. Nevertheless, the Boldrini study is currently unique in that it is the only stereological human *post mortem* study published on the subject (Boldrini et al. 2009). This is important as random, systematic sampling is generally regarded as essential for obtaining unbiased and quantitative data (West and Gundersen 1990).

Besides the above evidence of a possible link between adult neurogenesis in the human hippocampus and major depression, adult neurogenesis has also been investigated in other psychiatric and neurological disorders, including neurodegenerative disorders. In our above-mentioned study of Reif et al. (2006) data indicated that proliferation of cells in the DG was significantly reduced in schizophrenic patients, and this finding was offered as a possible contributory explanation for the cognitive deficits of this disorder. In addition to this study, hippocampal neurogenesis has been reported to be increased in epilepsy (Crespel et al. 2005) and Alzheimer's disease (Jin et al. 2004; Marlatt and Lucassen 2010). In the latter case, the expression of immature neuronal marker proteins that signal the birth of new neurons was measured in the hippocampus of senile subjects with Alzheimer's disease (Jin et al. 2004). Compared to controls, Alzheimer brains showed increased expression of DCX and the immature neuronal marker TUC-4 in cells of the SGZ of the hippocampal DG. Another study focussing on younger (pre-senile) Alzheimer's patients found increased proliferation in layers CA1-3, possibly reflecting glial and vascular-associated changes, but not adult neurogenesis (Boekhoorn et al. 2006). In contrast to the human *post mortem* tissue results of Jin et al. (2004), adult neurogenesis was found to be decreased in a genetic mouse model of Alzheimer's disease (TgCRND8 mice), which overexpresses human amyloid ß precursor protein (APP) (Chishti et al. 2001; Herring et al. 2009; Lewejohann et al. 2009). The latter transgenic mice can be distinguished from other mouse models of Alzheimer's disease through the early exhibition (3 months) of ß-amyloid plaque load accompanied by astrogliosis/microgliosis and cognitive deficits (Chishti et al. 2001). Interestingly, enriched environment enhanced neuroplasticity in these mice, and also increased the diminished number of newborn hippocampal cells to wildtype levels (Herring et al. 2009).

Bringing together these above studies, it is possible to postulate that the adult neurogenesis hypothesis of depression should be widened to include a role for adult neurogenesis in a range of neuropsychiatric disorders, which includes neurological and neurodegenerative disorders such as Alzheimer's disease. However it needs to be born in mind that altered adult neurogenesis in these conditions could be an epiphenomenon and not causally involved. Potentially there are multiple mediators of changes in adult neurogenesis including altered availability of cytokines and neurotrophic factors, to name just a few. As an example, in rodents BDNF is known to be increased in the DG and hippocampal CA subfields after chronic antidepressant administration (Nibuya et al. 1995; Dias et al. 2003), and BDNF has been shown to promote the differentiation and survival of progenitor cells in the adult rat brain (Zigova et al. 1998). Another mechanism that may change levels of adult neurogenesis is suggested in an in vitro study using the multipotent, human hippocampal progenitor cell line HPC03A/07 (Anacker et al. 2011). The latter study identifies a critical role for glucocorticoid receptors in the antidepressant effects on neurogenesis. Thus, the study found that treatment of these cells (for 3–10 days) with the SSRI sertraline, increased neuronal differentiation via a GR-dependent mechanism, possibly mediated through altered cAMP/protein kinase A signaling (Anacker et al. 2011).

6 Summary and Conclusion

The data presented in this review highlight the possibility that altered neurogenesis in the adult hippocampus is integrally involved in the pathophysiology of major depression. Primarily based on data from experimental studies with rodents, there is strong evidence that hippocampal adult neurogenesis is downregulated by stressful conditions and upregulated by antidepressant drugs and some other, but not all, antidepressant treatments. Some experiments have even suggested that neurogenesis is necessary for the behavioral effects of antidepressants (Santarelli et al. 2003), but this finding was not replicated using a different experimental approach (Meshi et al. 2006). Clinical evidence supporting this hypothesis includes reports of reduced hippocampal volume in MRI and *post mortem* studies of depressed patients, findings which could also be associated with cognitive deficits observed in these patients. However, reduced hippocampal volume and cognitive deficits could also be shown in schizophrenic patients as well as in patients with neurodegenerative disorders such as Alzheimer‘s disease.

To date, studies of *post mortem* human hippocampal tissue of patients with various types of neuropsychiatric disorders have not been able to find consistent evidence of altered adult neurogenesis in major depression. Three independent studies have set out to test the hypotheses that antidepressant treatment increases NPC number and proliferation rates, and that adult neurogenesis is disturbed in depressed patients compared to controls, but have produced contradictory findings (Reif et al. 2006; Boldrini et al. 2009; Lucassen et al. 2010b). However, decreased NPC proliferation could be linked to schizophrenia (Reif et al. 2006) and increased adult neurogenesis could be shown in patients with Alzheimer‘s disease (Jin et al. 2004).

Animal studies implicate the involvement of adult neurogenesis in hippocampus-dependent learning and memory processes, such as spatial learning, contextual fear learning and extinction, as well as the process of pattern separation. Recent studies suggest an additional role of hippocampal adult neurogenesis in emotional memory formation, which is an important issue as impaired regulation of emotional memory is a feature of several affective disorders. Early studies of the mechanism by which newborn cells become an integral part of learning and memory processes, suggest that the role of newborn cells depends on their differentiation stage. Moreover, the work of Kitamura et al. (2009) suggests that the level of adult neurogenesis determines the hippocampus-dependent period of memory formation.

As adult neurogenesis is much more prominent in rodents than human and non-human primates, it can be questioned whether hippocampal adult neurogenesis in different species is regulated in the same way, and has the same functional relevance. Considering this, we highlighted the possible differences of adult neurogenesis levels and distribution patterns of progenitor cells in the DG of humans and rodents. Primates, and humans in particular, show low levels of adult neurogenesis compared to rodents (Eriksson et al. 1998; Gould 1999; Lee et al. 2010), but all species show a clear decline of adult neurogenesis with aging (Seki and Arai 1995; Kuhn et al. 1996; Kempermann et al. 1998; Amrein et al. 2011). In addition, in humans adult neurogenesis-related cells are not predominantly localized in the SGZ from a certain postnatal age (Lee et al. 2010) as in the case of rodents (Namba et al. 2005). Rather, in the human hippocampus most of the NPCs seem to be "stuck" in the hilus (Lee et al. 2010).

Despite much that has been learned from the above-mentioned studies as well as others, one of the burning questions in the field that has not yet been satisfactorily answered is whether changes in adult neurogenesis associated with depression (or other neuropsychiatric disorders) and its treatment, are causally connected. Indeed, increasing preclinical and clinical data give researchers reason to doubt the causal nature of changes in adult neurogenesis and the pathophysiology of neuropsychiatric disorders and their treatment. As a consequence one might wonder whether altered adult neurogenesis is merely an epiphenomenon triggered by various external manipulations. Against this background of many findings, links and suggestions, the real functional significance of adult neurogenesis in general, and, especially the functional relevance of adult neurogenesis in different species, needs further detailed investigation.

References

Altman J (1962) Are new neurons formed in the brains of adult mammals? Science 135:1127–1128

Altman J, Das GD (1965) Autoradiographic and histological evidence of postnatal hippocampal neurogenesis in rats. J Comp Neurol 124:319–335

Altman J, Das GD (1967) Postnatal neurogenesis in the guinea-pig. Nature 214:1098–1101

Amrein I, Isler K, Lipp HP (2011) Comparing adult hippocampal neurogenesis in mammalian species and orders: influence of chronological age and life history stage. Eur J Neurosci 34:978–987

Anacker C, Zunszain PA, Cattaneo A, Carvalho LA, Garabedian MJ, Thuret S, Price J, Pariante CM (2011) Antidepressants increase human hippocampal neurogenesis by activating the glucocorticoid receptor. Mol Psychiatry 16:738–750

Banasr M, Hery M, Printemps R, Daszuta A (2004) Serotonin-induced increases in adult cell proliferation and neurogenesis are mediated through different and common 5-HT receptor subtypes in the dentate gyrus and the subventricular zone. Neuropsychopharmacology 29:450–460

Bannerman DM, Deacon RM, Offen S, Friswell J, Grubb M, Rawlins JN (2002) Double dissociation of function within the hippocampus: spatial memory and hyponeophagia. Behav Neurosci 116:884–901

Barnea A, Nottebohm F (1996) Recruitment and replacement of hippocampal neurons in young and adult chickadees: an addition to the theory of hippocampal learning. Proc Natl Acad Sci U S A 93:714–718

Bekinschtein P, Oomen CA, Saksida LM, Bussey TJ (2011) Effects of environmental enrichment and voluntary exercise on neurogenesis, learning and memory, and pattern separation: BDNF as a critical variable? Semin Cell Dev Biol 22:536–542

Boekhoorn K, Joels M, Lucassen PJ (2006) Increased proliferation reflects glial and vascular-associated changes, but not neurogenesis in the presenile Alzheimer hippocampus. Neurobiol Dis 24:1–14

Boldrini M, Arango V (2010) Antidepressants, age, and neuroprogenitors. Neuropsychopharmacology 35:351–352

Boldrini M, Underwood MD, Hen R, Rosoklija GB, Dwork AJ, John Mann J, Arango V (2009) Antidepressants increase neural progenitor cells in the human hippocampus. Neuropsychopharmacology 34:2376–2389

Bremner JD, Narayan M, Anderson ER, Staib LH, Miller HL, Charney DS (2000) Hippocampal volume reduction in major depression. Am J Psychiatry 157:115–118

Brezun JM, Daszuta A (2000) Serotonin may stimulate granule cell proliferation in the adult hippocampus, as observed in rats grafted with foetal raphe neurons. Eur J Neurosci 12: 391–396

Broadbent NJ, Squire LR, Clark RE (2004) Spatial memory, recognition memory, and the hippocampus. Proc Natl Acad Sci U S A 101:14515–14520

Burghardt NS, Park EH, Hen R, Fenton AA (2012) Adult-born hippocampal neurons promote cognitive flexibility in mice. Hippocampus 31(6):739–751

Cahill L, McGaugh JL (1998) Mechanisms of emotional arousal and lasting declarative memory. Trends Neurosci 21:294–299

Campbell S, Marriott M, Nahmias C, MacQueen GM (2004) Lower hippocampal volume in patients suffering from depression: a meta-analysis. Am J Psychiatry 161:598–607

Caspi A, Sugden K, Moffitt TE, Taylor A, Craig IW, Harrington H, McClay J, Mill J, Martin J, Braithwaite A, Poulton R (2003) Influence of life stress on depression: moderation by a polymorphism in the 5-HTT gene. Science 301:386–389

Chishti MA, Yang DS, Janus C, Phinney AL, Horne P, Pearson J, Strome R, Zuker N, Loukides J, French J, Turner S, Lozza G, Grilli M, Kunicki S, Morissette C, Paquette J, Gervais F, Bergeron C, Fraser PE, Carlson GA, George-Hyslop PS, Westaway D (2001) Early-onset amyloid deposition and cognitive deficits in transgenic mice expressing a double mutant form of amyloid precursor protein 695. J Biol Chem 276:21562–21570

Clelland CD, Choi M, Romberg C, Clemenson GD Jr, Fragniere A, Tyers P, Jessberger S, Saksida LM, Barker RA, Gage FH, Bussey TJ (2009) A functional role for adult hippocampal neurogenesis in spatial pattern separation. Science 325:210–213

Coras R, Siebzehnrubl FA, Pauli E, Huttner HB, Njunting M, Kobow K, Villmann C, Hahnen E, Neuhuber W, Weigel D, Buchfelder M, Stefan H, Beck H, Steindler DA, Blumcke I (2010) Low proliferation and differentiation capacities of adult hippocampal stem cells correlate with memory dysfunction in humans. Brain 133:3359–3372

Crespel A, Rigau V, Coubes P, Rousset MC, de Bock F, Okano H, Baldy-Moulinier M, Bockaert J, Lerner-Natoli M (2005) Increased number of neural progenitors in human temporal lobe epilepsy. Neurobiol Dis 19:436–450

Czeh B, Lucassen PJ (2007) What causes the hippocampal volume decrease in depression? Are neurogenesis, glial changes and apoptosis implicated? Eur Arch Psychiatry Clin Neurosci 257:250–260

Czeh B, Michaelis T, Watanabe T, Frahm J, de Biurrun G, van Kampen M, Bartolomucci A, Fuchs E (2001) Stress-induced changes in cerebral metabolites, hippocampal volume, and cell proliferation are prevented by antidepressant treatment with tianeptine. Proc Natl Acad Sci U S A 98:12796–12801

Czeh B, Welt T, Fischer AK, Erhardt A, Schmitt W, Muller MB, Toschi N, Fuchs E, Keck ME (2002) Chronic psychosocial stress and concomitant repetitive transcranial magnetic stimulation: effects on stress hormone levels and adult hippocampal neurogenesis. Biol Psychiatry 52:1057–1065

Davachi L, Wagner AD (2002) Hippocampal contributions to episodic encoding: insights from relational and item-based learning. J Neurophysiol 88:982–990

Dawirs RR, Hildebrandt K, Teuchert-Noodt G (1998) Adult treatment with haloperidol increases dentate granule cell proliferation in the gerbil hippocampus. J Neural Transm 105:317–327

Dayer AG, Ford AA, Cleaver KM, Yassaee M, Cameron HA (2003) Short-term and long-term survival of new neurons in the rat dentate gyrus. J Comp Neurol 460:563–572

Deng W, Aimone JB, Gage FH (2010) New neurons and new memories: how does adult hippocampal neurogenesis affect learning and memory? Nat Rev Neurosci 11:339–350

Deng W, Saxe MD, Gallina IS, Gage FH (2009) Adult-born hippocampal dentate granule cells undergoing maturation modulate learning and memory in the brain. J Neurosci 29: 13532–13542

Dias BG, Banerjee SB, Duman RS, Vaidya VA (2003) Differential regulation of brain derived neurotrophic factor transcripts by antidepressant treatments in the adult rat brain. Neuropharmacology 45:553–563

Drapeau E, Mayo W, Aurousseau C, Le Moal M, Piazza PV, Abrous DN (2003) Spatial memory performances of aged rats in the water maze predict levels of hippocampal neurogenesis. Proc Natl Acad Sci U S A 100:14385–14390

Dupret D, Fabre A, Dobrossy MD, Panatier A, Rodriguez JJ, Lamarque S, Lemaire V, Oliet SH, Piazza PV, Abrous DN (2007) Spatial learning depends on both the addition and removal of new hippocampal neurons. PLoS Biol 5:e214

Dupret D, Revest JM, Koehl M, Ichas F, De Giorgi F, Costet P, Abrous DN, Piazza PV (2008) Spatial relational memory requires hippocampal adult neurogenesis. PLoS ONE 3:e1959

Encinas JM, Vaahtokari A, Enikolopov G (2006) Fluoxetine targets early progenitor cells in the adult brain. Proc Natl Acad Sci U S A 103:8233–8238

Eriksson PS, Perfilieva E, Bjork-Eriksson T, Alborn AM, Nordborg C, Peterson DA, Gage FH (1998) Neurogenesis in the adult human hippocampus. Nat Med 4:1313–1317

Fahrner A, Kann G, Flubacher A, Heinrich C, Freiman TM, Zentner J, Frotscher M, Haas CA (2007) Granule cell dispersion is not accompanied by enhanced neurogenesis in temporal lobe epilepsy patients. Exp Neurol 203:320–332

Fanselow MS, Dong HW (2010) Are the dorsal and ventral hippocampus functionally distinct structures? Neuron 65:7–19

Frodl T, Jager M, Smajstrlova I, Born C, Bottlender R, Palladino T, Reiser M, Moller HJ, Meisenzahl EM (2008a) Effect of hippocampal and amygdala volumes on clinical outcomes in major depression: a 3-year prospective magnetic resonance imaging study. J Psychiatry Neurosci 33:423–430

Frodl TS, Koutsouleris N, Bottlender R, Born C, Jager M, Scupin I, Reiser M, Moller HJ, Meisenzahl EM (2008b) Depression-related variation in brain morphology over 3 years: effects of stress? Arch Gen Psychiatry 65:1156–1165

Galea LA (2008) Gonadal hormone modulation of neurogenesis in the dentate gyrus of adult male and female rodents. Brain Res Rev 57:332–341

Gaspar P, Cases O, Maroteaux L (2003) The developmental role of serotonin: news from mouse molecular genetics. Nat Rev Neurosci 4:1002–1012

Gerdes J, Li L, Schlueter C, Duchrow M, Wohlenberg C, Gerlach C, Stahmer I, Kloth S, Brandt E, Flad HD (1991) Immunobiochemical and molecular biologic characterization of the cell proliferation-associated nuclear antigen that is defined by monoclonal antibody Ki-67. Am J Pathol 138:867–873

Geuze E, Vermetten E, Bremner JD (2005) MR-based in vivo hippocampal volumetrics: 2 Findings in neuropsychiatric disorders. Mol Psychiatry 10:160–184

Glasper ER, Schoenfeld TJ, Gould E (2012) Adult neurogenesis: optimizing hippocampal function to suit the environment. Behav Brain Res 227:380–383
Goldman SA, Nottebohm F (1983) Neuronal production, migration, and differentiation in a vocal control nucleus of the adult female canary brain. Proc Natl Acad Sci U S A 80:2390–2394
Gould E (1999) Serotonin and hippocampal neurogenesis. Neuropsychopharmacology 21: 46S–51S
Gould E, Beylin A, Tanapat P, Reeves A, Shors TJ (1999a) Learning enhances adult neurogenesis in the hippocampal formation. Nat Neurosci 2:260–265
Gould E, Reeves AJ, Graziano MS, Gross CG (1999b) Neurogenesis in the neocortex of adult primates. Science 286:548–552
Gould E, Tanapat P (1999) Stress and hippocampal neurogenesis. Biol Psychiatry 46:1472–1479
Gould E, Tanapat P, McEwen BS, Flugge G, Fuchs E (1998) Proliferation of granule cell precursors in the dentate gyrus of adult monkeys is diminished by stress. Proc Natl Acad Sci U S A 95:3168–3171
Grassi Zucconi G, Cipriani S, Balgkouranidou I, Scattoni R (2006) 'One night' sleep deprivation stimulates hippocampal neurogenesis. Brain Res Bull 69:375–381
Gratzner HG (1982) Monoclonal antibody to 5-bromo- and 5-iododeoxyuridine: A new reagent for detection of DNA replication. Science 218:474–475
Halim ND, Weickert CS, McClintock BW, Weinberger DR, Lipska BK (2004) Effects of chronic haloperidol and clozapine treatment on neurogenesis in the adult rat hippocampus. Neuropsychopharmacology 29:1063–1069
Hanson ND, Owens MJ, Nemeroff CB (2011) Depression, antidepressants, and neurogenesis: a critical reappraisal. Neuropsychopharmacology 36:2589–2602
Hellsten J, Wennstrom M, Mohapel P, Ekdahl CT, Bengzon J, Tingstrom A (2002) Electroconvulsive seizures increase hippocampal neurogenesis after chronic corticosterone treatment. Eur J Neurosci 16:283–290
Henke PG (1990) Hippocampal pathway to the amygdala and stress ulcer development. Brain Res Bull 25:691–695
Herman JP, Dolgas CM, Carlson SL (1998) Ventral subiculum regulates hypothalamo-pituitary-adrenocortical and behavioural responses to cognitive stressors. Neuroscience 86:449–459
Herring A, Ambree O, Tomm M, Habermann H, Sachser N, Paulus W, Keyvani K (2009) Environmental enrichment enhances cellular plasticity in transgenic mice with Alzheimer-like pathology. Exp Neurol 216:184–192
Heuser I, Yassouridis A, Holsboer F (1994) The combined dexamethasone/CRH test: a refined laboratory test for psychiatric disorders. J Psychiatr Res 28:341–356
Holmes A (2008) Genetic variation in cortico-amygdala serotonin function and risk for stress-related disease. Neurosci Biobehav Rev 32:1293–1314
Holsboer F (2001) Stress, hypercortisolism and corticosteroid receptors in depression: implications for therapy. J Affect Disord 62:77–91
Honea R, Crow TJ, Passingham D, Mackay CE (2005) Regional deficits in brain volume in schizophrenia: a meta-analysis of voxel-based morphometry studies. Am J Psychiatry 162:2233–2245
Imayoshi I, Sakamoto M, Ohtsuka T, Takao K, Miyakawa T, Yamaguchi M, Mori K, Ikeda T, Itohara S, Kageyama R (2008) Roles of continuous neurogenesis in the structural and functional integrity of the adult forebrain. Nat Neurosci 11:1153–1161
Jacobs BL (2002) Adult brain neurogenesis and depression. Brain Behav Immun 16:602–609
Jacobson L, Sapolsky R (1991) The role of the hippocampus in feedback regulation of the hypothalamic-pituitary-adrenocortical axis. Endocr Rev 12:118–134
Jin K, Peel AL, Mao XO, Xie L, Cottrell BA, Henshall DC, Greenberg DA (2004) Increased hippocampal neurogenesis in Alzheimer's disease. Proc Natl Acad Sci U S A 101:343–347
Kee N, Sivalingam S, Boonstra R, Wojtowicz JM (2002) The utility of Ki-67 and BrdU as proliferative markers of adult neurogenesis. J Neurosci Methods 115:97–105
Kempermann G (2002) Why new neurons? Possible functions for adult hippocampal neurogenesis. J Neurosci 22:635–638

Kempermann G (2008) The neurogenic reserve hypothesis: what is adult hippocampal neurogenesis good for? Trends Neurosci 31:163–169

Kempermann G, Gage FH (1999) Experience-dependent regulation of adult hippocampal neurogenesis: effects of long-term stimulation and stimulus withdrawal. Hippocampus 9: 321–332

Kempermann G, Gast D, Gage FH (2002) Neuroplasticity in old age: sustained fivefold induction of hippocampal neurogenesis by long-term environmental enrichment. Ann Neurol 52: 135–143

Kempermann G, Jessberger S, Steiner B, Kronenberg G (2004) Milestones of neuronal development in the adult hippocampus. Trends Neurosci 27:447–452

Kempermann G, Kuhn HG, Gage FH (1997) More hippocampal neurons in adult mice living in an enriched environment. Nature 386:493–495

Kempermann G, Kuhn HG, Gage FH (1998) Experience-induced neurogenesis in the senescent dentate gyrus. J Neurosci 18:3206–3212

Kirby ED, Friedman AR, Covarrubias D, Ying C, Sun WG, Goosens KA, Sapolsky RM, Kaufer D (2012) Basolateral amygdala regulation of adult hippocampal neurogenesis and fear-related activation of newborn neurons. Mol Psychiatry 17:527–536

Kitamura T, Saitoh Y, Takashima N, Murayama A, Niibori Y, Ageta H, Sekiguchi M, Sugiyama H, Inokuchi K (2009) Adult neurogenesis modulates the hippocampus-dependent period of associative fear memory. Cell 139:814–827

Knoth R, Singec I, Ditter M, Pantazis G, Capetian P, Meyer RP, Horvat V, Volk B, Kempermann G (2010) Murine features of neurogenesis in the human hippocampus across the lifespan from 0 to 100 years. PLoS ONE 5:e8809

Kobayashi K, Ikeda Y, Sakai A, Yamasaki N, Haneda E, Miyakawa T, Suzuki H (2010) Reversal of hippocampal neuronal maturation by serotonergic antidepressants. Proc Natl Acad Sci U S A 107:8434–8439

Kodama M, Fujioka T, Duman RS (2004) Chronic olanzapine or fluoxetine administration increases cell proliferation in hippocampus and prefrontal cortex of adult rat. Biol Psychiatry 56:570–580

Koolhaas JM, Bartolomucci A, Buwalda B, de Boer SF, Flugge G, Korte SM, Meerlo P, Murison R, Olivier B, Palanza P, Richter-Levin G, Sgoifo A, Steimer T, Stiedl O, van Dijk G, Wohr M, Fuchs E (2011) Stress revisited: a critical evaluation of the stress concept. Neurosci Biobehav Rev 35:1291–1301

Kronenberg G, Bick-Sander A, Bunk E, Wolf C, Ehninger D, Kempermann G (2006) Physical exercise prevents age-related decline in precursor cell activity in the mouse dentate gyrus. Neurobiol Aging 27:1505–1513

Kronenberg G, Kirste I, Inta D, Chourbaji S, Heuser I, Endres M, Gass P (2009) Reduced hippocampal neurogenesis in the GR (±) genetic mouse model of depression. Eur Arch Psychiatry Clin Neurosci 259:499–504

Kuhn HG, Dickinson-Anson H, Gage FH (1996) Neurogenesis in the dentate gyrus of the adult rat: age-related decrease of neuronal progenitor proliferation. J Neurosci 16:2027–2033

Lanfumey L, Mongeau R, Cohen-Salmon C, Hamon M (2008) Corticosteroid-serotonin interactions in the neurobiological mechanisms of stress-related disorders. Neurosci Biobehav Rev 32:1174–1184

Lee M, Olmes DG, Nietzer SL, Winnig M, Lesch KP, Schmitt AG (2010) Hippocampal neurogenesis during murine and human ontogenesis: same same but different?. FENS Abstr. vol 5, 011.39

Lewejohann L, Reefmann N, Widmann P, Ambree O, Herring A, Keyvani K, Paulus W, Sachser N (2009) Transgenic Alzheimer mice in a semi-naturalistic environment: more plaques, yet not compromised in daily life. Behav Brain Res 201:99–102

Lowry CA, Johnson PL, Hay-Schmidt A, Mikkelsen J, Shekhar A (2005) Modulation of anxiety circuits by serotonergic systems. Stress 8:233–246

Lucassen PJ, Meerlo P, Naylor AS, van Dam AM, Dayer AG, Fuchs E, Oomen CA, Czeh B (2010a) Regulation of adult neurogenesis by stress, sleep disruption, exercise

and inflammation: Implications for depression and antidepressant action. Eur Neuropsychopharmacol 20:1–17
Lucassen PJ, Stumpel MW, Wang Q, Aronica E (2010b) Decreased numbers of progenitor cells but no response to antidepressant drugs in the hippocampus of elderly depressed patients. Neuropharmacology 58:940–949
Lyons DM, Buckmaster PS, Lee AG, Wu C, Mitra R, Duffey LM, Buckmaster CL, Her S, Patel PD, Schatzberg AF (2010) Stress coping stimulates hippocampal neurogenesis in adult monkeys. Proc Natl Acad Sci U S A 107:14823–14827
Madsen TM, Treschow A, Bengzon J, Bolwig TG, Lindvall O, Tingstrom A (2000) Increased neurogenesis in a model of electroconvulsive therapy. Biol Psychiatry 47:1043–1049
Malberg JE, Eisch AJ, Nestler EJ, Duman RS (2000) Chronic antidepressant treatment increases neurogenesis in adult rat hippocampus. J Neurosci 20:9104–9110
Marlatt MW, Lucassen PJ (2010) Neurogenesis and Alzheimer's disease: biology and pathophysiology in mice and men. Curr Alzheimer Res 7:113–125
Maya Vetencourt JF, Sale A, Viegi A, Baroncelli L, De Pasquale R, O'Leary OF, Castren E, Maffei L (2008) The antidepressant fluoxetine restores plasticity in the adult visual cortex. Science 320:385–388
Meshi D, Drew MR, Saxe M, Ansorge MS, David D, Santarelli L, Malapani C, Moore H, Hen R (2006) Hippocampal neurogenesis is not required for behavioral effects of environmental enrichment. Nat Neurosci 9:729–731
Ming GL, Song H (2005) Adult neurogenesis in the mammalian central nervous system. Annu Rev Neurosci 28:223–250
Ming GL, Song H (2011) Adult neurogenesis in the mammalian brain: significant answers and significant questions. Neuron 70:687–702
Mirescu C, Peters JD, Gould E (2004) Early life experience alters response of adult neurogenesis to stress. Nat Neurosci 7:841–846
Moghaddam B, Bolinao ML, Stein-Behrens B, Sapolsky R (1994) Glucocorticoids mediate the stress-induced extracellular accumulation of glutamate. Brain Res 655:251–254
Morris SA, Eaves DW, Smith AR, Nixon K (2010) Alcohol inhibition of neurogenesis: a mechanism of hippocampal neurodegeneration in an adolescent alcohol abuse model. Hippocampus 20:596–607
Moser MB, Moser EI (1998) Functional differentiation in the hippocampus. Hippocampus 8: 608–619
Moser MB, Moser EI, Forrest E, Andersen P, Morris RG (1995) Spatial learning with a minislab in the dorsal hippocampus. Proc Natl Acad Sci U S A 92:9697–9701
Namba T, Mochizuki H, Onodera M, Mizuno Y, Namiki H, Seki T (2005) The fate of neural progenitor cells expressing astrocytic and radial glial markers in the postnatal rat dentate gyrus. Eur J Neurosci 22:1928–1941
Namestkova K, Simonova Z, Sykova E (2005) Decreased proliferation in the adult rat hippocampus after exposure to the Morris water maze and its reversal by fluoxetine. Behav Brain Res 163:26–32
Nibuya M, Morinobu S, Duman RS (1995) Regulation of BDNF and trkB mRNA in rat brain by chronic electroconvulsive seizure and antidepressant drug treatments. J Neurosci 15: 7539–7547
Nilsson M, Perfilieva E, Johansson U, Orwar O, Eriksson PS (1999) Enriched environment increases neurogenesis in the adult rat dentate gyrus and improves spatial memory. J Neurobiol 39:569–578
O'Keefe J, Nadel L (1978) The hippocampus as a cognitive map. Oxford University Press, Oxford
Overstreet DH, Pucilowski O, Retton MC, Delagrange P, Guardiola-Lemaitre B (1998) Effects of melatonin receptor ligands on swim test immobility. NeuroReport 9:249–253
Paizanis E, Hamon M, Lanfumey L (2007) Hippocampal neurogenesis, depressive disorders, and antidepressant therapy. Neural Plast 2007:73754

Papp M, Gruca P, Boyer PA, Mocaer E (2003) Effect of agomelatine in the chronic mild stress model of depression in the rat. Neuropsychopharmacology 28:694–703
Pariante CM, Lightman SL (2008) The HPA axis in major depression: classical theories and new developments. Trends Neurosci 31:464–468
Perera TD, Coplan JD, Lisanby SH, Lipira CM, Arif M, Carpio C, Spitzer G, Santarelli L, Scharf B, Hen R, Rosoklija G, Sackeim HA, Dwork AJ (2007) Antidepressant-induced neurogenesis in the hippocampus of adult nonhuman primates. J Neurosci 27:4894–4901
Pham K, McEwen BS, Ledoux JE, Nader K (2005) Fear learning transiently impairs hippocampal cell proliferation. Neuroscience 130:17–24
Pittenger C, Duman RS (2008) Stress, depression, and neuroplasticity: a convergence of mechanisms. Neuropsychopharmacology 33:88–109
Pollak DD, Monje FJ, Zuckerman L, Denny CA, Drew MR, Kandel ER (2008) An animal model of a behavioral intervention for depression. Neuron 60:149–161
Ramirez-Amaya V, Marrone DF, Gage FH, Worley PF, Barnes CA (2006) Integration of new neurons into functional neural networks. J Neurosci 26:12237–12241
Ramirez-Rodriguez G, Klempin F, Babu H, Benitez-King G, Kempermann G (2009) Melatonin modulates cell survival of new neurons in the hippocampus of adult mice. Neuropsychopharmacology 34:2180–2191
Reif A, Fritzen S, Finger M, Strobel A, Lauer M, Schmitt A, Lesch KP (2006) Neural stem cell proliferation is decreased in schizophrenia, but not in depression. Mol Psychiatry 11:514–522
Reif A, Schmitt A, Fritzen S, Lesch KP (2007) Neurogenesis and schizophrenia: dividing neurons in a divided mind? Eur Arch Psychiatry Clin Neurosci 257:290–299
Revest JM, Dupret D, Koehl M, Funk-Reiter C, Grosjean N, Piazza PV, Abrous DN (2009) Adult hippocampal neurogenesis is involved in anxiety-related behaviors. Mol Psychiatry 14: 959–967
Ridder S, Chourbaji S, Hellweg R, Urani A, Zacher C, Schmid W, Zink M, Hortnagl H, Flor H, Henn FA, Schutz G, Gass P (2005) Mice with genetically altered glucocorticoid receptor expression show altered sensitivity for stress-induced depressive reactions. J Neurosci 25:6243–6250
Sahay A, Hen R (2007) Adult hippocampal neurogenesis in depression. Nat Neurosci 10: 1110–1115
Sahay A, Scobie KN, Hill AS, O'Carroll CM, Kheirbek MA, Burghardt NS, Fenton AA, Dranovsky A, Hen R (2011) Increasing adult hippocampal neurogenesis is sufficient to improve pattern separation. Nature 472:466–470
Santarelli L, Saxe M, Gross C, Surget A, Battaglia F, Dulawa S, Weisstaub N, Lee J, Duman R, Arancio O, Belzung C, Hen R (2003) Requirement of hippocampal neurogenesis for the behavioral effects of antidepressants. Science 301:805–809
Saveanu RV, Nemeroff CB (2012) Etiology of depression: genetic and environmental factors. Psychiatr Clin North Am 35:51–71
Saxe MD, Battaglia F, Wang JW, Malleret G, David DJ, Monckton JE, Garcia AD, Sofroniew MV, Kandel ER, Santarelli L, Hen R, Drew MR (2006) Ablation of hippocampal neurogenesis impairs contextual fear conditioning and synaptic plasticity in the dentate gyrus. Proc Natl Acad Sci U S A 103:17501–17506
Schloesser RJ, Lehmann M, Martinowich K, Manji HK, Herkenham M (2010) Environmental enrichment requires adult neurogenesis to facilitate the recovery from psychosocial stress. Mol Psychiatry 15:1152–1163
Schmitt A, Weber S, Jatzko A, Braus DF, Henn FA (2004) Hippocampal volume and cell proliferation after acute and chronic clozapine or haloperidol treatment. J Neural Transm 111:91–100
Scholzen T, Gerdes J (2000) The Ki-67 protein: from the known and the unknown. J Cell Physiol 182:311–322
Segal M, Richter-Levin G, Maggio N (2010) Stress-induced dynamic routing of hippocampal connectivity: a hypothesis. Hippocampus 20:1332–1338

Seki T, Arai Y (1995) Age-related production of new granule cells in the adult dentate gyrus. NeuroReport 6:2479–2482
Sheline YI (1996) Hippocampal atrophy in major depression: a result of depression-induced neurotoxicity? Mol Psychiatry 1:298–299
Smith ME (2005) Bilateral hippocampal volume reduction in adults with post-traumatic stress disorder: a meta-analysis of structural MRI studies. Hippocampus 15:798–807
Snyder JS, Cameron HA (2012) Could adult hippocampal neurogenesis be relevant for human behavior? Behav Brain Res 227:384–390
Snyder JS, Soumier A, Brewer M, Pickel J, Cameron HA (2011) Adult hippocampal neurogenesis buffers stress responses and depressive behaviour. Nature 476:458–461
Spritzer MD, Galea LA (2007) Testosterone and dihydrotestosterone, but not estradiol, enhance survival of new hippocampal neurons in adult male rats. Dev Neurobiol 67:1321–1333
Spritzer MD, Ibler E, Inglis W, Curtis MG (2011) Testosterone and social isolation influence adult neurogenesis in the dentate gyrus of male rats. Neuroscience 195:180–190
Squire LR, Knowlton B, Musen G (1993) The structure and organization of memory. Annu Rev Psychol 44:453–495
Squire LR, Ojemann JG, Miezin FM, Petersen SE, Videen TO, Raichle ME (1992) Activation of the hippocampus in normal humans: a functional anatomical study of memory. Proc Natl Acad Sci U S A 89:1837–1841
Steen RG, Mull C, McClure R, Hamer RM, Lieberman JA (2006) Brain volume in first-episode schizophrenia: systematic review and meta-analysis of magnetic resonance imaging studies. Br J Psychiatry 188:510–518
Stockmeier CA, Mahajan GJ, Konick LC, Overholser JC, Jurjus GJ, Meltzer HY, Uylings HB, Friedman L, Rajkowska G (2004) Cellular changes in the postmortem hippocampus in major depression. Biol Psychiatry 56:640–650
Swanson LW, Cowan WM (1977) An autoradiographic study of the organization of the efferent connections of the hippocampal formation in the rat. J Comp Neurol 172:49–84
Tanapat P, Hastings NB, Reeves AJ, Gould E (1999) Estrogen stimulates a transient increase in the number of new neurons in the dentate gyrus of the adult female rat. J Neurosci 19: 5792–5801
Tata DA, Anderson BJ (2010) The effects of chronic glucocorticoid exposure on dendritic length, synapse numbers and glial volume in animal models: implications for hippocampal volume reductions in depression. Physiol Behav 99:186–193
Tata DA, Marciano VA, Anderson BJ (2006) Synapse loss from chronically elevated glucocorticoids: relationship to neuropil volume and cell number in hippocampal area CA3. J Comp Neurol 498:363–374
Taupin P (2010) A dual activity of ROS and oxidative stress on adult neurogenesis and Alzheimer's disease. Cent Nerv Syst Agents Med Chem 10:16–21
Teicher MH, Anderson CM, Polcari A (2012) Childhood maltreatment is associated with reduced volume in the hippocampal subfields CA3, dentate gyrus, and subiculum. Proc Natl Acad Sci U S A 109:E563–E572
Thomas RM, Hotsenpiller G, Peterson DA (2007) Acute psychosocial stress reduces cell survival in adult hippocampal neurogenesis without altering proliferation. J Neurosci 27:2734–2743
van Groen T, Wyss JM (1990) Extrinsic projections from area CA1 of the rat hippocampus: olfactory, cortical, subcortical, and bilateral hippocampal formation projections. J Comp Neurol 302:515–528
van Os J, Kenis G, Rutten BP (2010) The environment and schizophrenia. Nature 468:203–212
van Praag H, Christie BR, Sejnowski TJ, Gage FH (1999) Running enhances neurogenesis, learning, and long-term potentiation in mice. Proc Natl Acad Sci U S A 96:13427–13431
van Praag H, Shubert T, Zhao C, Gage FH (2005) Exercise enhances learning and hippocampal neurogenesis in aged mice. J Neurosci 25:8680–8685
van Praag HM (2004) Can stress cause depression? Prog Neuropsychopharmacol Biol Psychiatry 28:891–907

Veena J, Srikumar BN, Raju TR, Shankaranarayana Rao BS (2009) Exposure to enriched environment restores the survival and differentiation of new born cells in the hippocampus and ameliorates depressive symptoms in chronically stressed rats. Neurosci Lett 455:178–182

Videbech P, Ravnkilde B (2004) Hippocampal volume and depression: a meta-analysis of MRI studies. Am J Psychiatry 161:1957–1966

von Bohlen and Halbach O (2011) Immunohistological markers for proliferative events, gliogenesis, and neurogenesis within the adult hippocampus. Cell Tissue Res 345:1–19

Wahlberg KE, Wynne LC, Oja H, Keskitalo P, Pykalainen L, Lahti I, Moring J, Naarala M, Sorri A, Seitamaa M, Laksy K, Kolassa J, Tienari P (1997) Gene-environment interaction in vulnerability to schizophrenia: findings from the finnish adoptive family study of Schizophrenia. Am J Psychiatry 154:355–362

Wakade CG, Mahadik SP, Waller JL, Chiu FC (2002) Atypical neuroleptics stimulate neurogenesis in adult rat brain. J Neurosci Res 69:72–79

Wang HD, Dunnavant FD, Jarman T, Deutch AY (2004) Effects of antipsychotic drugs on neurogenesis in the forebrain of the adult rat. Neuropsychopharmacology 29:1230–1238

Watson DR, Bai F, Barrett SL, Turkington A, Rushe TM, Mulholland CC, Cooper SJ (2012) Structural changes in the hippocampus and amygdala at first episode of psychosis. Brain Imaging Behav 6:49–60

West MJ, Gundersen HJ (1990) Unbiased stereological estimation of the number of neurons in the human hippocampus. J Comp Neurol 296:1–22

Zhang CL, Zou Y, He W, Gage FH, Evans RM (2008) A role for adult TLX-positive neural stem cells in learning and behaviour. Nature 451:1004–1007

Zigova T, Pencea V, Wiegand SJ, Luskin MB (1998) Intraventricular administration of BDNF increases the number of newly generated neurons in the adult olfactory bulb. Mol Cell Neurosci 11:234–245

Neurobiological Markers of Familial Risk for Depression

Lara C. Foland-Ross, Michael G. Hardin and Ian H. Gotlib

Abstract Major depression is associated with a wide range of neurobiological disturbances, including anomalies in the structure and function of cortical and subcortical gray matter and dysregulation of the HPA axis. In this chapter, we review research demonstrating that many of these abnormalities are also present in never-depressed offspring of adults with recurrent depression and discuss how such findings might reflect dysfunctional neuroregulatory systems that precede the onset of this disorder. We also briefly discuss candidate genes and environmental factors that have been posited to be directly involved in the transmission of neural and HPA-axis abnormalities from depressed parents to their offspring, and we review how, by obtaining a better understanding of the neurobiological markers of depression risk, we can facilitate the development of targeted strategies for the prevention and treatment of major depression.

Keywords Depression · Risk · Neuroimaging · Cortisol · Hippocampus · DLPFC · Amygdala · Genetics

Contents

L. C. Foland-Ross · M. G. Hardin · I. H. Gotlib (✉)
Department of Psychology, Stanford University, Stanford, CA, USA
e-mail: ian.gotlib@stanford.edu

Curr Topics Behav Neurosci (2013) 14: 181–206
DOI: 10.1007/7854_2012_213

Published Online: 10 May 2012

Major depressive disorder (MDD) is among the most prevalent of all psychiatric disorders; almost 20 % of the American population (or more than 30 million adults) will experience a clinically significant episode of depression in their lifetime (Kessler and Wang 2009). Not surprisingly given this high prevalence, MDD is associated with disturbingly high societal costs; in fact, the World Health Organization has ranked MDD as the leading cause of years lived with disability and the single most burdensome disease in the world in terms of total disability-adjusted years among people in the middle years of life (Murray and Lobez 1996).

Given these alarming statistics, researchers have worked to understand factors that are involved in the onset and maintenance of MDD. In this context, scientists have described a complex interplay of genetic and environmental factors as playing an important role in contributing to vulnerability for the development of MDD. Indeed, the strongest and most reliable risk factor for the development of MDD is a family history of the disorder (Williamson et al. 2004): having a parent with MDD is associated with a three- to fivefold increase in the risk for developing a depressive episode (Beardslee et al. 1998; Williamson et al. 2004). Despite this fact, we know little about the mechanisms involved in this transmission of risk. Although investigators have identified a number of depression-related physiological and neural abnormalities, we do not know if these anomalies are epiphenomena of MDD, consequences of having been depressed, or familially transmitted aberrations that increase vulnerability to the disorder. By examining individuals who have not yet experienced MDD themselves, but who are at high risk for developing the disorder by virtue of having a family history of MDD, we can begin to elucidate these issues.

In this chapter, we describe research that has been conducted with individuals who are at familial risk for the development of depression, with a specific focus on neurobiological factors. We begin by reviewing findings from neuroimaging studies documenting abnormalities in brain structure and function in never-depressed offspring of parents with recurrent depression, and then review evidence implicating the specific involvement of hypothalamic–pituitary–adrenal (HPA) axis dysfunction in the intergenerational transmission of risk for depression. Finally, we present a brief discussion of specific genes and environmental factors that have been posited to be directly involved in the transmission of neural and HPA-axis abnormalities from depressed parents to their offspring.

1 Neuroimaging of Individuals at Familial Risk for Depression

Over the past 20 years, considerable advances in non-invasive neuroimaging technologies, particularly magnetic resonance imaging (MRI), have contributed significantly to our knowledge of anomalies in brain structure and function that are associated with MDD. While some of these abnormalities may be a function of the depressive state at the time of assessment, other anomalies may reflect stable predisposing factors that place individuals at increased risk for the onset of depression. In the following section, we present findings from neuroimaging studies designed to identify structural and/or functional aberrations associated with an elevated risk for depression.

1.1 Structural Findings

Given that depression is primarily a disorder of the experience and regulation of emotion, it is perhaps not surprising that structural neuroimaging studies of this disorder have revealed volumetric abnormalities in brain regions that subserve the processing of emotional stimuli and information. Indeed, recent meta-analyses of depression-related structural neuroimaging studies indicate that depressed individuals reliably exhibit abnormal gray matter volumes of several structures, including the hippocampus, amygdala, anterior cingulate cortex (ACC), and dorsolateral prefrontal cortex (DLPFC) (Videbech and Ravnkilde 2004; Hamilton et al. 2008; Koolschijn et al. 2009; Kempton et al. 2011). While it is possible that structural abnormalities in these regions are a consequence of major depression, it is also possible that these aberrations are present prior to the onset of the disorder and contribute to an individual's likelihood of developing MDD. To examine this issue, it is necessary that investigators determine whether neural abnormalities in depressed adults are also present in individuals who are at high risk for developing depression but have not yet experienced clinically significant symptoms. Although no structural neuroimaging studies have taken a prospective longitudinal approach to identifying whether neural abnormalities in never-depressed individuals at high risk for depression predict the eventual onset of MDD, over the past 5 years, a number of reports have indicated that abnormalities in brain structure that have previously been found to be associated with depression in adulthood are also present in younger, vulnerable individuals.

Given the involvement of the hippocampus in mood (Critchley et al. 2000; Eisenberger et al. 2007; Fusar-Poli et al. 2009) and physiological reactions (e.g., cortisol response) associated with psychosocial stress (Herman et al. 1989; Mizoguchi et al. 2003), several structural neuroimaging investigations of individuals at high familial risk for MDD have focused on this structure. For example, Chen et al. (2010) examined differences in hippocampal volume between never-depressed adolescent daughters of mothers with a history of MDD (high risk for depression) and never-depressed adolescent daughters of mothers with no history of psychopathology (low risk for depression). Results of this investigation revealed that high-risk girls had smaller hippocampal volumes than did their low-risk counterparts, suggesting that neuroanatomic anomalies of the hippocampus that are associated with depression (Videbech and Ravnkilde 2004; Koolschijn et al. 2009; Kempton et al. 2011) precede the onset of the disorder. This finding is consistent with the results of two other studies that examined hippocampal abnormalities in never-depressed adults with a first-degree relative who was experiencing depression. In the first study, Baaré et al. (2010) found reduced hippocampal volume in never-depressed adult siblings of adults with recurrent depression compared to never-depressed siblings of never-depressed adults. In the second investigation, Amico et al. (2011) found that adults who had either a first- or a second-degree relative with depression had smaller hippocampal volumes than did both adults without a positive family history of depression and adults who were diagnosed with current MDD. This latter finding is particularly interesting in that it involves a comparison of hippocampal volumes between never-depressed high-risk individuals and diagnosed depressed persons. While it may, at first, seem counterintuitive that high-risk individuals had smaller hippocampal volume than did diagnosed depressed persons, it is important to note that this finding is consistent with a growing body of evidence indicating that hippocampal volume increases with pharmacological treatment of MDD (Santarelli et al. 2003; Vermetten et al. 2003; Lai et al. 2010), likely as a result of hippocampal neurogenesis (Malberg et al. 2000). In this context, therefore, Amico et al.'s findings not only point to reduced hippocampal volume as a risk factor for depression, but also suggest that depression-associated volumetric reductions of the hippocampus can be reversed with medication.

Adding to this body of work, Rao et al. (2010), like Chen et al. (2010), found adolescents at high familial risk for depression to exhibit volumetric decreases in the hippocampus compared to never-depressed adolescents with no family history of depression. Moreover, Rao et al. examined an additional group of adolescents with current depression and found these participants to exhibit deficits in hippocampal volume similar to those found in the high-risk group. Although this latter finding conflicts with the results of Amico et al. (2011), such a discrepancy may be due to effects of medication; whereas Amico et al. included mainly medicated participants in their depressed sample, Rao et al. assessed primarily medication-na individuals. Importantly, Rao et al.'s study of high-risk adolescents also extends previous findings obtained with high-risk populations by linking structural anomalies in the hippocampus to childhood stress. More specifically, using mediation analyses, Rao et al. found that early life adversity during childhood increased the likelihood of having smaller hippocampi at initial assessment, and that this association was present only in offspring of depressed parents, and not in children of never-depressed parents. These findings are intriguing in suggesting that depressed parents transmit a heightened sensitivity to stress to their offspring, and that this sensitivity directly influences hippocampal development. Alternatively, depressed parents may be more likely to select themselves, and therefore their children, into high-stress environments, and this form of chronic childhood stress may combine with other early life adversities to result in the volumetric reductions. Finally, Rao et al. followed their participants longitudinally and found that depressive symptoms at a follow-up assessment were partially mediated by smaller hippocampal volumes at initial assessment. Thus, a positive family history of depression in this study interacted with life stress to influence hippocampal development, and these factors collectively contributed to a greater likelihood of developing depression. These findings are particularly notable because they are the first to suggest that abnormalities in brain structure, present in individuals at initial assessment, can predict which high-risk adolescents are most likely to develop depression.

Recent structural neuroimaging investigations of youths at high familial risk for depression suggest, in addition to the hippocampus, that volumetric abnormalities in other brain regions may also precede the onset of MDD. In the same study in which Amico et al. (2011) examined differences in hippocampal volume between adults at low and high familial risk for depression, group differences were also reported in the gray matter of the frontal lobe and striatum. More specifically, individuals at high familial risk for depression were found to exhibit volumetric reductions in the right DLPFC (rDLPFC) and putamen compared with low-risk individuals. These findings are particularly interesting given the respective roles of the rDLPFC and putamen in the processing of behavior and affect. For example, the rDLPFC has been implicated in cognitive and inhibitory control processes (e.g., Aron 2011; Ridderinkhoff et al. 2004), and is an integral component of a frontal system that is involved in the regulation of emotion through inhibitory control over limbic regions (e.g., Hariri et al. 2000; Beauregard et al. 2001). In turn, the putamen, as part of the striatum and the cortico-striatal-pallidal-thalamic (CSPT) loop (Alexander et al. 1986), has been posited to be involved in reward-seeking behavior (Kawagoe et al. 1998; Gold 2003). Volumetric reductions in each of these regions have also been reported in structural neuroimaging studies of depressed adults (see Kempton et al. 2011). Thus, Amico et al.'s finding of reduced volume of these structures in nondepressed individuals at familial risk for MDD suggests that these anomalies may precede the onset of depression.

Other investigators conducting structural and functional neuroimaging studies of depression have focused on the amygdala, given its critical involvement in the perception and memory of emotional information (Kensinger and Corkin 2004; Adolphs 2008; Pessoa and Adolphs 2010). Lupien et al. (2011) followed women with recurrent depression and never-depressed controls who had given birth to healthy offspring over a 10-year period and assessed their depressive symptoms at regular intervals. At the end of this 10-year period, the children were scanned to examine whether children of depressed mothers were characterized by structural abnormalities in subcortical limbic areas. Results of this study showed that children of depressed mothers had larger amygdala volumes than did their low-risk peers and, further, that these volumes were significantly positively correlated with the mothers' mean depressive scores assessed over the 10-year interval. These findings suggest that children's neural development is affected not only by whether their parent is depressed, but also by the severity and duration of the mother's depression during the child's lifetime. The authors compared their findings to results of studies that examined amygdala volume in children reared in orphanages—a model of early maternal separation (Mehta et al. 2009; Tottenham et al. 2010)—and interpreted the greater amygdala volume in the high-risk children to be a consequence of poor parental care caused by the mothers' depression. Indeed, maternal depression has been found in other investigations to decrease the quality of maternal care (Field 1984, 1994; Murray and Cooper 2003); the influence of depression on parenting and the consequent effects on healthy neurobiological development in the child are topics that warrant discussion and are addressed in later sections of this chapter.

Finally, in a study examining structural abnormalities of cortical gray matter using measurements of cortical thickness, Peterson et al. (2009) found significant cortical thinning in never-depressed individuals with first- or second-degree relatives with depression in the right dorsal and inferior frontal gyri, somatosensory and motor cortices, dorsal and inferior regions of parietal cortex, and posterior regions of temporal cortex, compared to individuals who did not have a family history of depression. Follow-up correlational analyses of these data indicated that thinner cortical gray matter in these regions of high-risk individuals was associated with higher levels of inattention and poorer performance on tests measuring immediate and delayed visual memory. Risk-related cortical thinning in these brain regions may therefore directly underlie poor performance on measures of inattention and lead, in turn, to maladaptive emotion processing and subsequent difficulties in the ability to regulate emotional arousal, which may contribute to the onset of MDD. Indeed, mediator analyses confirmed that cortical thickness abnormalities mediated the association of familial risk with performance on cognitive tests. These findings both provide insight into how neurobiological abnormalities might mediate the intergenerational transmission of depression and highlight the need for future neuroimaging studies of samples at familial risk for depression to incorporate and integrate measures of behavior and cognition.

While the results of these studies suggest that structural differences within certain brain regions are involved in the intergenerational transmission of risk for depression, other research implicates the neural connections among brain regions in this process. For example, depression-related abnormalities have been documented in the white matter tracts that connect proximal and distal brain regions to facilitate complex behaviors (e.g. Sexton et al. 2009; Kieseppä et al. 2010). Using diffusion tensor imaging (DTI) to examine the possibility that these white matter abnormalities precede the onset of depression and represent a vulnerability marker for MDD, Huang et al. (2011) examined fractional anisotropy (FA, a measure that reflects aspects of membrane integrity and myelin thickness) in healthy adolescents who were at high or low familial risk for depression. Results of this investigation showed that high-risk adolescents exhibited reduced FA in the cingulum, which connects gray matter of the cingulate with that of the hippocampus and perihippocampus, as well as reduced FA in tracts that connect rostral portions of the temporal lobe (e.g., amygdala and hippocampus) with the inferior portions of the frontal lobe that have been posited to be involved in social and emotional regulation (Schmahmann et al. 2007; Sexton et al. 2009).

Taken together, these structural neuroimaging findings suggest that both subcortical and cortical gray matter structure, as well as FA of white matter, are aberrant in individuals who are at familial risk for depression but who have not yet experienced a depressive episode themselves. While the precise causes of these structural anomalies are not yet known, genetic and/or environmental factors are likely to be involved. We will address the possible influence of a specific set of genes and environmental variables later in this chapter. It is also unclear how abnormalities in brain structure might affect function in overlapping and connected areas. Although we review findings below from studies examining brain function in individuals at familial risk for depression, experiments are required that address the direct consequences of changes in gray matter structure on brain function and behavior.

1.2 Functional Findings

Like the research described above examining risk-associated anomalies in brain structure, functional neuroimaging of individuals at familial risk for depression is still in its infancy. Given that depression is primarily a disorder of emotional processing, and that most neural functional abnormalities that have been found in depression are reported in emotion-related circuitry (Price and Drevets 2010), research conducted to date on populations at high risk for depression has focused almost entirely on neural activity associated with the processing of emotional stimuli. In support of such a focus, behavioral studies have demonstrated that, like depressed persons, never-depressed individuals at familial risk for depression are characterized by negatively biased processing of emotional information; high-risk individuals have been found behaviorally to exhibit negative biases in the interpretation (Dearing and Gotlib 2009) and identification (Joormann et al. 2010) of stimuli, in attentional processing (Joormann et al. 2007), and in the categorization of emotional information (Mannie et al. 2007a).

Results of the few emotion-related functional neuroimaging studies that have been conducted with individuals at high risk for MDD are generally consistent with findings from neuroimaging studies of depressed adults indicating that strong bottom-up neural responses in emotion-processing regions (such as the amygdala and limbic circuits), coupled with hypo-responsivity in top-down regulatory regions (such as the DLPFC), may contribute to these negatively based responses to emotional information (Browning et al. 2010; De Raedt et al. 2010). For example, compared to their low-risk peers, adolescents at high risk for MDD have been found to have a greater neural response in brain regions that mediate emotional reactivity to affective stimuli, such as the amygdala and insula, as they view negative emotion faces (Monk et al. 2008) or sad film clips (Lévesque et al. 2011; Joormann et al. 2012). High-risk young adults have also been found to have a smaller left DLPFC response than do their low-risk counterparts as they view negative emotion faces during an emotion-matching task (Mannie et al. 2011) or attempt to repair sad mood through recalling positive memories (Joormann et al. 2012). While these findings suggest that limbic and frontal functioning serve as markers for increased vulnerability to depression, it is clear that further investigation is warranted. For example, Monk et al. (2008) observed increased limbic response by high-risk adolescents only when their attention was unconstrained, not in a constrained-attention condition, suggesting that attention-related processes mediate this effect. Similarly, Mannie et al. (2011) found no differences between young low- and high-risk adults in subcortical limbic response, highlighting the importance of considering the role of development in emotion processing.

Given the diminished experience of pleasure and the high levels of anhedonia that often characterize individuals diagnosed with MDD, investigators studying intergenerational risk for depression have also started to examine the possibility that functional anomalies related to the processing of reward information may serve as markers for increased vulnerability to MDD. Building on evidence that both depressed adults (Steele et al. 2007; Kumar et al. 2008; Pizzagalli et al. 2009) and adolescents (Forbes et al. 2007, 2009) show attenuated striatal activity in response to reward information, Gotlib et al. (2010) scanned low- and high-risk adolescent girls as they performed a task designed to study reward anticipation and consumption. These investigators found that high-risk adolescents demonstrated attenuated striatal response during the processing of reward, suggesting that reward-related neural function is a marker of increased risk for depression. While these data are encouraging, this is the only study to date to examine reward-related neural function in high-risk individuals. Therefore, it is critical that investigators extend this finding so that we can gain a better understanding of the role of reward-related processing in contributing to vulnerability for depression.

2 HPA-Axis Dysregulation in Individuals at Familial Risk for Depression

The observations that patients with Cushing's syndrome often experience severe depression (Gallagher et al. 2009) and that levels of cortisol increase in response to stress (Dickerson and Kemeny 2004) have contributed to the refinement of stress-diathesis hypotheses of depression (e.g., Monroe and Simons 1991), in which high levels of cortisol are posited to play a significant pathophysiologic role in the etiology of MDD. Indeed, one of the more consistently reported physiological abnormalities in major depression is abnormally elevated levels of cortisol (Burke et al. 2005; Knorr et al. 2010; Stetler and Miller 2011). This glucocorticoid hormone is released into systemic circulation by the HPA axis following awakening or during times of stress. More specifically, corticotropin-releasing hormone (CRH) is secreted from the hypothalamus, which causes the neighboring pituitary gland to secrete adrenocorticotropin hormone (ACTH). ACTH then acts on the adrenal glands to induce the release of cortisol, in addition to a number of other hormonal compounds, which help the body respond adaptively to awakening or to a stressful event by increasing heart rate, boosting vascular tone, and speeding reaction time. Elevated levels of cortisol also typically function through negative feedback mechanisms in the hippocampus to inhibit the further production and secretion of cortisol by the HPA system (Sapolsky et al. 1986). Thus, in normal functioning, cortisol facilitates individuals' ability to adapt to, and recover from, stress.

Previous studies that have examined HPA responses to stress have typically used laboratory tasks that measure the temporal dynamics of cortisol release following exposure to a psychosocial stressor. Such a sampling approach typically consists of three phases: (1) a basal activity phase, involving baseline cortisol levels in the absence of a stressor; (2) a 'stress reactivity' phase, in which cortisol increases from baseline (i.e., pre-stressor) levels following the onset of a stressor; and (3) a 'stress recovery' phase, in which cortisol levels return to baseline following the offset of the stressor (McEwen 1998). Although reports of cortisol release in depressed individuals in response to a laboratory stressor are mixed (Gotthardt et al. 1995; Heim et al. 2000; Young et al. 2000), a recent meta-analysis showed that, controlling for baseline levels, elevations in cortisol in response to laboratory stressors are typically attenuated compared to levels in never-depressed control subjects (Burke et al. 2005). This meta-analysis also reported abnormalities during the stress recovery phase; across studies, depressed individuals generally take longer than do nondepressed individuals for their cortisol to return to baseline levels following the removal or termination of the stressor (Burke et al. 2005). Thus, whereas never-depressed individuals show reliable and dynamic patterns of neuroendocrine responses to laboratory stress, with greater stress reactivity and rapid recovery following stress, individuals diagnosed with MDD exhibit a relatively flat and unresponsive pattern of cortisol secretion. The meaning of this pattern in depression is unclear, although it has been posited that abnormal cortisol responses to stress may reflect difficulties in the ability of depressed individuals to adapt to, and recover from stressful events.

A second, more widely replicated finding in MDD concerns diurnal fluctuations in salivary cortisol. Typically, cortisol levels rise in the early morning, peaking approximately 30 min after awakening, and then diminish over the course of the day, reaching the lowest points in the evening (Posener et al. 1996). Although some investigators have reported a blunted cortisol awakening response in depression (Stetler and Miller 2005; Huber et al. 2006), most studies have documented abnormally high levels of morning salivary cortisol in MDD (Pruessner et al. 2003; Bhagwagar et al. 2005; Vreeburg et al. 2009). Consistent with this pattern of results, two recent meta-analyses concluded that morning salivary cortisol levels are higher in individuals diagnosed with MDD than they are in never-depressed participants (Knorr et al. 2010; Stetler and Miller 2011). Abnormally elevated levels of cortisol following awakening have also been found in studies of individuals who report chronic stress or worrying (Wust et al. 2000; Schlotz et al. 2004) and work overload (Steptoe et al. 2000). Thus, elevations in waking cortisol in depression could be a consequence of chronic stress, or could be related to the anticipation of more burdensome demands in daily life (Fries et al. 2009). Meta-analyses examining diurnal fluctuations of cortisol secretion in depression have also found, in addition to higher morning cortisol levels, that depressed individuals exhibit elevated cortisol levels before bedtime. This finding provides empirical support for the suggestion that baseline levels of cortisol are chronically elevated in MDD, which may contribute to a ceiling effect observed in some laboratory studies of depressed individuals' cortisol response to stress.

Although anomalous functioning of the HPA axis could be the consequence, or "scar," of having a depressive illness, or of having experienced traumatic life events, it is also possible that dysregulation of the HPA-axis represents a vulnerability marker for depression. Indeed, investigators have found that depressed individuals who exhibit no change in HPA-axis activity following pharmacotherapy for MDD have poorer outcome than do depressed individuals who show medication-related normalization of the HPA axis (Heuser et al. 1994). Researchers have also found that alterations in HPA-axis activity can predict subsequent episodes of depression in remitted individuals, independent of the type of treatment they receive (Harris et al. 2000; Wust et al. 2000; Zobel et al. 2001).

Whether dysregulation of the HPA-axis might also predict the initial onset of MDD is a focus of ongoing research. Because, as we noted earlier, one of the strongest risk factors for depression is a family history of MDD, studies of young offspring of depressed parents are particularly promising in elucidating this issue. Interestingly, investigators who have examined HPA-axis function have found that children and adolescents who have a depressed parent demonstrate both increased basal (Lupien et al. 2000; Young et al. 2006; Lupien et al. 2011) and increased waking (Mannie et al. 2007b; Vreeburg et al. 2010) levels of salivary cortisol compared to offspring of never-depressed parents. These patterns are in striking overlap with findings of studies of currently depressed adults (Pruessner et al. 2003; Bhagwagar et al. 2005; Vreeburg et al. 2009; Knorr et al. 2010; Stetler and Miller 2011), and provide support for the formulation that aberrant HPA-axis function precedes the onset of MDD.

The gold-standard approach, however, to determining whether HPA-axis dysfunction precedes the onset of depression involves measuring cortisol secretion in participants prior to the onset of the first depressive episode and then following the same individuals longitudinally to determine whether anomalies in HPA-axis regulation predict higher rates of depression over time. Few studies have used this methodological approach. Of these few, however, there is a striking consistency in their findings. For example, Adam et al. (2010) found higher levels of morning salivary cortisol never-depressed adolescents who developed an episode of major depression over the following year than in adolescents who did not develop depression. Consistent with this finding, Rao et al (2009) examined nocturnal urinary-free cortisol (NUFC) excretion (i.e., basal levels of cortisol) in adolescents with or without a family history of depression. These investigators found that adolescents at familial risk for depression exhibited higher NUFC excretion than did their low-risk counterparts and, further, that levels of NUFC excretion were highest in those adolescents who developed MDD by a 5-year follow-up assessment. Goodyer et al. (2000) also followed adolescents who were at high risk for depression by virtue of either having a depressed parent or of experiencing psychosocial adversities. These adolescents were assessed at baseline for morning levels of salivary cortisol, and were assessed again for depressive symptoms 12 months later. Goodyer et al. found that high-risk adolescents who met diagnostic criteria for MDD at follow-up exhibited significantly higher morning cortisol levels at initial assessment a year earlier than did high-risk adolescents who did not develop depression. Interestingly, subsequent analyses of these data showed that the association between cortisol and risk for depression was moderated by genetic factors (Goodyer et al. 2009). More specifically, of the adolescents at familial or psychosocial risk for depression, those who carried at least one copy of the short (s) allele in the promoter region of the serotonin transporter gene (5-HTTLPR) exhibited significant higher levels of morning salivary cortisol than did high-risk individuals who were homozygous for the long (l) allele. Moreover, the risk for depression was highest in individuals who both were s-carriers and exhibited abnormally elevated waking cortisol levels at the initial assessment. Thus, genetic factors appear to influence both HPA-axis function and subsequent risk for depression. We discuss the role of genetic factors in the intergenerational transmission of neurobiological abnormalities below.

3 Mechanisms of Transmission of Neurobiological Abnormalities

One challenge facing researchers interested in elucidating intergenerational neurobiological markers of depression risk involves understanding the multiple pathways through which neurobiological abnormalities are transmitted from depressed parents to their children. Although several mechanisms have been proposed, in this section, we focus on the roles of genes and maternal environment.

3.1 Genetic Factors

Genetic factors are likely to contribute to associations that are observed between maternal depression and the emergence of depression-related neurobiological abnormalities in their children. One relatively straightforward first step in elucidating the specific genes that are involved in the intergenerational transmission of these anomalies is to study healthy individuals at genetic risk for depression to examine whether their brain structure and function resembles that of depressed persons. Although different genes have been proposed to moderate individuals' vulnerability to depression (Levinson 2006), investigators in this area have focused on two genes, in particular.

First, due in part to the efficacy of selective serotonin reuptake inhibitors (SSRIs) in the treatment of depression, researchers have examined neural characteristics associated with functional polymorphisms of the serotonin transporter (5-HTT) gene. More specifically, the short (s) and long (l) alleles of the promoter region of the serotonin transporter gene (5-HTTLPR) have been found to decrease and increase transcription rates of 5-HTT, respectively. Recent investigations of 5-HTTLPR suggest that functional polymorphisms of this gene contribute to the onset of depression by influencing individuals' sensitivity to stress. 5-HTTLPR s-allele carriers have been found to exhibit a greater number of, and more severe, depressive symptoms in response to stressful life events than do individuals who are homozygous for the long allele (Caspi et al. 2003; Kaufman et al. 2004). Evidence supporting this formulation comes from two studies reporting that increases in cortisol in response to stress are mediated by 5-HTTLPR polymorphisms, with s-allele homozygotes showing greater cortisol increases, as well as prolonged recovery from stress, compared to l-allele carriers (Gotlib et al. 2008; Way and Taylor 2010). Similarly, Chen et al. (2010) found that never-depressed girls who were s-allele carriers exhibited higher waking, but not afternoon or evening, cortisol levels than did l-allele homozygotes. Moreover, in a study we described earlier, Goodyer et al. (2009) found that s-allele carriers who were at familial or psychosocial risk for depression, but were not themselves depressed, exhibited both higher waking cortisol levels and higher rates of depression at a follow-up assessment than did high-risk individuals who were homozygous for the l-allele. Thus, a consistent picture is emerging across studies indicating that the s-allele genotype of 5-HTTLPR is associated with higher cortisol in response to awakening and psychosocial stress, and that this may serve to increase vulnerability for the development of depression.

Other investigators have focused on the involvement of 5-HTTLPR polymorphisms in the emergence of alterations in the structure and/or function of neural gray matter. In one study of never-depressed individuals, Hariri et al. (2005) found carriers of the s-allele to exhibit significantly greater activation in the amygdala during the performance of an affective faces task than did l-allele homozygotes. The association between amygdala response and s-allele carrier status has been replicated by other researchers (Dannlowski et al. 2010; Furman et al. 2010; von dem Hagen et al. 2011), adding support to the hypothesis that broad measures of neural system engagement are sensitive to alterations in serotonergic function influenced by 5-HTTLPR polymorphisms. Interestingly, the amygdala is well-known to provide positive feedback to the HPA axis (Feldman et al. 1982); therefore, 5-HTTLPR-related hyperreactivity of this region may contribute to the higher cortisol responses to stress that have been documented in s-allele carriers.

Researchers have also examined the effects of 5-HTTLPR polymorphisms on the structure of, and functional connectivity between, brain regions involved in the processing of emotional information. Pezawas et al. (2005) found that relative to l-allele homozygotes, s-allele homozygotes are characterized by reduced gray matter volume in, and abnormal functional connectivity between, the amygdala and the peri- and subgenual anterior cingulate cortices. Moreover, decreased cortical gray matter volume of the inferior frontal gyrus has been reported in s-allele carriers compared to l-allele homozygotes (Selvaraj et al. 2011). As we noted above, these areas have been implicated consistently in neuroimaging studies of depressed adults (Gotlib and Hamilton 2008; Koolschijn et al. 2009) and their never-depressed offspring (Monk et al. 2008; Peterson et al. 2009). It is tempting to speculate, therefore, that depression-related abnormalities in brain structure and function (and, consequently, risk for MDD) may be transmitted directly from depressed parent to child through allelic variations in 5-HTTLPR.

Another gene that has been implicated in depression is one that encodes brain-derived neurotrophic factor (BDNF), a protein that aids in the survival and plasticity of neurons (Murer et al. 2001), the growth and differentiation of new neurons and synapses (especially in the hippocampus, e.g., Huang and Reichardt 2001), and the protection of neurons against stress-induced damage (Radecki et al. 2005). BDNF polymorphisms involving the substitution of Valine (Val) for Methionine (Met) at codon 66, result in an increase of activity-dependent secretion of BDNF (Egan et al. 2003). The amount of available BDNF protein, therefore, is significantly decreased in carriers of the Met allele.

Interestingly, recent research has documented that low levels of the BDNF protein are present in individuals with MDD (Karege et al. 2002), and that polymorphisms of the BDNF gene are associated with increased risk for depression (Schumacher et al. 2005). One hypothesis concerning the etiology of MDD is that lower levels of the BDNF protein decrease protection against stress-induced damage to neuronal gray matter (Duman and Monteggia 2006). Indeed, at least one study examining the association between MDD and the BDNF gene has supported this formulation. Frodl et al. (2007) found that currently depressed individuals exhibited reduced hippocampal volume compared with never-depressed individuals; interestingly, the greatest reductions were found in individuals who carried the BDNF Met allele. Importantly, however, this study, like others (Pezawas et al. 2004; Bueller et al. 2006) also found associations between BDNF and hippocampus volume in the nondepressed group, suggesting that the BDNF gene alone does not account for MDD-related abnormalities in brain structure. In this context, the results of more recent studies point to a better-defined role of BDNF polymorphisms in the pathogenesis of depression. For example, Gatt et al. (2009) found that the combination of Met-allele carrier status and exposure to childhood trauma predicted a reduction in gray matter in hippocampus and lateral prefrontal cortex and, in turn, higher rates of depression. Similarly, Gerritsen and colleagues (in press) reported that Met-allele carriers with a history of childhood trauma had significantly less gray matter volume in the subgenual ACC than did both Met-allele carriers without trauma and Val/Val homozygotes with trauma. Therefore, similar to the role of 5-HTTLPR in the emergence of depression and its related neurobiological abnormalities, the likelihood that neural markers for depression are transmitted from parent to child could also be a function of the interaction of the BDNF genotype and other personal or environmental factors.

Finally, a growing body of research suggests that interactions among various genes play a significant role in the development of both depression and depression-related abnormalities in neural structure and function. For example, in a study of healthy preschoolers, HPA-axis reactivity to a laboratory stressor was found to be higher in s-allele homozygotes who were also carriers of the BDNF Met-allele than it was in s-allele homozygotes who carried two copies of the BDNF Val allele

(Dougherty et al. 2010). This finding is consistent with the results of at least one study linking an interaction of the BDNF Met-allele and 5-HTTLPR s-allele with elevated levels of depressive symptoms in children (Kaufman et al. 2006), and extends prior suggestions of a protective effect of the BDNF Val-allele. However, the results of research examining the protective effects of the BDNF gene are equivocal. Pezawas et al. (2008), for example, found significant volume reductions in subgenual ACC in non-disordered individuals who were homozygous for the 5-HTTLPR s-allele compared to carriers of the l-allele, with the most severe deficits present in individuals who were also homozygous for the BDNF Val allele.

Indeed, a recurring difficulty in this field involves the non-replication of initially promising findings (e.g., see Scharinger et al. 2010 for a review). While this may be due in part to small sample sizes, another issue that serves to complicate matters is gene–environment correlations (rGE), which occur when specific genetic factors influence individuals' exposure to certain environments. It therefore remains possible for example, that reductions in neural function and structure of 5-HTTLPR s-allele homozygotes could be directly related to heritable increases in the exposure to high-risk environments. Indeed, relevant to the focus of this review, a growing body of research provides evidence to support the notion that environmental risk factors for depression, including aspects of the social environment, such as the experience of negative life events and negative parent–child relationships, are heritable (Thapar et al. 1998; Rice et al. 2003; Lau et al. 2006; Lau and Eley 2008; Wichers et al. 2008). These findings, taken in context with the knowledge that environmental factors have a profound effect on the development of depression and its neurobiological hallmarks (as we discuss in greater detail below), make it clear that additional studies are needed, with sample sizes large enough to be able to test the interaction of BDNF and 5-HTTLPR polymorphisms, to help clarify epistatic relations between these genes and their interaction with other neural and environmental risk markers for depression.

3.2 Environmental Factors

It is now well-documented that long periods of separation of neonatal rodents and non-human primates from their mothers elicit changes in HPA-axis function that persist into adulthood and that resemble characteristics of depressed adults (Sanchez et al. 2001). Paralleling these findings are reports showing that HPA-axis dysfunction (Heim et al. 2002, 2008) and volumetric reductions in cortical (Hanson et al. 2010; van Harmelen et al. 2010) and subcortical gray matter (Vythilingam et al. 2002) are present in individuals who have experienced childhood trauma.

Given these findings, researchers seeking to identify the factors that are involved in the intergenerational transmission of depression risk have also focused on whether environmental factors, such as depression-related alterations in parental care, might influence both risk for depression, and abnormal development of the HPA axis. For example, Halligan et al. (2004) found in a longitudinal study that the presence of postnatal depression in mothers was significantly associated 13 years later with higher morning waking cortisol levels in their offspring, suggesting long-term associations between early exposure to maternal depression and HPA-axis dysfunction. Interestingly, subsequent analyses of these data indicated that cortisol abnormalities in these adolescent offspring were significantly associated with the quality of maternal care received during the first year of life, but not 5 years later when the majority of cases of mothers' depression had remitted (Murray et al. 2010). Additional data collected from this sample indicate that the level of cortisol abnormalities at age 13 significantly predicted the severity of depressive symptoms an average of 3 years later (Halligan et al. 2007). Thus, depression-related disturbances in maternal care may contribute to the risk for depressive disorder through long-term alterations of HPA-axis function in the first year of life.

Indeed, researchers are documenting that depression in mothers is associated with disturbed parenting processes, including a decreased sensitivity to infant needs and increased rates of withdrawn and disengaged behaviors (Field 1984, 1994; Murray and Cooper 2003). These types of behaviors may be particularly stressful to the child (Langrock et al. 2002), and researchers have suggested that this type of chronic stress mediates the association between maternal depression and depression in the offspring. For example, Hammen et al. (2003) found that whereas the onset of depression in offspring of depressed mothers was more frequently associated with chronic interpersonal stress (i.e., stress resulting from social interactions between the child/adolescent and others in his or her environment), the onset of depression in offspring of never depressed mothers was associated more frequently with an episodic stressor (i.e., a stressful life event). Moreover, although there are not yet studies similar to that conducted by Murray et al. (2010), who examined neuroendocrine systems while taking into account depression-related alterations in maternal care, it is becoming increasingly clear that poor maternal care is associated with alterations in structure of the amygdala (Caldji et al. 1998) and disturbances in HPA-axis function (Liu et al. 2000; Huot et al. 2004). Moreover, periods of maternal separation lasting as little as 24 hours have been documented in animal studies to lead to permanent reductions in hippocampal BDNF expression (Roceri et al. 2002). It is clear, therefore, that we need more investigations that integrate the construct of maternal care into models testing the mediating effects of this variable on the association between maternal depression and neurobiological abnormalities in the children.

Finally, maternal depression may contribute to neurobiological abnormalities in the child through prenatal exposure to biological factors present during pregnancy (see Goodman and Gotlib 1999, for a model and review). Although depression during pregnancy is at least as common as, if not more common than postnatal or lifetime depression (O'Hara et al. 1990; Evans et al. 2001), relatively few studies have examined this issue. Of these few however, women who experience depression during pregnancy have been found to demonstrate higher basal cortisol levels than do pregnant women who are not depressed (Lundy et al. 1999; Field et al. 2006a, b). Moreover, these elevations in cortisol have been linked with higher levels of fetal (Glover 1997; Gitau et al. 2001) and newborn cortisol (Lundy et al. 1999; Diego et al. 2004). Newborns of mothers who were depressed during pregnancy also exhibit greater relative right frontal electroencephalography (EEG) asymmetries, as well as more negative and "withdrawal" emotions compared to newborns of mothers who were not depressed during pregnancy (Lundy et al. 1999; Diego et al. 2004). Thus, neurobiological abnormalities in offspring of depressed parents may emerge not only as a consequence of genetic factors and/or depression-related alterations in parental care, but may result from abnormal development of neural systems caused by exposure to aberrant biological factors in utero.

4 Conclusion and Future Directions

Major depression is associated with a wide range of neurobiological disturbances, including dysregulation of the HPA axis and disruptions in the structure and function of cortical and subcortical gray matter. In this chapter, we have reviewed research demonstrating that many of these same abnormalities are also present in the never-depressed offspring of adults with recurrent depression, suggesting that these aberrations are not necessarily correlates or consequences of the experience of depression and/or its treatment, but could reflect dysfunctional neuroregulatory systems that precede the onset of this disorder. Indeed, literature in this area has demonstrated that disruptions in the structure and/or function of the hippocampus, amygdala, prefrontal cortex, basal ganglia, and striatum are present in individuals who are at familial risk for MDD, as are abnormalities in the regulation of the HPA axis. Studies using a prospective approach have extended these findings to show that high-risk offspring who later go on to develop depression have more severe abnormalities in HPA-axis function and hippocampal structure than do individuals who do not develop depression. These findings indicate that anomalies in neurobiological functioning may be useful in predicting which high-risk individuals are at even greater risk for developing MDD. Whether neural function and structure of other brain regions may be similarly useful in predicting which individuals will develop depression is not known, and clearly is an important area for future research.

We also presented and discussed different mechanisms by which abnormalities in neural and neuroendocrine function may be transmitted from depressed parent to child. Genetic factors (e.g., 5-HTTLPR, BDNF) are likely to be important in explaining associations between maternal depression and neural anomalies in the offspring. In addition, environmental factors, particularly those involving stress and/or disruptions in prenatal, infant, or childhood surroundings, likely also influence the emergence of neurobiological markers of risk. It is imperative that researchers continue to try to understand how these and other factors may be involved in the transmission of neurobiological markers of risk; such studies will be critical to elucidating the mechanisms that underlie the intergenerational transmission of risk for depression. Complicating matters, however, are findings that allelic variants of certain genes are correlated with environmental risk factors for depression (Lau and Eley 2008). Moreover, it is likely that there are several other variables that we did not discuss in this chapter but that are also involved in the intergenerational transmission of neurobiological abnormalities, and that these other variables interact with genetic and environmental risk factors for depression in complex ways, creating diverse pathways to MDD and depression-related patterns of neural anomaly.

In this context, future studies examining the intergenerational transmission of neural abnormalities should also elucidate the potential causal relations among different neurobiological markers of risk. The hippocampus and PFC, for example, both provide negative feedback to the HPA axis (Herman et al. 2005), and both are structurally and/or functionally abnormal in depressed individuals and their never-depressed offspring (Videbech and Ravnkilde 2004; Hamilton et al. 2008; Koolschijn et al. 2009; Kempton et al. 2011). Altered HPA-axis activity in depressed and high-risk individuals, therefore, could be secondary to damage to hippocampal or PFC gray matter. Chronically elevated levels of cortisol, however, can have adverse morphologic effects on gray matter (Sapolsky 2000), particularly in brain areas that contain a high density of cortisol receptors, such as the hippocampus and PFC (Chao et al. 1989; Sapolsky 2000; Furay et al. 2008). Thus, increased systemic levels of this hormone may cause the volumetric reductions that have been found in these brain regions. Future studies that integrate multiple measures across several time points would be helpful in determining whether and how these neurobiological markers of risk are causally related to each other in forming a vicious cycle of neurobiological abnormality.

Finally, although the research discussed in this chapter serves to increase our understanding of how depression in parents can influence neurobiological abnormalities in their never-depressed offspring, these findings could also contribute to the development of effective prevention strategies. Certainly, it is clear that there is still much to be learned about what leads to the development of neurobiological abnormalities in offspring of depressed parents; it is also clear that we do not yet understand precisely how these anomalies might mediate the appearance of depressive symptoms. Nevertheless, we are beginning to link the study of neurobiological anomalies with the elucidation of possibilities for prevention efforts. For example, Dozier et al. (2008) found that HPA-axis dysfunction becomes normalized in foster children, a population posited to share adversities similar to those of animals studied in early-separation paradigms, whose foster parents participate in behavior intervention paradigms that promote nurturing care and attachment security. These findings are consistent with "cross-fostering" experiments with animals, in which researchers have found that neurobiological changes associated with early life trauma can be altered by subsequent "optimal" caregiving experience (Maccari et al. 1995; Barbazanges et al. 1996). Other investigators have found that specific pharmacologic agents are capable of reversing neurobiological alterations caused by early life stress (Suomi 1991; Lopez et al. 1999). And, importantly, other classes of medications have been found to *prevent* abnormalities in neural structure (McEwen et al. 1997; Magariños et al. 1999) that have been documented in previous studies to be caused by trauma or stress. Thus, pharmacologic agents and optimal caregiving environments may be useful in protecting against further neurobiological abnormality in individuals at familial risk for depression.

Findings from prevention and treatment studies are intriguing because they suggest that the results of these investigations, when combined with research examining aspects of parental depression that may contribute to the presence of neurobiological abnormalities in the children, could facilitate the development of targeted strategies for prevention and intervention, and thereby reduce the socioeconomic burden of MDD. Moreover, by continuing to study and elucidate neurobiological abnormalities in individuals who are at high risk for depression but who have not yet developed depression themselves, we can begin to understand whether and how these abnormalities, once inherited, can influence the subsequent onset of clinically significant symptoms of MDD.

Acknowledgments Preparation of this chapter was facilitated by NIMH Grants MH59259 and MH74849 to Ian H. Gotlib and MH090617 to Lara Foland-Ross, and by a Hope for Depression Research Foundation Grant awarded to Ian H. Gotlib and Lara Foland-Ross.

References

Adam EK, Doane LD, Zinbarg RE, Mineka S, Craske MG, Griffith JW (2010) Prospective prediction of major depressive disorder from cortisol awakening responses in adolescence. Psychoneuroendocrinology 35(6):921–931

Adolphs R (2008) Fear, faces, and the human amygdala. Curr Opin Neurobiol 18(2):166–172

Alexander GE, DeLong MR, Strick PL (1986) Parallel organization of functionally segregated circuits linking basal ganglia and cortex. Annu Rev Neurosci 9:357–381

Amico F, Meisenzahl E, Koutsouleris N, Reiser M, Moller HJ, Frodl T (2011) Structural MRI correlates for vulnerability and resilience to major depressive disorder. J Psychiatry Neurosci 36(1):15–22

Aron AR (2011) From reactive to proactive and selective control: developing a richer model for stopping inappropriate responses. Biol Psychiatry 69(12):e55–e68

Baaré WF, Vinberg M, Knudsen GM, Paulson OB, Langkilde AR, Jernigan TL, Kessing LV (2010) Hippocampal volume changes in healthy subjects at risk of unipolar depression. J Psychiatr Res 44(10):655–662

Barbazanges A, Vallée M, Mayo W, Day J, Simon H, Le Moal M, Maccari S (1996) Early and later adoptions have different long-term effects on male rat offspring. J Neurosci 16(23):7783–7790

Beardslee WR, Versage EM, Gladstone TR (1998) Children of affectively ill parents: a review of the last 10 years. J Am Acad Child Adolesc Psychiatry 37(11):11134–11141

Beauregard M, Levesque J, Bourgouin P (2001) Neural correlates of conscious self-regulation of emotion. J Neurosci 21(18):RC165

Bhagwagar Z, Hafizi S, Cowen PJ (2005) Increased salivary cortisol after waking in depression. Psychopharmacologia 182(1):54–57

Browning M, Holmes EA, Harmer CJ (2010) The modification of attentional bias to emotional information: a review of the techniques, mechanisms, and relevance to emotional disorders. Cogn Affect Behav Neurosci 10(1):8–20

Bueller JA, Aftab M, Sen S, Gomez-Hassan D, Burmeister M, Zubieta JK (2006) BDNF Val66Met allele is associated with reduced hippocampal volume in healthy subjects. Biol Psychiatry 59(9):812–815

Burke HM, Davis MC, Otte C, Mohr DC (2005) Depression and cortisol responses to psychological stress: a meta-analysis. Psychoneuroendocrinology 30(9):846–856

Caldji C, Tannenbaum B, Sharma S, Francis D, Plotsky PM, Meaney MJ (1998) Maternal care during infancy regulates the development of neural systems mediating the expression of fearfulness in the rat. Proc Nat Acad Sci USA 95(9):5335–5340

Caspi A, Sugden K, Moffitt TE, Taylor A, Craig IW, Harrington H, McClay J, Mill J, Martin J, Braithwaite A, Poulton R (2003) Influence of life stress on depression: moderation by a polymorphism in the 5-HTT gene. Science 301(5631):386–389

Chen MC, Hamilton JP, Gotlib IH (2010) Decreased hippocampal volume in healthy girls at risk of depression. Arch Gen Psychiatry 67(3):270–276

Chao HM, Choo PH, McEwen BS (1989) Glucocorticoid and mineralocorticoid receptor mRNA expression in rat brain. Neuroendocrinology 50(4):365–371

Critchley H, Daly E, Phillips M, Brammer M, Bullmore E, Williams S, Van Amelsvoort T, Robertson D, David A, Murphy D (2000) Explicit and implicit neural mechanisms for processing of social information from facial expressions: a functional magnetic resonance imaging study. Hum Brain Mapp 9(2):93–105

Dannlowski U, Konrad C, Kugel H, Zwitserlood P, Domschke K, Schöning S, Ohrmann P, Bauer J, Pyka M, Hohoff C, Zhang W, Baune BT, Heindel W, Arolt V, Suslow T (2010) Emotion specific modulation of automatic amygdala responses by 5-HTTLPR genotype. Neuroimage 53(3):893–898

De Raedt R, Koster EH, Joormann J (2010) Attentional control in depression: a translational affective neuroscience approach. Cogn Affect Behav Neurosci 10(1):1–7

Dearing KF, Gotlib IH (2009) Interpretation of ambiguous information in girls at risk for depression. J Abnorm Child Psychol 37(1):79–91

Dickerson SS, Kemeny ME (2004) Acute stressors and cortisol responses: a theoretical integration and synthesis of laboratory research. Psychol Bull 130(3):355–391

Diego MA, Field T, Hernandez-Reif M, Cullen C, Schanberg S, Kuhn C (2004) Prepartum, postpartum, and chronic depression effects on newborns. Psychiatry 67(1):63–80

Dougherty LR, Klein DN, Congdon E, Canli T, Hayden EP (2010) Interaction between 5-HTTLPR and BDNF Val66Met polymorphisms on HPA axis reactivity in preschoolers. Biol Psychiatry 83(2):93–100

Dozier M, Peloso E, Lewis E, Laurenceau JP, Levine S (2008) Effects of an attachment-based intervention on the cortisol production of infants and toddlers in foster care. Dev Psychopathol 20(3):845–859

Duman RS, Monteggia LM (2006) A neurotrophic model for stress-related mood disorders. Biol Psychiatry 59(12):1116–1127

Egan MF, Kojima M, Callicott JH, Goldberg TE, Kolachana BS, Bertolino A, Zaitsev E, Gold B, Goldman D, Dean M, Lu B, Weinberger DR (2003) The BDNF val66met polymorphism affects activity-dependent secretion of BDNF and human memory and hippocampal function. Cell 112(2):257–269

Eisenberger NI, Gable SL, Lieberman MD (2007) Functional magnetic resonance imaging responses relate to differences in real-world social experience. Emotion 7(4):745–754

Evans J, Heron J, Francomb H, Oke S, Golding J (2001) Cohort study of depressed mood during pregnancy and after childbirth. BMJ 323(7307):257–260

Feldman S, Conforti N, Siegel RA (1982) Adrenocortical responses following limbic stimulation in rats with hypothalamic deafferentations. Neuroendocrinology 35:205

Field T (1984) Early interactions between infants and their postpartum depressed mothers. Infant Behav Dev 7(4):537–540

Field T (1994) The effects of mother's physical and emotional unavailability on emotion regulation. The development of emotion regulation. N. A. Fox Chicago, University of Chicago Press

Field T, Diego M, Dieter J, Hernandez-Reif M, Schanberg S, Kuhn C, Yando R, Bendell D (2006a) Prenatal depression effects on the fetus and newborn: a review. Infant Behav Dev 29(3):445–455
Field T, Hernandez-Reif M, Diego M, Figueiredo B, Schanberg S, Kuhn C (2006b) Prenatal cortisol, prematurity and low birthweight. Infant Behav Dev 29(2):268–275
Forbes EE, Hariri AR, Martin SL, Silk JS, Moyles DL, Fisher PM, Brown SM, Ryan ND, Birmaher B, Axelson DA, Dahl RE (2009) Altered striatal activation predicting real-world positive affect in adolescent major depressive disorder. Am J Psychiatry 166(1):64–73
Forbes EE, Shaw DS, Dahl RE (2007) Alterations in reward-related decision making in boys with recent and future depression. Biol Psychiatry 61(5):633–639
Fries E, Dettenborn L, Kirschbaum C (2009) The cortisol awakening response (CAR): facts and future directions. Int J Psychophysiol 72(1):67–73
Frodl T, Schüle C, Schmitt G, Born C, Baghai T, Zill P, Bottlender R, Rupprecht R, Bondy B, Reiser M, Möller HJ, Meisenzahl EM (2007) Association of the brain-derived neurotrophic factor Val66Met polymorphism with reduced hippocampal volumes in major depression. Arch Gen Psychiatry 64(4):410–416
Furay AR, Bruestle AE, Herman JP (2008) The role of the forebrain glucocorticoid receptor in acute and chronic stress. Endocrinology 149(11):5482–5490
Furman DJ, Hamilton JP, Joormann J, Gotlib IH (2010) Altered timing of amygdala activation during sad mood elaboration as a function of 5-HTTLPR. Soc Cogn Affect Neurosci 6(3):270–276
Fusar-Poli P, Placentino A, Carletti F, Landi P, Allen P, Surguladze S, Benedetti F, Abbamonte M, Gasparotti R, Barale F, Perez J, McGuire P, Politi P (2009) Functional atlas of emotional faces processing: a voxel-based meta-analysis of 105 functional magnetic resonance imaging studies. J Psychiatry Neurosci 34(6):418–432
Gallagher P, Reid KS, Ferrier IN (2009) Neuropsychological functioning in health and mood disorder: modulation by glucocorticoids and their receptors. Psychoneuroendocrinology 34(Suppl 1):S196–S207
Gatt JM, Nemeroff CB, Dobson-Stone C, Paul RH, Bryant RA, Schofield PR, Gordon E, Kemp AH, Williams LM (2009) Interactions between BDNF Val66Met polymorphism and early life stress predict brain and arousal pathways to syndromal depression and anxiety. Mol Psychiatry 14(7):681–695
Gerritsen L, Tendolkar I, Franke B, Vasquez AA, Kooijman S, Buitelaar J, Fernández G, Rijpkema M (in press) BDNF Val66Met genotype modulates the effect of childhood adversity on subgenual anterior cingulate cortex volume in healthy subjects. Mol Psychiatry
Gitau R, Fisk NM, Teixeira JM, Cameron A, Glover V (2001) Fetal hypothalamic-pituitaryadrenal stress responses to invasive procedures are independent of maternal responses. J Clin Endocrinol Metab 86(1):104–109
Glover V (1997) Maternal stress or anxiety in pregnancy and emotional development of the child. Br J Psychiatry 171:105–106
Gold GI (2003) Linking reward expectation to behavior in the basal ganglia. Trends Neurosci 26(1):12–14
Goodman SH, Gotlib IH (1999) Risk for psychopathology in the children of depressed mothers: a developmental model for understanding mechanisms of transmission. Psychol Rev 106(3): 458–490
Goodyer IM, Bacon A, Ban M, Croudace T, Herbert J (2009) Serotonin transporter genotype, morning cortisol and subsequent depression in adolescents. Br J Psychiatry 195(1):39–45
Goodyer IM, Tamplin A, Herbert J, Altham PM (2000) Recent life events, cortisol, dehydroepiandrosterone and the onset of major depression in high-risk adolescents. Br J Psychiatry 177:499–504
Gotlib IH, Hamilton JP (2008) Neuroimaging and depression: current status and unresolved issues. Curr Dir Psychol Sci 17(2):159–163
Gotlib IH, Hamilton JP, Cooney RE, Singh MK, Henry ML, Joormann J (2010) Neural processing of reward and loss in girls at risk for major depression. Arch Gen Psychiatry 67(4):380–387

Gotlib IH, Joormann J, Minor KL, Hallmayer J (2008) HPA-axis reactivity: a mechanism underlying the associations among 5-HTTLPR, stress, and depression. Biol Psychiatry 63(9):847–851

Gotthardt U, Schweiger U, Fahrenberg J, Lauer CJ, Holsboer F, Heuser I (1995) Cortisol, ACTH, and cardiovascular response to a cognitive challenge paradigm in aging and depression. Am J Physiol 268(Pt 4):R865–R873

Halligan SL, Herbert J, Goodyer I, Murray L (2007) Disturbances in morning cortisol secretion in association with maternal postnatal depression predict subsequent depressive symptomatology in adolescents. Biol Psychiatry 62(1):40–46

Halligan SL, Herbert J, Goodyer IM, Murray L (2004) Exposure to postnatal depression predicts elevated cortisol in adolescent offspring. Biol Psychiatry 55(4):376–381

Hamilton JP, Siemer M, Gotlib IH (2008) Amygdala volume in major depressive disorder: a meta-analysis of magnetic resonance imaging studies. Mol Psychiatry 13(11):993–1000

Hammen C, Shih J, Altman T, Brennan PA (2003) Interpersonal impairment and the prediction of depressive symptoms in adolescent children of depressed and nondepressed mothers. J Am Acad Child Adolesc Psychiatry 42(5):571–577

Hanson JL, Chung MK, Avants BB, Shirtcliff EA, Gee JC, Davidson RJ, Pollak SD (2010) Early stress is associated with alterations in the orbitofrontal cortex: a tensor-based morphometry investigation of brain structure and behavioral risk. J Neurosci 30(22):7466–7472

Hariri AR, Bookheimer SY, Mazziotta JC (2000) Modulating emotional responses: effects of a neocortical network on the limbic system. NeuroReport 11(1):43–48

Hariri AR, Drabant EM, Munoz KE, Kolachana BS, Mattay VS, Egan MF, Weinberger DR (2005) A susceptibility gene for affective disorders and the response of the human amygdala. Arch Gen Psychiatry 62(2):146–152

Harris TO, Borsanyi S, Messari S, Stanford K, Brown GW, Cleary SE, Shiers HM, Herbert J (2000) Morning cortisol as a risk factor for subsequent major depressive disorder in adult women. Br J Psychiatry 177:505–510

Heim C, Mletzko T, Purselle D, Musselman DL, Nemeroff CB (2008) The dexamethasone/corticotropin-releasing factor test in men with major depression: role of childhood trauma. Biol Psychiatry 63(4):398–405

Heim C, Newport DJ, Heit S, Graham YP, Wilcox M, Bonsall R, Miller AH, Nemeroff CB (2000) Pituitary-adrenal and autonomic responses to stress in women after sexual and physical abuse in childhood. J Am Med Assoc 284(5):592–597

Heim C, Newport DJ, Wagner D, Wilcox MM, Miller AH, Nemeroff CB (2002) The role of early adverse experience and adulthood stress in the prediction of neuroendocrine stress reactivity in women: a multiple regression analysis. Depress Anxiety 15(3):117–125

Herman JP, Ostrander MM, Mueller NK, Figueiredo H (2005) Limbic system mechanisms of stress regulation: hypothalamo-pituitary-adrenocortical axis. Prog Neuropsychopharmacol Biol Psychiatry 29(8):1201–1213

Herman JP, Schäfer MK, Young EA, Thompson R, Douglass J, Akil H, Watson SJ (1989) Evidence for hippocampal regulation of neuroendocrine neurons of the hypothalamo-pituitary-adrenocortical axis. J Neurosci 9(9):3072–3082

Heuser I, Yassouridis A, Holsboer F (1994) The combined dexamethasone/CRH test: a refined laboratory test for psychiatric disorders. J Psychiatr Res 28(4):341–356

Huang EJ, Reichardt LF (2001) Neurotrophins: roles in neuronal development and function. Annu Rev Neurosci 24:677–736

Huang H, Fan X, Williamson DE, Rao U (2011) White matter changes in healthy adolescents at familial risk for unipolar depression: a diffusion tensor imaging study. Neuropsychopharmacology 36(3):684–691

Huber TJ, Issa K, Schik G, Wolf OT (2006) The cortisol awakening response is blunted in psychotherapy inpatients suffering from depression. Psychoneuroendocrinology 31(7):900–904

Huot RL, Gonzalez ME, Ladd CO, Thrivikraman KV, Plotsky PM (2004) Foster litters prevent hypothalamic-pituitary-adrenal axis sensitization mediated by neonatal maternal separation. Psychoneuroendocrinology 29(2):279–289

Joormann J, Cooney RE, Henry ML, Gotlib IH (2012) Neural correlates of automatic mood regulation in girls at high risk for depression. J Abnorm Psychol 121(1):61–72

Joormann J, Gilbert K, Gotlib IH (2010) Emotion identification in girls at high risk for depression. J Child Psychol Psychiatry 51(5):575–582

Joormann J, Talbot L, Gotlib IH (2007) Biased processing of emotional information in girls at risk for depression. J Abnorm Psychol 116(1):135–143

Karege F, Perret G, Bondolfi G, Schwald M, Bertschy G, Aubry JM (2002) Decreased serum brain-derived neurotrophic factor levels in major depressed patients. Psychiatric Research 109(2):143–148

Kaufman J, Yang BZ, Douglas-Palumberi H, Grasso D, Lipschitz D, Houshyar S, Krystal JH, Gelernter J (2006) Brain-derived neurotrophic factor-5-HTTLPR gene interactions and environmental modifiers of depression in children. Biol Psychiatry 59(8):673–680

Kaufman J, Yang BZ, Douglas-Palumberi H, Houshyar S, Lipschitz D, Krystal JH, Gelernter J (2004) Social supports and serotonin transporter gene moderate depression in maltreated children. Proc Nat Acad Sci USA 101(49):17316–17321

Kawagoe R, Takikawa Y, Hikosaka O (1998) Expectation of reward modulates cognitive signals in the basal ganglia. Nat Neurosci 1(5):411–416

Kempton MJ, Salvador Z, Munafò MR, Geddes JR, Simmons A, Frangou S, Williams SC (2011) Structural neuroimaging studies in major depressive disorder: meta-analysis and comparison with bipolar disorder. Arch Gen Psychiatry 68(7):675–690

Kensinger EA, Corkin S (2004) Two routes to emotional memory: distinct neural processes for valence and arousal. Proc Nat Acad Sci USA 101(9):3310–3315

Kessler RC, Wang PS (2009) Epidemiology of depression. In: Gotlib IH, Hammen CL (eds) Handbook of depression, 2nd edn. Guilford, New York

Kieseppä T, Eerola M, Mäntylä R, Neuvonen T, Poutanen V, Luoma K, Tuulio-Henriksson A, Jylhä P, Mantere O, Melartin T, Rytsälä H, Vuorilehto M, Isometsä E (2010) Major depressive disorder and white matter abnormalities: a diffusion tensor imaging study with tract-based spatial statistics. J Affect Disord 120(1–3):240–244

Knorr U, Vinberg M, Kessing LV, Wetterslev J (2010) Salivary cortisol in depressed patients versus control persons: a systematic review and meta-analysis. Psychoneuroendocrinology 35(9):1275–1286

Koolschijn PC, van Haren NE, Lensvelt-Mulders GJ, Hulshoff Pol HE, Kahn RS (2009) Brain volume abnormalities in major depressive disorder: a meta-analysis of magnetic resonance imaging studies. Hum Brain Mapp 30(11):3719–3735

Kumar P, Waiter G, Ahearn T, Milders M, Reid I, Steele JD (2008) Abnormal temporal difference reward-learning signals in major depression. Brain 131(Pt 8):2084–2093

Lai CH, Hsu YY, Wu YT (2010) First episode drug-naive major depressive disorder with panic disorder: gray matter deficits in limbic and default network structures. Eur Neuropsychopharmacol 20(10):676–682

Langrock AM, Compas BE, Keller G, Merchant MJ, Copeland ME (2002) Coping with the stress of parental depression: parents' reports of children's coping, emotional, and behavioral problems. J Clin Child Adolesc Psychol 31(3):312–324

Lau JY, Eley TC (2008) Disentangling gene-environment correlations and interactions on adolescent depressive symptoms. J Child Adolesc Psychiatry 49(2):142–150

Lau JYF, Rijsdijk FV, Eley TC (2006) I think, therefore I am: a twin study of attributional style in adolescents. J Child Psychol Psychiatry 47:696–703

Lévesque ML, Beauregard M, Ottenhof KW, Fortier E, Tremblay RE, Brendgen M, Pérusse D, Dionne G, Robaey P, Vitaro F, Boivin M, Booij L (2011) Altered patterns of brain activity during transient sadness in children at familial risk for major depression. J Affect Disord 135(1–3):410–413

Levinson DF (2006) The genetics of depression: a review. Biol Psychiatry 60(2):84–92

Liu D, Diorio J, Day JC, Francis DD, Meaney MJ (2000) Maternal care, hippocampal synaptogenesis and cognitive development in rats. Nat Neurosci 3(8):799–806

Lopez JF, Akil H, Watson SJ (1999) Neural circuits mediating stress. Biol Psychiatry 46(11):1461–1471

Lundy BL, Jones NA, Field T, Nearing G, Davalos M, Pietro PA, Schanberg S, Kuhn C (1999) Prenatal depression effects on neonates. Infant Behav Dev 22(1):119–129

Lupien SJ, King S, Meaney MJ, McEwen BS (2000) Child's stress hormone levels correlate with mother's socioeconomic status and depressive state. Biol Psychiatry 48(10):976–980

Lupien SJ, Parent S, Evans AC, Tremblay RE, Zelazo PD, Corbo V, Pruessner JC, Séguin JR (2011) Larger amygdala but no change in hippocampal volume in 10-year-old children exposed to maternal depressive symptomatology since birth. Proc Nat Acad Sci USA 108(34):14324–14329

Maccari S, Piazza PV, Kabbaj M, Barbazanges A, Simon H, Le Moal M (1995) Adoption reverses the long-term impairment in glucocorticoid feedback induced by prenatal stress. J Neurosci 15(1):110–116

Magariños AM, Deslandes A, McEwen BS (1999) Effects of antidepressants and benzodiazepine treatments on the dendritic structure of CA3 pyramidal neurons after chronic stress. Eur J Pharmacol 371(2–3):113–122

Malberg JE, Eisch AJ, Nestler EJ, Duman RS (2000) Chronic antidepressant treatment increases neurogenesis in adult rat hippocampus. J Neurosci 20:9104–9110

Mannie ZN, Bristow GC, Harmer CJ, Cowen PJ (2007a) Impaired emotional categorisation in young people at increased familial risk of depression. Neuropsychologia 45(13):2975–2980

Mannie ZN, Harmer CJ, Cowen PJ (2007b) Increased waking salivary cortisol levels in young people at familial risk of depression. Am J Psychiatry 164(4):617–621

Mannie ZN, Taylor MJ, Harmer CJ, Cowen PJ, Norbury R (2011) Frontolimbic responses to emotional faces in young people at familial risk of depression. J Affect Disord 130(1–2): 127–132

McEwen BS (1998) Protective and damaging effects of stress mediators. N Engl J Med 338(3):171–179

McEwen BS, Conrad CD, Kuroda Y, Frankfurt M, Magarinos AM, McKittrick C (1997) Prevention of stress-induced morphological and cognitive consequences. Eur Neuropsychopharmacol 7(3):S323–S328

Mehta MA, Golembo NI, Nosarti C, Colvert E, Mota A, Williams SC, Rutter M, Sonuga-Barke EJ (2009) Amygdala, hippocampal and corpus callosum size following severe early institutional deprivation: the English and Romanian adoptees study pilot. J Child Psychol Psychiatry 50(8):943–951

Mizoguchi K, Ishige A, Aburada M, Tabira T (2003) Chronic stress attenuates glucocorticoid negative feedback: involvement of the prefrontal cortex and hippocampus. Neuroscience 119(3):887–897

Monk CS, Klein RG, Telzer EH, Schroth EA, Mannuzza S, Moulton JL III, Guardino M, Masten CL, McClure-Tone EB, Fromm S, Blair RJ, Pine DS, Ernst M (2008) Amygdala and nucleus accumbens activation to emotional facial expressions in children and adolescents at risk for major depression. Am J Psychiatry 165(1):90–98

Monroe SM, Simons AD (1991) Diathesis-stress theories in the context of life stress research: implications for the depressive disorders. Psychol Bull 110(3):406–425

Murer MG, Yan Q, Raisman-Vozari R (2001) Brain-derived neurotrophic factor in the control human brain, and in Alzheimer's disease and Parkinson's disease. Prog Neurobiol 63(1):71–124

Murray C, Lobez A (1996) The global burden of disease. Harvard School of Public Health, Cambridge

Murray L, Cooper P (2003) Intergenerational transmission of affective and cognitive processes asociated with depression: infancy and the pre-school years. Oxford University Press, Oxford

Murray L, Halligan SL, Goodyer I, Herbert J (2010) Disturbances in early parenting of depressed mothers and cortisol secretion in offspring: a preliminary study. J Affect Disord 122(3):218–223

O'Hara MW, Zekoski EM, Philipps LH, Wright EJ (1990) Controlled prospective study of postpartum mood disorders: comparison of childbearing and non childbearing women. J Abnorm Psychol 99(1):3–15

Pessoa L, Adolphs R (2010) Emotion processing and the amygdala: from a 'low road' to 'many roads' of evaluating biological significance. Nat Rev Neurosci 11(11):773–783
Peterson BS, Warner V, Bansal R, Zhu H, Hao X, Liu J, Durkin K, Adams PB, Wickramaratne P, Weissman MM (2009) Cortical thinning in persons at increased familial risk for major depression. Proc Nat Acad Sci USA 106(15):6273–6278
Pezawas L, Meyer-Lindenberg A, Drabant EM, Verchinski BA, Munoz KE, Kolachana BS, Egan MF, Mattay VS, Hariri AR, Weinberger DR (2005) 5-HTTLPR polymorphism impacts human cingulate–amygdala interactions: a genetic susceptibility mechanism for depression. Nat Neurosci 8(6):828–834
Pezawas L, Meyer-Lindenberg A, Goldman AL, Verchinski BA, Chen G, Kolachana BS, Egan MF, Mattay VS, Hariri AR, Weinberger DR (2008) Evidence of biologic epistasis between BDNF and SLC6A4 and implications for depression. Mol Psychiatry 13(7):709–716
Pezawas L, Verchinski BA, Mattay VS, Callicott JH, Kolachana BS, Straub RE, Egan MF, Meyer-Lindenberg A, Weinberger DR (2004) The brain-derived neurotrophic factor val66met polymorphism and variation in human cortical morphology. J Neurosci 24(45):10099–10102
Pizzagalli DA, Holmes AJ, Dillon DG, Goetz EL, Birk JL, Bogdan R, Dougherty DD, Iosifescu DV, Rauch SL, Fava M (2009) Reduced caudate and nucleus accumbens response to rewards in unmedicated individuals with major depressive disorder. Am J Psychiatry 166(6):702–710
Posener JA, Schildkraut JJ, Samson JA, Schatzberg AF (1996) Diurnal variation of plasma cortisol and homovanillic acid in healthy subjects. Psychoneuroendocrinology 21(1):33–38
Price JL, Drevets WC (2010) Neurocircuitry of mood disorders. Neuropsychopharmacology 35(1):192–216
Pruessner M, Hellhammer DH, Pruessner JC, Lupien SJ (2003) Self-reported depressive symptoms and stress levels in healthy young men: associations with the cortisol response to awakening. Psychosom Med 65(1):92–99
Radecki DT, Brown LM, Martinez J, Teyler TJ (2005) BDNF protects against stress-induced impairments in spatial learning and memory and LTP. Hippocampus 15(2):246–253
Rao U, Chen LA, Bidesi AS, Shad MU, Thomas MA, Hammen CL (2010) Hippocampal changes associated with early-life adversity and vulnerability to depression. Biol Psychiatry 67(4):357–364
Rao U, Hammen CL, Poland RE (2009) Risk markers for depression in adolescents: sleep and HPA measures. Neuropsychopharmacology 34(8):1936–1945
Ridderinkhoff KR, Ullsperger M, Crone EA, Nieuwenhuis S (2004) The role of the medial frontal cortex in cognitive control. Science 306(5695):443–447
Rice F, Harold GT, Thapar A (2003) Negative life events as an account of age-related differences in the genetic aetiology of depression in childhood and adolescence. J Child Psychol Psychiatry 44:977–987
Roceri M, Hendriks W, Racagni G, Ellenbroek BA, Riva MA (2002) Early maternal deprivation reduces the expression of BDNF and NMDA receptor subunits in rat hippocampus. Mol Psychiatry 7(6):609–616
Sanchez MM, Ladd CO, Plotsky PM (2001) Early adverse experience as a developmental risk factor for later psychopathology: evidence from rodent and primate models. Dev Psychopathol 13(3):419–449
Santarelli L, Saxe M, Gross C, Surget A, Battaglia F, Dulawa S, Weisstaub N, Lee J, Duman R, Arancio O, Belzung C, Hen R (2003) Requirement of hippocampal neurogenesis for the behavioral effects of antidepressants. Science 301(5634):805–809
Sapolsky RM (2000) The possibility of neurotoxicity in the hippocampus in major depression: a primer on neuron death. Biol Psychiatry 48(8):755–765
Sapolsky RM, Krey LC, McEwen BS (1986) The neuroendocrinology of stress and aging: the glucocorticoid cascade hypothesis. Endocr Rev 7(3):284–301
Scharinger C, Rabl U, Sitte HH, Pezawas L (2010) Imaging genetics of mood disorders. Neuroimage 53(3):810–821
Schlotz W, Hellhammer J, Schulz P, Stone AA (2004) Perceived work overload and chronic worrying predict weekend-weekday differences in the cortisol awakening response. Psychosom Med 66(2):207–214

Schmahmann JD, Pandya DN, Wang R, Dai G, D'Arceuil HE, de Crespigny AJ, Wedeen VJ (2007) Association fibre pathways of the brain: parallel observations from diffusion spectrum imaging and autoradiography. Brain 130(3):630–653

Schumacher J, Jamra RA, Becker T, Ohlraun S, Klopp N, Binder EB, Schulze TG, Deschner M, Schmäl C, Höfels S, Zobel A, Illig T, Propping P, Holsboer F, Rietschel M, Nöthen MM, Cichon S (2005) Evidence for a relationship between genetic variants at the brain-derived neurotrophic factor (BDNF) locus and major depression. Biol Psychiatry 58(4):307–314

Selvaraj S, Godlewska BR, Norbury R, Bose S, Turkheimer F, Stokes P, Rhodes R, Howes O, Cowen PJ (2011) Decreased regional gray matter volume in S' allele carriers of the 5-HTTLPR triallelic polymorphism. Mol Psychiatry 16(5):472–473

Sexton CE, Mackay CE, Ebmeier KP (2009) A systematic review of diffusion tensor imaging studies in affective disorders. Biol Psychiatry 66(9):814–823

Steele JD, Kumar P, Ebmeier KP (2007) Blunted response to feedback information in depressive illness. Brain 130(Pt 9):2367–2374

Steptoe A, Cropley M, Griffith J, Kirschbaum C (2000) Job strain and anger expression predict early morning elevations in salivary cortisol. Psychosom Med 62(2):286–292

Stetler C, Miller GE (2005) Blunted cortisol response to awakening in mild to moderate depression: regulatory influences of sleep patterns and social contacts. J Abnorm Psychol 114(4):697–705

Stetler C, Miller GE (2011) Depression and hypothalamic-pituitary-adrenal activation: a quantitative summary of four decades of research. Psychosom Med 73(2):114–126

Suomi SJ (1991) Early stress and adult emotional reactivity in rhesus monkeys. Ciba Found Symp 156(171–183):183–188

Thapar A, Harold G, McGuffin P (1998) Life events and depressive symptoms in childhood-shared genes or shared adversity? A research note. J Child Psychol Psychiatry 39: 1153–1158

Tottenham N, Hare TA, Quinn BT, McCarry TW, Nurse M, Gilhooly T, Millner A, Galvan A, Davidson MC, Eigsti IM, Thomas KM, Freed PJ, Booma ES, Gunnar MR, Altemus M, Aronson J, Casey BJ (2010) Prolonged institutional rearing is associated with atypically large amygdala volume and difficulties in emotion regulation. Dev Sci 13(1):46–61

van Harmelen AL, van Tol MJ, van der Wee NJ, Veltman DJ, Aleman A, Spinhoven P, van Buchem MA, Zitman FG, Penninx BW, Elzinga BM (2010) Reduced medial prefrontal cortex volume in adults reporting childhood emotional maltreatment. Biol Psychiatry 68(9):832–838

Vermetten E, Vythilingam M, Southwick SM, Charney DS, Bremner JD (2003) Long-term treatment with paroxetine increases verbal declarative memory and hippocampal volume in posttraumatic stress disorder. Biol Psychiatry 54(7):693–702

Videbech P, Ravnkilde B (2004) Hippocampal volume and depression: a meta-analysis of MRI studies. Am J Psychiatry 161(11):1957–1966

von dem Hagen EA, Passamonti L, Nutland S, Sambrook J, Calder AJ (2011) The serotonin transporter gene polymorphism and the effect of baseline on amygdala response to emotional faces. Neuropsychologia 49(4):674–680

Vreeburg SA, Hartman CA, Hoogendijk WJ, van Dyck R, Zitman FG, Ormel J, Penninx BW (2010) Parental history of depression or anxiety and the cortisol awakening response. Br J Psychiatry 197(3):180–185

Vreeburg SA, Hoogendijk WJ, van Pelt J, Derijk RH, Verhagen JC, van Dyck R, Smit JH, Zitman FG, Penninx BW (2009) Major depressive disorder and hypothalamic-pituitary-adrenal axis activity: results from a large cohort study. Arch Gen Psychiatry 66(6):617–626

Vythilingam M, Heim C, Newport J, Miller AH, Anderson E, Bronen R, Brummer M, Staib L, Vermetten E, Charney DS, Nemeroff CB, Bremner JD (2002) Childhood trauma associated with smaller hippocampal volume in women with major depression. Am J Psychiatry 159(12):2072–2080

Way BM, Taylor SE (2010) The serotonin transporter promoter polymorphism is associated with cortisol response to psychosocial stress. Biol Psychiatry 67(5):487–492

Wichers M, Kenis G, Jacobs N, Mengelers R, Derom C, Vlietinck R, van Os J (2008) The BDNF Val(66)Met x 5-HTTLPR x child adversity interaction and depressive symptoms: an attempt at replication. Am J Med Genet 147B(1):120–123
Williamson DE, Birmaher B, Axelson DA, Ryan ND, Dahl RE (2004) First episode of depression in children at low and high familial risk for depression. J Am Acad Child Adolesc Psychiatry 43(3):291–297
Wust S, Federenko I, Hellhammer DH, Kirschbaum C (2000) Genetic factors, perceived chronic stress, and the free cortisol response to awakening. Psychoneuroendocrinology 25(7):707–720
Young EA, Aggen SH, Prescott CA, Kendler KS (2000) Similarity in saliva cortisol measures in monozygotic twins and the influence of past major depression. Biol Psychiatry 48(1):70–74
Young EA, Vazquez D, Jiang H, Pfeffer CR (2006) Saliva cortisol and response to dexamethasone in children of depressed parents. Biol Psychiatry 60(8):831–836
Zobel AW, Nickel T, Sonntag A, Uhr M, Holsboer F, Ising M (2001) Cortisol response in the combined dexamethasone/CRH test as predictor of relapse in patients with remitted depression: a prospective study. J Psychiatr Res 35(2):83–94

Part III
Cognitive Accounts

Emotional Processing and Antidepressant Action

Catherine J. Harmer

Abstract Negative affective schema and associated biases in information processing have long been associated with clinical depression. Such an approach has guided the development of successful psychological therapies for this and other emotional disorders. However, until quite recently, there has been a large chasm between the practitioners and scientists working with this approach and those working on the neurobiological basis of depression and its treatment. Recent research, however, has started to bridge this gap and our understanding of the neural processes underpinning these cognitive processes has progressed markedly over the past decade. Moreover, rather than representing separate targets for psychological and biological treatments, novel findings suggest that pharmacological interventions for depression also modify these psychological maintaining factors early in treatment and may be involved in the later emergence of clinically relevant change. Such findings offer the possibility of greater integration between psychological and pharmacological conceptualisations of psychiatric illness and provide an experimental medicine model to generate and test specific predictions. Such a model could be applied to improve treatment development, stratification and combination approaches for patients with depression and provide a framework for considering and overcoming treatment nonresponse.

Keywords Emotional processing · fMRI · Depression · Antidepressant drugs · Serotonin

---

C. J. Harmer (✉)
Warneford Hospital, University Department of Psychiatry,
Headington, Oxford, OX3 7JX, UK
e-mail: Catherine.Harmer@psych.ox.ac.uk

Curr Topics Behav Neurosci (2013) 14: 209–222
DOI: 10.1007/7854_2012_210

Published Online: 8 May 2012

Contents

1 Introduction

The first antidepressant drug treatments were discovered by chance over 50 years ago. Since then, intense research effort both in academic and industrial laboratories has explored the key processes underlying antidepressant drug action and in using this information to refine and develop more effective and targeted antidepressant agents. This characterisation has largely explored the neurophysiological actions of drug treatments on neurochemical, cellular and molecular processes. By contrast, there has been little work exploring how these neural changes act to reverse the psychological experiences seen in depression and little integration between pharmacological and psychological theories and treatment for depression (see Harmer et al. 2009a).

However, it is likely that these different approaches, rather than being mediated by opposing processes, simply reflect different levels of analysis. Indeed, recent work reveals that antidepressant drug treatments normalise key psychological processes in depression very early on in treatment and that these actions may mediate later therapeutic change (Harmer et al. 2009a, 2011a). This chapter will review the role of psychological emotional biases in depression and the evidence that antidepressant treatments may work via early correction of emotional biases.

2 Negative Schema and Emotional Processing Bias in Depression

The original cognitive model of depression was developed by Beck (1967) and provided a revolutionary framework for the empirical identification and characterisation of factors that lead to and maintain depression. This cognitive model has stimulated research and clinical practice since then and has served as the foundation for the development of cognitive therapy. In this model, risk for depression is associated with the existence of negative self-referent schema's (such as the underlying belief that 'I will never succeed') which become activated by stressors. It is well documented in cognitive psychology that our own beliefs tend to shape the kind of information we receive and remember, since with limited capacity at all stages of cognitive processing and a world full of potential cues and sources of information, we preferentially process the stimuli which make most sense to us (Bartlett 1932). For the individual at risk of depression then, information processing becomes biased towards negative self-referent information at the expense of more positive cues, potentially reinforcing and maintaining negative self beliefs and low mood (Disner et al. 2011).

Since Beck's original formulation, the existence of these kinds of negative biases in processing have been extensively researched and described. These kinds of biases affect processes of emotional perception, elements of attention, memory and response to feedback (Roiser et al. 2012; Elliott et al. 2011) and have been linked both to the risk of depression and risk of relapse. For example, depressed patients are more likely to label facial stimuli as showing sad expressions and are typically less able to detect mild facial expressions of happiness (Harmer et al. 2009b; Gur et al. 1992; Surguladze et al. 2005). This kind of bias has been found to predict depression levels 3 and 6 months later (Hale 1998) and subsequent relapse to depression (Bouhuys et al. 1999). Further support for the hypothesis that these biases play a causal role in depression comes from studies where emotional biases have been manipulated experimentally using implicit training paradigms ("cognitive bias modification") and which have shown subsequent effects on mood responses to stressors. That is, artificially manufacturing negative biases in healthy people can mimic some of the symptoms seen in depression (Macleod et al. 2002).

Recent neurobiological formulations of the cognitive negative bias seen in depression emphasise the role of aberrant responses and interplay between limbic, frontal and striatal areas (Disner et al. 2011; Roiser et al. 2012; Elliott et al. 2011; Harmer et al. 2011a). An imbalance in processing stimuli across this network has been suggested to lead to increased salience of negative versus positive stimuli, inefficient attentional disengagement and control processes and reduced reward response. Evidence in favour of these formulations is provided by research findings, predominantly using functional magnetic resonance imaging in combination with the presentation of emotional stimuli. Results suggest that depression is associated with increased amygdala reactivity to negative stimuli that is both more intense and longer lasting (Fu et al. 2004; Sheline et al. 2001; Siegle et al. 2002). Such effects are seen, for example, during presentation of negative pictorial stimuli such as fearful or sad facial expressions and have been related (through interactions with the hippocampus) to enhanced memory for negative stimuli in depression (Hamilton and Gotlib 2008). It has also been suggested that the ongoing perception and sustained attention towards negative stimuli may occur because of reduced inhibition from cognitive control networks including the dorsolateral prefrontal cortex, dorsal anterior cingulate and ventrolateral prefrontal cortex (Disner et al. 2011; Roiser et al. 2012). Indeed, altered responses in these key areas, along with altered connectivity with the amygdala, have been observed in depression during the processing of emotional stimuli (Disner et al. 2011;

Roiser et al. 2012). Exaggerated responses to negative vs positive cues has also been reported in sensory areas such as the visual cortex, which receives back-projections from the amygdala, and may function to modify visual attention to stimuli depending on their salience or relevance to the individual (Vuilleumier and Driver 2007). The responses to anticipation and receipt of reinforcing stimuli (such as money or chocolate) in the nucleus accumbens and medial orbitofrontal cortex have also been reported to be blunted in depression (Knutson et al. 2008; Pizzagalli et al. 2009) even during remission (McCabe et al. 2009) and has been suggested to underpin the experience of anhedonia (Wacker et al. 2009).

Together, these results suggest that the preferential processing of negative affective information is a key process in depression which can be indexed using both behavioural and neuroimaging methods. It is likely that the root cause of these negative biases varies across individuals as a function of genetic and environmental factors. However, it is a key hypothesis that targeting these kinds of emotional biases may be an effective way of treating depression (Harmer et al. 2009b, 2011a; Pringle et al. 2011; Roiser et al. 2012).

3 Targeting Emotional Biases in the Treatment of Depression

Cognitive behavioural therapy (CBT) explicitly aims to reduce negative emotional biases in the treatment of depression and anxiety and the magnitude of these effects have been associated with therapeutic response (Mogg et al. 1995). However, it is only recently that the potential importance of antidepressant drug effects on these processes has begun to be recognised. These effects have largely been characterised using a healthy volunteer model to allow direct actions of the drug treatment on emotional processing to be assessed in a tightly controlled laboratory setting and independently from shifts in clinical symptoms. Such an approach allows the refinement of the design and hypotheses for subsequent clinical studies. Using this translational strategy, we have established that antidepressant administration affects emotional processing across a range of paradigms, early in treatment and independently from changes in mood (see Harmer et al. 2011a). For example, 7 days administration of the selective reuptake inhibitor (SSRI) antidepressant citalopram reduced the perception of facial expressions of fear, anger and disgust and decreased negative affective memory recall compared to placebo treatment in healthy volunteers (Harmer et al. 2004).

Such effects on emotional processing are characteristic of a number of different effective antidepressants across different pharmacological classes (including citalopram, reboxetine, duloxetine, agomelatine and mirtazapine) (Arnone et al. 2009; Harmer et al. 2003, 2008, 2011b). Each of these quite different antidepressants decreased interpretation and memory for negative compared to positive information in healthy volunteers early in treatment in double blind randomised controlled studies. Such effects would be expected to reverse negative biases seen in depression and reduce the influence of this important maintaining factor. Critically, a recent study suggests that depressed patients also show these early effects of antidepressants on emotional processing, before subjective changes are seen (Harmer et al. 2009b). Thus, a single dose of reboxetine (4 mg) was able to increase positive affective responses in facial expression recognition, emotional categorisation and emotional memory in depressed patients with a similar magnitude of effects to those seen in a sample of healthy controls. Such actions suggest that effects on emotional processing are apparent within the first few hours after the first antidepressant drug dose and could be an important mechanism for emergent therapeutic effects.

4 The Neural Correlates of Early Changes in Emotional Bias

The behavioural changes seen in emotional processing with antidepressant administration are associated with altered patterns of neural response across a network of areas including the amygdala, extra striate cortex and medial prefrontal cortex (see Harmer et al. 2009a, 2011a). The amygdala overactivity seen in depression to negative affective stimuli (Fu et al. 2004; Sheline et al. 2001; Siegle et al. 2002; Victor et al. 2010) is normalised following SSRI drug treatment after 1 week (Godlewska et al. 2012) and after longer term treatment (8 weeks or more) (Sheline et al. 2001; Fu et al. 2004; Victor et al. 2010). We have also shown similar effects in healthy volunteers early in treatment. Thus, administration of SSRIs (Harmer et al. 2006; Murphy et al. 2009a), noradrenaline reuptake inhibitors (Norbury et al. 2007) and receptor blocking antidepressants (Rawlings et al. 2010) decreased amygdala response to negative vs positive facial expressions. Similar effects have now been replicated in a number of independent research groups (Windischberger et al. 2010; Arce et al. 2008). In addition to these actions within the amygdala, emotionally specific drug effects are also seen within the medial prefrontal cortex (Di Simplicio et al. 2011) involved in self-referential processing and visual areas important for the translation of emotional salience into increased visual processing and attention (Norbury et al. 2007; Rawlings et al. 2010).

It is notable that the early effects of antidepressant drug treatments largely impact on "lower-level" responses to emotional information and there is a relative absence of reported effects on prefrontal and other higher order areas thought to play a role in regulation and control (see Harmer et al. 2011a). Although it is possible that this is an artefact of the kind of emotional paradigms used in these studies, it could indicate that pharmacological treatments are particularly working via this bottom-up route (i.e., decreasing automatic responses to emotional information rather than increasing more strategic control processes) (Harmer et al. 2011a; Roiser et al. 2012). Longer term antidepressant treatment has been reported to increase connectivity between frontal cortex and amygdala responses (Anand et al. 2005, 2007; Chen et al. 2008) though the direction of this action is not clear and may also occur through reduced amygdala responsively. These findings raise interesting questions about the initial locus of antidepressant drug action and whether efficacy could be improved by agents targeting the integrity and function of strategic control pathways (Roiser et al. 2012).

There is also increasing interest in the effect of drug treatment on resting-state connectivity patterns. Sheline et al. (2010) reported increased levels of connectivity between resting state networks involving the medial PFC ("Dorsal Nexus") in depression, suggesting aberrant interconnection between cognitive and emotional processes. The effect of treatment on network function is currently being investigated, but an initial study found that a 1-week treatment of citalopram was able to reduce aberrant connectivity between the dorsal nexus and hippocampus in healthy controls (McCabe et al. 2011). Given the hypothesised role for emotional modulation of hippocampal memory encoding (Hamilton and Gotlib 2008), this may be an intriguing corollary of the reduced negative recall bias seen with antidepressant treatment (Harmer et al. 2009b, 2004). Again, further studies are needed to explore early effects in depression and in relation to early changes in emotional processing and to later therapeutic action.

Together, these results suggest that antidepressants can affect limbic areas underpinning emotional processing very early on in treatment. Such effects reinforce the findings seen in the behavioural models of cognitive bias and highlight a previously undiscovered action of antidepressants which appears to precede clinical response. Such results, however, raise the question of why the clinical therapeutic actions of antidepressants are not seen earlier in treatment.

5 Understanding the Delayed Onset of Clinical Therapeutic Antidepressant Effects

A key hypothesis of the cognitive neuropsychological model is that the time-lag in clinical therapeutic effects of antidepressant treatments occur because environmental and social interaction is needed before a change in emotional bias will be able to affect mood and behaviour (Harmer et al. 2009a). A similar explanation has been proposed to account for the delayed effects of other manipulations which target emotional bias on subjective mood and anxiety responses. In particular, cognitive bias modification, which experimentally manipulates emotional bias using a psychological training technique (such as repeatedly pairing a probe to which the volunteer has to respond quickly with the location of a threatening stimulus (Macleod et al. 2002) also affects mood and anxiety with a delayed onset of action. However, although subjective state is not immediately affected under baseline conditions, this cognitive bias modification does modulate emotional reactions to a stressful task situation. For example, volunteers who have been trained to attend to a threatening stimulus show an increased negative mood response to a laboratory stressor relative to those volunteers who had been trained to selectively attend away from threatening stimuli (MacLeod et al. 2002). These results reinforce the idea that environmental interactions are critically important in determining how changes in emotional bias will affect mood. In depressed patients receiving antidepressant treatment, such interactions are likely to happen naturally with every day challenges, social contact and life events. However, it is also possible to probe such environmental effects experimentally using a laboratory stressor.

To test the idea that antidepressant-induced changes in emotional bias impact on mood only in interaction with stress and emotional events, we investigated the effects of 7 days administration of citalopram on the response to a negative mood challenge in healthy volunteers. As expected, short-term citalopram administration did not affect mood or subjective state under baseline conditions, but was able to protect healthy volunteers against the mood lowering effects of a negative mood induction compared to double blind administration of placebo (Browning et al. 2011). Furthermore, as hypothesised by the cognitive neuropsychological model, the magnitude of this protective action on mood response was predicted by the effect of citalopram on emotional memory. That is, those volunteers showing the largest reduction in negative bias were least susceptible to the negative mood induction.

These findings support the idea that although changes in emotional bias may occur quickly with antidepressant administration, such treatment would still be expected to have a delayed effect on clinical symptoms such as depressed mood. As such, this approach can provide a psychological explanation for the clinical delay in antidepressant drug treatment efficacy and also suggests a plausible mechanism whereby changes in neurochemical function induced by antidepressant drug treatment become translated into clinical improvement. Consistent with this delayed onset of action, results from a clinical study suggested that the change in the recognition of happy facial expressions with antidepressant use at 2 weeks correlated with therapeutic change 4 weeks later in depressed patients (Tranter et al. 2009).

While other models have been proposed to account for the delay in clinical efficacy of antidepressant drug treatment, these hypotheses usually emphasise the importance of delayed neurobiological actions in treatment effects. For example, the neurotrophic theory suggests that antidepressant effects on brain plasticity are required for successful antidepressant response (Duman and Monteggia 2006). These delayed antidepressant effects on plasticity are believed to include increased elaboration of neurotrophic factors (such as BDNF), neurogenesis, neuronal maturation and synaptic plasticity and at a behavioural level have been linked to changes in learning (Nissen et al. 2010; Normann et al. 2007). In isolation, it is difficult to explain how these improvements in plasticity and learning are linked to reduction in the different clinical symptoms seen in depression. However, such effects may well be critical for the translation of increased positive emotional bias described above into sustained improvements in the affect and social function of depressed patients. Hence, it may be predicted that the therapeutic effects of antidepressant agents which positively bias emotional processing could benefit from additional effects on neuroplasticity, particularly in view of the known deleterious effects of depression on learning and memory performance (Elliott 1998).

6 Applications

The surprising observation that antidepressants have early effects on emotional processing both in depressed and healthy volunteers has a number of theoretical and practical applications. This approach provides an experimental medicine model which can be used to generate specific predictions regarding treatment development, nonresponse and combination therapy.

6.1 Drug Development

Despite innovations in the process of drug development, new treatments for central nervous disorder conditions often fail relatively late in development because of problems in efficacy (Kola and Landis 2004). Typically, these candidate treatments are assessed for safety in Phase 1 healthy volunteer studies before efficacy is tested in Phase 2 and 3 studies in depressed patients. Phase 2 and 3 clinical trials are expensive, time-consuming and expensive but are often embarked on in the absence of any indication of efficacy in humans. It is clear that the inclusion of efficacy markers in Phase 1 studies may provide information for decision making and refinement of the compound for development but until recently there has been the absence of these kinds of biomarker models of efficacy which could be used in healthy volunteers.

The data reviewed above, however, suggests that early effects of drug treatments on emotional processing may provide a possible biomarker. These measures of emotional processing are affected early on in treatment, in healthy volunteers as well as depressed patients and target a key psychological process known to be important in depression. It is also notable that these effects are seen across antidepressant classes and in the absence of more overt changes in mood. Such results suggest that characterising novel candidate treatments early in development may provide some key information concerning effective dose, clinical profile and mechanism of action. Consistent with this, we have recently found that these human models were maximally sensitive to the same dose of the novel antidepressant agomelatine identified as effective in RCTs (Harmer et al. 2011b). Furthermore, these measures of emotional processing were unaffected, or not consistently affected, by failed antidepressant candidates such as the NK1 antagonist aprepitant (Chandra et al. 2010) which were positive in preclinical animal models of depression. These preliminary findings support the use of this human experimental medicine approach for the characterisation and screening of new agents for the treatment of depression and anxiety. Further work is required to assess the ability of these models to predict wanted and unwanted actions of novel drug treatments and differentiate between different candidates at this early stage of development.

6.2 Prediction of Treatment Nonresponse

While around 30 % of patients achieve remission with a single course of an antidepressant, this figure can be effectively doubled by trying different treatment iterations (Rush et al. 2007). However, such a trial and error approach for the treatment of a patient, coupled with the slow onset of clinical antidepressant drug effects, can lead to significant delay in effective management and increase risk of long-term morbidity and disability. Recent data suggest that early change in emotional processing with antidepressant administration may predict therapeutic response seen 4 weeks later (Tranter et al. 2009).

We are currently investigating whether the initial emotional processing response with treatment has sufficient sensitivity and specificity to inform whether a change in drug or dose is necessary early in treatment. This framework also provides a model for understanding why a particular treatment may fail for an individual. Thus, a treatment would be predicted to fail if the drug was unsuccessful in shifting cognitive bias. However, a treatment may also be unsuccessful if, despite a shift in cognitive bias, the patient is not receiving sufficient exposure to emotional and social stimuli to allow the change in processing to affect mood. This may be a particular problem for depression which is often characterised by social isolation and withdrawal and would be predicted to limit the ability of the antidepressants to improve depressed mood. For other patients, cognitive biases may be too entrenched to shift with standard antidepressant treatment. It is an intriguing idea that it may be possible to boost the effects of pharmacological treatment with specific psychological intervention though further investigation is needed to assess the best methods for this combination. For example, it is interesting to consider whether boosting exposure to and interpretation of the environment and social situations early in the treatment with antidepressants could build on these early positive effects. Such hypotheses need to be explored systematically using these experimental medicine models but could have far reaching implications for clinical practice.

6.3 Treatment Stratification

While different antidepressants with different pharmacological effects have common actions on the processing of emotional information which may be important in their therapeutic actions, there also seems to be some specific differences between agents which may be relevant to their clinical profiles. For example, negative biases in emotional memory recall have been suggested to be particularly relevant to depression, with less clear-cut effects emerging in the anxiety disorders (Bradley et al. 1997; Williams et al. 1997). Drug treatments for depression also have very consistent effects on emotional memory. Thus, SSRIs like citalopram, novel antidepressants like agomelatine and receptor blocking antidepressants such as mirtazapine all decreased relative recall for negative compared to positive emotional information in healthy volunteers in double blind placebo controlled studies (see Harmer et al. 2011b). Such effects have also been described for non-drug treatments for depression such as vagus nerve stimulation (Critchley et al. 2007), negative ion administration (Malcolm et al. 2009; Harmer et al. 2012) and cognitive bias modification (Browning et al. 2011). These actions suggest that remediation of negatively biased memory processes may be important in different therapeutic strategies for the treatment of depression, perhaps representing a final common pathway important for therapeutic effects.

By contrast, anxiety and anxiolytic drug treatments have been particularly associated with changes in attentional bias and physiological reactivity to threatening stimuli (Grillon and Charney 2011; MacLeod et al. 1986). The visual dot-probe task is able to provide a snapshot of attention to threat across time by varying the duration of stimuli presentations (MacLeod et al. 1986). Anxiety disorders and high trait levels of anxiety have been associated with increased vigilance to threatening information which is most consistently evident with short stimulus durations of 150 ms or less i.e., reflecting increased early orienting to threat rather than increased maintenance of attention (see Mogg and Bradley 1998). While attentional biases in depression have also been reported, such effects are usually seen with longer exposure durations in the order of 1,000 ms, possibly reflecting increased sustained attention to negative information rather than fast attentional orienting (Bradley et al. 1997). Consistent with this, we have observed key effects of drug treatments used for anxiety on the early attentional processing of threat in healthy volunteers (Murphy et al. 2008, 2009b). Thus, both the SSRI citalopram (also used in the treatment of anxiety) and the benzodiazepine diazepam reduced attentional vigilance to fearful facial expressions at exposure durations of 100 ms or less in healthy volunteers. These effects also seem to be relatively specific since we found that diazepam did not affect emotional memory and an antidepressant without profound anxiolytic action, reboxetine, did not affect attentional dot-probe performance (Murphy et al. 2008, 2009b). Hence, these results suggest that correction of initial hypervigilance to threat may be a common effect of drug treatments for anxiety, whereas remediation of biases in emotional memory may be particularly relevant for antidepressant agents. These differences between drug treatments imply that it may be possible to predict who is likely to benefit from different antidepressants from particular patterns of emotional bias shown before treatment. In other words, a treatment may work best for an individual if drug action is matched to the particular emotional processing bias seen in that person. These speculations remain to be tested, however, in large-scale randomised clinical trials.

7 Conclusions

It has often been assumed that antidepressants have no clinically relevant effects until they have been given over days or weeks of treatment. However, recent evidence suggests that key psychological processes which are important in depression, are affected early in treatment and may over time and experience be an important mechanism of therapeutic change in depression. This approach refocuses research on key processes that are occurring shortly after treatment initiation and provides a model by which psychological and pharmacological processes in depression can be integrated and understood. This approach also provides an experimental medicine model for the generation of specific predictions which may overcome some limitations in drug development, treatment nonresponse and the application of psychological–pharmacological combination strategies. There remains the need for large-scale clinical translation of this work, where we examine to what extent these biases predict therapeutic response, whether this could be used to improve clinical practice and treatment allocation and provide a testbed for the development and screening of novel treatment strategies.

Conflict of Interest Catherine serves on the advisory board of P1vital Ltd and owns shares in the company. She is also a company director of Oxford Psychologists Ltd and has received consultancy fees over the last 3 years from Eli-Lilly, P1vital, GSK and Servier.

References

Anand A, Li Y, Wang Y, Wu J, Gao S, Bukhari L, Mathews VP, Kalnin A, Lowe MJ (2005) Activity and connectivity of brain mood regulating circuit in depression: a functional magnetic resonance study. Biol Psychiatry 57(10):1079–1088

Anand A, Li Y, Wang Y, Gardner K, Lowe MJ (2007) Reciprocal effects of antidepressant treatment on activity and connectivity of the mood regulating circuit: an FMRI study. J Neuropsychiatry Clin Neurosci 19(3):274–282

Arce E, Simmons AN, Lovero KL, Stein MB, Paulus MP (2008) Escitalopram effects on insula and amygdala BOLD activation during emotional processing. Psychopharmacology 196(4): 661–672

Arnone D, Horder J, Cowen PJ, Harmer CJ (2009) Early effects of mirtazapine on emotional processing. Psychopharmacology 203(4):685–691

Bartlett FC (1932) Remembering: a study in experimental and social psychology. Cambridge University Press, Cambridge

Beck AT (1967) Depression: clinical, experimental and theoretical aspects. Harper & Row, New York

Bouhuys AL, Geerts E, Gordijn MC (1999) Depressed patients' perceptions of facial emotions in depressed and remitted states are associated with relapse: a longitudinal study. J Nerv Ment Dis 187:595–602

Bradley BP, Mogg K, Lee SC (1997) Attentional biases for negative information in induced and naturally occurring dysphoria. Behav Res Ther 35:911–927

Browning M, Grol M, Ly V, Goodwin GM, Holmes EA, Harmer CJ (2011) Using an experimental medicine model to explore combination effects of pharmacological and cognitive interventions for depression and anxiety. Neuropsychopharmacology. doi:10.1038/npp.2011.15

Chandra P, Hafizi S, Massey-Chase RM, Goodwin GM, Cowen PJ, Harmer CJ (2010) NK1 receptor antagonism and emotional processing in healthy volunteers. J Psychopharmacol 24(4):481–487 (Epub 2009 Apr 7)

Chen CH, Suckling J, Ooi C, Fu CH, Williams SC, Walsh ND, Mitterschiffthaler MT, Pich EM, Bullmore E (2008) Functional coupling of the amygdala in depressed patients treated with antidepressant medication. Neuropsychopharmacology 33(8):1909–1918

Critchley HD, Lewis PA, Orth M, Josephs O, Deichmann R, Trimble MR, Dolan RJ (2007) Vagus nerve stimulation for treatment-resistant depression: behavioral and neural effects on encoding negative material. Psychosom Med 69(1):17–22

Di Simplicio M, Norbury R, Harmer CJ (2011) Short-term antidepressant administration reduces negative self-referential processing in the medial prefrontal cortex in subjects at risk for depression. Mol Psychiatry. Mar 1 (Epub ahead of print)

Disner SG, Beevers CG, Haigh EA, Beck AT (2011) Neural mechanisms of the cognitive model of depression. Nat Rev Neurosci 12(8):467–477. doi:10.1038/nrn3027

Duman RS, Monteggia LM (2006) A neurotrophic model for stress-related mood disorders. Biol Psychiatry 59(12):1116–1127

Elliott R (1998) The neuropsychological profile in unipolar depression. Trends Cogn Sci 2(11):447–454

Elliott R, Zahn R, Deakin JF, Anderson IM (2011) Affective cognition and its disruption in mood disorders. Neuropsychopharmacology 36(1):153–182

Fu CH, Williams SC, Cleare AJ, Brammer MJ, Walsh ND, Kim J, Andrew CM, Pich EM, Williams PM, Reed LJ, Mitterschiffthaler MT, Suckling J, Bullmore ET (2004) Attenuation of the neural response to sad faces in major depression by antidepressant treatment: a prospective, event-related functional magnetic resonance imaging study. Arch Gen Psychiatry 61(9):877–889

Godlewska BR, Norbury R, Selvaraj S, Cowen PJ, Harmer CJ (2012) Effect of short term SSRI treatment on amygdala hyperactivity in depressed patients: a placebo controlled study. Psychol Med (in press)

Grillon C, Charney DR (2011) In the face of fear: anxiety sensitizes defensive responses to fearful faces. Psychophysiology 48:1745–1752

Gur RC, Erwin RJ, Gur RE, Zwil AS, Heimberg C, Kraemer HC (1992) Facial emotion discrimination: II. Behavioral findings in depression. Psychiatry Res 42:241–251

Hale WW III (1998) Judgment of facial expressions and depression persistence. Psychiatry Res 80:265–274

Hamilton JP, Gotlib IH (2008) Neural substrates of increased memory sensitivity for negative stimuli in major depression. Biol Psychiatry 63:1155–1162

Harmer CJ, Hill SA, Taylor MJ, Cowen PJ, Goodwin GM (2003) Toward a neuropsychological theory of antidepressant drug action: increase in positive emotional bias after potentiation of norepinephrine activity. Am J Psychiatry 160:990–992

Harmer CJ, Shelley NC, Cowen PJ, Goodwin GM (2004) Increased positive versus negative affective perception and memory in healthy volunteers following selective serotonin and norepinephrine reuptake inhibition. Am J Psychiatry 161:1256–1263

Harmer CJ, Mackay CE, Reid CB, Cowen PJ, Goodwin GM (2006) Antidepressant drug treatment modifies the neural processing of nonconscious threat cues. Biol Psychiatry 59:816–820

Harmer CJ, Heinzen J, O'Sullivan U, Ayres RA, Cowen PJ (2008) Dissociable effects of acute antidepressant drug administration on subjective and emotional processing measures in healthy volunteers. Psychopharmacology 199:495–502

Harmer CJ, Goodwin GM, Cowen PJ (2009a) Why do antidepressants take so long to work? A cognitive neuropsychological model of antidepressant drug action. Br J Psychiatry 195:102–108

Harmer CJ, O'Sullivan U, Favaron E, Massey-Chase R, Ayres R, Reinecke A, Goodwin GM, Cowen PJ (2009b) Effect of acute antidepressant administration on negative affective bias in depressed patients. Am J Psychiatry 166:1178–1184

Harmer CJ, Cowen PJ, Goodwin GM (2011a) Efficacy markers in depression. J Psychopharmacol 25:1148–1158

Harmer CJ, de Bodinat C, Dawson GR, Dourish CT, Waldenmaier L, Adams S, Cowen PJ, Goodwin GM (2011b) Agomelatine facilitates positive versus negative affective processing in healthy volunteer models. J Psychopharmacol 25:1159–1167

Harmer CJ, Charles M, McTavish S, Favaron E, Cowen PJ (2012) Negative ion treatment increases positive emotional processing in seasonal affective disorder. Psychol Med (Epub)

Knutson B, Bhanji JP, Cooney RE, Atlas LY, Gotlib IH (2008) Neural responses to monetary incentives in major depression. Biol Psychiatry 63(7):686–692

Kola I, Landis J (2004) Can the pharmaceutical industry reduce attrition rates? Nat Rev Drug Discov 3(8):711–715

MacLeod C, Mathews A, Tata P (1986) Attentional bias in emotional disorders. J Abnorm Psychol 95(1):15–20

MacLeod C, Rutherford E, Campbell L, Ebsworthy G, Holker L (2002) Selective attention and emotional vulnerability: assessing the causal basis of their association through the experimental manipulation of attentional bias. J Abnorm Psychol 111:107–123

Malcolm CP, Cowen PJ, Harmer CJ (2009) High-density negative ion treatment increases positive affective memory. Psychol Med 39(11):1930–1932

McCabe C, Cowen PJ, Harmer CJ (2009) Neural representation of reward in recovered depressed patients. Psychopharmacology 205(4):667–677

McCabe C, Mishor Z, Filippini N, Cowen PJ, Taylor MJ, Harmer CJ (2011) SSRI administration reduces resting state functional connectivity in dorso-medial prefrontal cortex. Mol Psychiatry 16(6):592–594

Mogg K, Bradley BP, Millar N, White J (1995) A follow-up study of cognitive bias in generalized anxiety disorder. Behav Res Ther 33(8):927–935

Mogg K, Bradley BP (1998) A cognitive-motivational analysis of anxiety. Behav Res Ther 36: 809–848

Murphy SE, Downham C, Cowen PJ, Harmer CJ (2008) Direct effects of diazepam on emotional processing in healthy volunteers. Psychopharmacology 199:503–513

Murphy SE, Norbury R, O'Sullivan U, Cowen PJ, Harmer CJ (2009a) Effect of a single dose of citalopram on amygdala response to emotional faces. Br J Psychiatry 194:535–540

Murphy SE, Yiend J, Lester KJ, Cowen PJ, Harmer CJ (2009b) Short-term serotonergic but not noradrenergic antidepressant administration reduces attentional vigilance to threat in healthy volunteers. Int J Neuropsychopharmacol 12:169–179

Nissen C, Holz J, Blechert J, Feige B, Riemann D, Voderholzer U, Normann U (2010) Learning as a model for neural plasticity in major depression. Biol Psychiatry 68(6):544–552

Norbury R, Mackay CE, Cowen PJ, Goodwin GM, Harmer CJ (2007) Short-term antidepressant treatment and facial processing. Functional magnetic resonance imaging study. Br J Psychiatry 190:531–532

Normann C, Schmitz D, Fürmaier A, Bach M, Döing C (2007) Long-term plasticity of visually evoked potentials in humans is altered in major depression. Biol Psychiatry 62(5):373–380

Pizzagalli DA, Holmes AJ, Dillon DG, Goetz EL, Birk JL, Bogdan R, Dougherty DD, Ilosifescu DV, Rauch SL, Fava M (2009) Reduced caudate and nucleus accumbens response to rewards in unmedicated individuals with major depressive disorder. Am J Psychiatry 166(6):702–710

Pringle A, Browning M, Cowen PJ, Harmer CJ (2011) A cognitive neuropsychological model of antidepressant drug action. Prog Neuropsychopharmacol Biol Psychiatry 35(7):1586–1592

Rawlings NB, Norbury R, Cowen PJ, Harmer CJ (2010) A single dose of mirtazapine modulates neural responses to emotional faces in healthy people. Psychopharmacology 212(4):625–634

Roiser JP, Elliott R, Sahakian BJ (2012) Cognitive mechanisms of treatment in depression. Neuropsychopharmacology 37(1):117–136

Rush AJ (2007) STAR*D: what have we learned? Am J Psychiatry 164:201–204

Sheline YI, Barch DM, Donnelly JM, Ollinger JM, Snyder AZ, Mintun MA (2001) Increased amygdala response to masked emotional faces in depressed subjects resolves with antidepressant treatment: an fMRI study. Biol Psychiatry 50(9):651–658

Sheline YI, Price JL, Yan Z, Mintun MA (2010) Resting-state functional MRI in depression unmasks increased connectivity between networks via the dorsal nexus. Proc Natl Acad Sci U S A 107(24):11020–11025

Siegle GJ, Steinhauer SR, Thase ME, Stenger VA, Carter CS (2002) Can't shake that feeling: event-related fMRI assessment of sustained amygdala activity in response to emotional information in depressed individuals. Biol Psychiatry 51(9):693–707

Surguladze S, Brammer MJ, Keedwell P, Giampietro V, Young AW, Travis MJ, Williams SC, Phillips ML (2005) A differential pattern of neural response toward sad versus happy facial expressions in major depressive disorder. Biol Psychiatry 57(3):201–209

Tranter R, Bell D, Gutting P, Harmer C, Healy D, Anderson IM (2009) The effect of serotonergic and noradrenergic antidepressants on face emotion processing in depressed patients. J Affect Disord 118:87–93

Victor TA, Furey ML, Fromm SJ, Ohman A, Drevets WC (2010) Relationship between amygdala responses to masked faces and mood state and treatment in major depressive disorder. Arch Gen Psychiatry 67(11):1128–1138

Vuilleumier P, Driver J (2007) Modulation of visual processing by attention and emotion: windows on causal interactions between human brain regions. Philos Trans R Soc Lond B Biol Sci 362(1481):837–855

Wacker J, Dillon DG, Pizzagalli DA (2009) The role of the nucleus accumbens and rostral anterior cingulate cortex in anhedonia: integration of resting EEG, fMRI, and volumetric techniques. Neuroimage 46(1):327–333

Williams JMG, Watts FN, MacLeod C, Mathews A (1997) Cognitive psychology and emotional disorders, 2nd edn. Wiley, Chichester

Windischberger C, Lanzenberger R, Holik A, Spindelegger C, Stein P, Moser U, Gerstl F, Fink M, Moser E, Kasper S (2010) Area-specific modulation of neural activation comparing escitalopram and citalopram revealed by pharmaco-fMRI: a randomized cross-over study. Neuroimage 49(2):1161–1170 (Epub 2009 Oct 13)

What Can Fear and Reward Learning Teach Us About Depression?

Katherine A. Collins and Daniela Schiller

Abstract The precise neural substrates of major depressive disorder (MDD) remain elusive, and FDA-approved antidepressants fail at least one-third of treatment-seeking patients. It is imperative, therefore, to identify novel research strategies to tackle the factors impeding progress. In this chapter we propose that the knowledge derived from computational investigations of associative learning might offer new insights into the neurobiology of MDD.

Keywords Associative learning · Reward learning · Fear learning/conditioning · Prediction error · Major Depressive Disorder (Mdd)

Contents

K. A. Collins · D. Schiller (✉)
Mt. Sinai School of Medicine, New York, NY, US
e-mail: daniela.schiller@mssm.edu

Curr Topics Behav Neurosci (2013) 14: 223–242
DOI: 10.1007/7854_2012_236

Published Online: 9 January 2013

1 Introduction

The origins of depressive illness have been the subject of documented intellectual inquiry since Hippocrates wrote about melancholic disorders in approximately 400 BC (Hippocrates 1849). Since the serendipitous discovery in 1951, that a drug developed to treat tuberculosis appeared to induce improvements in mood, basic scientists and clinicians have been continuously engaged in investigations of the biological correlates of depression, employing a range of techniques at multiple levels of analysis, such as behavior, genetics, neuronal etc. Here we focus on the neural systems underlying cognitive process in depression (see Tables 1 and 3).

Despite the efforts of so many, current antidepressant interventions fail at least one-third of treatment-seeking patients (Rush et al. 2006). Moreover, the precise neural substrates of major depressive disorder (MDD) remain elusive. Those individuals who continue to suffer from the illness are both burdened by their own pain and can constitute a burden to their families and society. MDD is a leading cause of worldwide disability and is estimated to cost the United States alone over 80 billion dollars annually in workplace, medical and suicide-related mortality expenditures (Greenberg et al. 1990; World Health Organization 2012).

In this chapter we propose that studies of associative learning constitute one important untapped resource available to depression researchers. We begin by highlighting the characteristics of MDD that make it an illness so difficult to understand. We then describe how insights from fear and reward learning research can facilitate the interpretation of abnormal neural activity patterns in depressed patients, and inform the design and hypotheses of future investigations. Our aim is to encourage interdisciplinary collaboration that can accelerate the application of new insights about the pathophysiology of MDD in clinical settings.

2 Why do the Neural Mechanisms of Depression Remain Elusive?

In 1937 Papez proposed that, “the hypothalamus, the anterior thalamic nuclei, the gyrus cinguli, the hippocampus, and their interconnections constitute a harmonious mechanism which may elaborate the functions of central emotion, as well as participate in emotional expression” (Papez 1937). By offering evidence that affect is generated in the brain and is not a “magic product”, Papez and his contemporaries set the stage for modern researchers to study mood disorders as a biological phenomenon. In the subsequent decades, scientists have tested several theories of depression’s etiology (Table 1) and identified countless behavioral, neural, molecular, and genetic correlates of the illness. Given the bulk of data acquired, why do the precise neural mechanisms of depression remain elusive?

Table 1 Major Theories of Depression's Pathophysiology

Hypothesis	Example of supporting evidence
1. Monoamine: Deficits in monoamine neurotransmission can cause depression (Lehmann et al. 1958; Bunney and Davis 1965)	Drugs that modulate monoamine neurotransmission have antidepressant effects (Schildkraut 1967)
2. Cognitive: Automatic negative thoughts about the self, the world, and the future can cause depression (Blackburn 1986)	Depression patients reliably report more automatic negative thoughts than healthy volunteers (Hollon 1965; Beck et al. 1963)
3. Neurotrophic: Decreased expression of neurotrophic factors including BDNF can cause depression (Smith et al. 1995; Nibuya et al. 1995)	Stress exposure decreases, but antidepressants increase, hippocampal BDNF levels in animals (Duman and Monteggia 2006; Duman et al. 1997)
4. Glutamatergic: Dysfunction of the glutamatergic system can cause depression (Trullas and Skolnick 1990)	NMDA antagonists function as antidepressants (Sanacora et al. 2012)

MDD's complexity is one barrier to progress. The diagnostic criteria for MDD (Table 2) (American Psychiatric Association 2000) were developed in order to achieve reliability across clinicians and are not grounded in biology. They are so broad as to encompass a diverse set of patients. One patient may describe sadness, insomnia, weight loss, fatigue, and feelings of worthlessness. Another will report hyper-phagia, agitation, poor concentration, and suicidal ideation, complaining that she no longer enjoys any of her activities. A single individual, in addition, may describe different constellations of symptoms during different episodes of depression.

The diversity of depressive phenotypes makes it difficult to create a laboratory model of depression. Laboratory manipulations do not provoke a syndrome as durable and multifaceted as MDD. In the Velten Mood Induction Procedure, for instance, participants read self-referential statements that describe emotional ("I am worthless") and/or physical states ("I am listless") associated with MDD (Emmett 1968; Riskind et al. 1982). After completion, subjects not only report feeling more depressed, they also evince mild psychomotor retardation and find it easier to recall negative versus positive autobiographical memories (David 1983). These effects are transient, however, and participants do not describe changes in neurovegetative symptoms such as sleep, appetite, or energy.

Pharmacological models of depression have similar limitations. Acute tryptophan depletion is a procedure originally used in humans to test the hypothesis that abnormally low levels of serotonin trigger depression (Table 1) (Reilly et al. 1997). In this protocol, a nutritional supplement is used to temporarily reduce the bioavailability of tryptophan, serotonin's amino-acid precursor (Reilly et al. 1997). Tryptophan depletion changes sleep architecture in healthy volunteers so that it more closely resembles that observed in depression patients (Zimmermann et al. 1993; Voderholzer et al. 1998; Arnulf et al. 2002). It does not, however, reliably

Table 2 Major Depressive Disorder: Diagnostic Criteria

*History of at least one *major depressive episode* in the absence of a separate psychotic disorder
*No history of manic, hypomanic, or mixed episodes
Major Depressive Episode: Diagnostic Criteria
**Prevasive depressed mood and/or anhedonia lasting at least two weeks*
**At least 4 of the following symptoms:*
*Change in appetite or weight
*Insomnia or hypersomnia
*Psychomotor agitation or retardation
*Fatigue or loss of energy
*Feelings of worthlessness or excessive/inappropriate guilt
*Impaired concentration or indecisiveness
*Suicidal ideation
* All symptoms are pervasive
**Symptoms cause clinically significant distress or impairment*
*Symptoms are not associated with substance use, a medical condition, or bereavement

influence mood in healthy volunteers (Ruhe et al. 2007) and any impact is short-lived (Reilly et al. 1997).

More intense psychological or pharmacological challenges might induce a quorum of persistent MDD symptoms. Yet such experimental conditions would likely be unethical. Hence, it remains impossible to dissociate the neural states that precipitate depression from those that are secondary or caused by the illness, using laboratory experiments in healthy volunteers.

Scientists have developed several animal models of depression in an effort to compensate for the weaknesses inherent in human models (Table 3). Researchers can subject animals to extreme stressors in order to produce lasting behavioral change and use invasive techniques to probe depression's neurobiology. Unfortunately, the utility of these protocols is still undermined by the internal nature of MDD symptoms. Phenomena such as depressed mood, anhedonia, suicidal ideation, and feelings of worthlessness or helplessness lack reliable physiological or behavioral signatures (Meaney 2001). It is impossible to confirm the presence of parallel "symptoms" in animals. Instead, models and measures of animal "depression" are primarily selected because traditionally, their effects can be reversed by acute administration of monoaminergic antidepressants (Porsolt 1978). In the forced swim test, for instance, animals are placed in a cylinder of water. The time that passes before they stop struggling and become immobile is the measurement used to quantify "depression" severity. Animals who struggle the least are considered the most "depressed" because antidepressants such as imipramine and fluoxetine increase latency to immobility (Porsolt 1978).

In recent years, clinical research confirmed that monoaminergic medications rarely elicit meaningful improvement in patients until after one week to several months of treatment (Rush et al. 2006). This understanding triggered a surge of more complex models of MDD, which involve more chronic sources of stress. In the social defeat model, for instance, male mice repeatedly confront larger

Table 3 Animal Models and Measures of Depression (Yan et al. 2010)

A. Models

Model	Protocol	Behavioral and physiological changes	Strengths	Weaknesses
Forced Swim (Connor et al. 1997; Porsolt 1978)	Subject suspended in cylinder of water, time swimming is assessed. Less = depressed	Transient increase in stress hormones, transient reduction in immune function	Predicts drug efficacy, easy to replicate	Face validity, no other behavioral symptoms elicited, no durable effects
Tail Suspension (Kanda et al. 1993; Steru et al. 1985)	Subject suspended by tail, time struggling to escape is assessed. Less = depressed	Transient increase in stress hormones, catecholamines	Predicts drug efficacy, easy to replicate	Face validity, no other behavioral symptoms elicited, no durable effects
Learned Helplessness (Seligman 1972)	Repeated exposure to intense, unpredictable, and inescapable electric foot shock	Altered sleep, reduced body weight, energy, sexual activity, and locomotion; elevated levels of stress hormones	Rapid, easy to replicate, predicts drug efficacy	Symptoms not durable, requires repeated exposures
Chronic Mild Stress (Katz et al. 1981; Katz and Baldrighi 1982; Willner 1987, 1997)	Exposure to multiple innocuous but unpredictable stressors (e.g. light/dark reversal, temperature change, food deprivation, restraint) over an extended period of time (3 weeks or longer)	Gradual attenuation of sucrose preference, and coat deterioration, decreased sexual, aggressive, exploratory, and general locomotive behavior	Face validity, elicits many symptoms Durability: behavior changes lasts several weeks	Time-intensive, labor intensive, and difficult to replicate across laboratories. No changes in elevated plus maze or social interaction
Social Defeat (Berton et al. 2006)	Repeated exposure to larger more aggressive breeding animals. Initial contact is direct and the aggressor attacks the subject. Subsequently separated from aggressor by translucent	Reduced social interaction, attenuated sucrose preference, changes in stress hormones	Social nature of induction	Only used in male animals Durability depends on number of defeat sessions

(continued)

	barrier but visual, olfactory, auditory contact maintained.			
Maternal Deprivation (Plotsky and Meaney 1993; Meaney 2001; Ruedi-Bettschen 2004)	Pup separated from mother for extended periods of time during first two weeks after birth	Changes in stress hormone and catelcolamine systems, increased startle, suppressed feeding during novelty exposure, changes in hippocampal development	Face validity, elicits many symptoms Durability: behavior changes persist into adulthood	Time-intensive, variability across laboratories

B. Measures

B. Paradigm	Protocol	Quantification
Sucrose Preference	Subject given access to water with high sucrose content	Proportion of sucrose solution consumed out of all liquid consumed Less = depressed
Novelty-suppressed feeding (Dulawa and Hen 2005)	Subject given access to food that is placed under a bright light	Time until feeding commences Longer = depressed
Social interaction	Subject put in cage with other animals	Time spent interacting with others Less = depressed
Latency to Avoid/Escape Aversive Stimuli	Animals are trained to perform a specific behavior to avoid or terminate exposure to aversive stimulus	Time until escape behavior Longer = depressed
Coat change	Coat appearance and grooming are observed	Grooming score given for 7 body parts Less = depressed

aggressor mice (Berton et al. 2006). In the chronic mild stress model, animals endure innocuous stressors daily for three weeks (Steru et al. 1985; Dulawa and Hen 2005) and in maternal deprivation young pups are separated from their mothers during the first three weeks of life (Seligman 1972).

Animals subjected to these protocols indeed evince lasting behavioral and physiological changes that are reversed only with chronic antidepressant treatment (Yan et al. 2010; Dulawa and Hen 2005; Ruedi-Bettschen et al. 2004). Yet the translational value of these "symptoms" is still unclear. Animals who have completed either the chronic stress (Willner et al. 1987) or social defeat (Berton et al. 2006) paradigm, for example, consume less sugar-water than controls when given free access to both pure water and sugar-water. Attenuated sucrose preference is conceptualized as evidence of anhedonia because the animals show a lack of motivation to seek out an innately rewarding sensation. While this interpretation seems reasonable, there is still no evidence that reduced consumption of sweet foods is associated with anhedonia or depression severity in humans (Amsterdam et al. 1987).

As a group, animal models of MDD have limited face validity. While useful in drug screening and in illuminating the neural mechanisms of antidepressant action, it remains unclear how much insight they offer into the pathophysiology of human depression.

3 The Neurocircuitry of Depression

Neuroimaging studies of MDD patients find abnormalities in brain regions involved in evaluating the salience of stimuli, executive function (control of cognition), encoding and storing complex mental representations of contextual information, and value representation. Although different theories of MDD emphasize the importance of different brain structures (Murray et al. 2011; Mayberg et al. 1997; Mayberg 2003; Hamani et al. 2011), there is a consensus that these highly interconnected regions interact in an affective network, and that persistent dysfunction in its hubs may cause MDD (Fig. 1).

Mayberg and colleagues (Mayberg 2003; Hamani et al. 2011) proposed the "Limbic-Cortical Dysregulation Model", which posits that persistent imbalance between subcortical and cortical regions, influencing mood and arousal, results in MDD. A key correlate of pathology in this model is hyperactivity in the subgenual anterior cingulate cortex, a prefrontal region residing under the genu of corpus callosum. Drevets, Price, and Murray (Price and Drevets 2010; Murray et al. 2011) proposed an alternative model whereby abnormal signaling in the amygdala and a set of interacting prefrontal networks could produce depressive illness. These models do make some predictions about relationships between structure and symptoms, but they lack specificity. Drevets and peers, for instance, discuss how orbitofrontal dysfunction could contribute both to feelings of worthlessness/helplessness (self-concept) and anhedonia (Price and Drevets 2010; Murray et al. 2011).

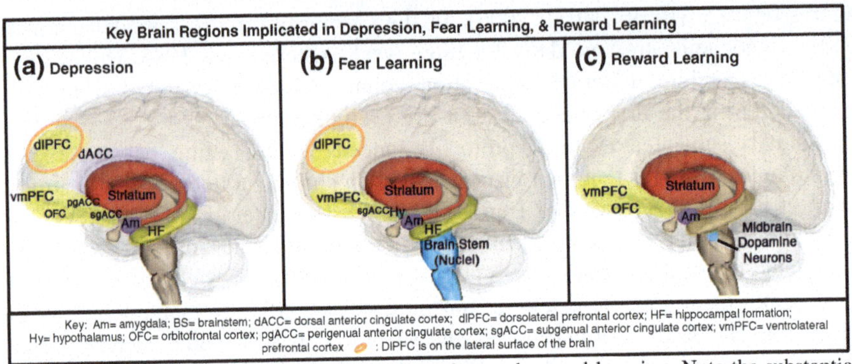

Fig. 1 Key brain areas implicated in depression, fear and reward learning. Note the substantial overlap

Below we describe knowledge hitherto of the neural mechanisms underlying core symptoms of MDD.

3.1 Salience Evaluation

Depressed individuals display excessive reactions to negative stimuli (e.g., sad faces) in the amygdala, insula, and dorsal anterior cingulate cortex (dACC) (Sheline et al. 2001; Siegle et al. 2002; Fu et al. 2004; Suslow et al. 2010; Victor et al. 2010; Hamilton et al. 2012). This triad is implicated in detection of salient events in the environment (Rangel and Hare 2010). In this framework, the amygdala orients the learner towards biologically relevant stimuli (Schoenbaum et al. 2011). The dACC activates the sympathetic nervous system in preparation for action (O'Doherty 2011). The insula mediates conscious awareness of changes in arousal and contributes to the intensity of the emotional experience (Schiller et al. 2008). With hyperactivity in these three regions, MDD patients may be more likely than healthy individuals to overestimate the salience of negative cues. They may also experience more robust physiological responses to aversive experiences. Network hypersensitivity to negative stimuli could underlie a bias towards perception and processing of negative information.

3.2 Executive Dysfunction

The dorsolateral prefrontal cortex (dlPFC) is involved in "executive" control of cognition, enabling deliberate regulation of attention. The dlPFC is recruited during activities that require working memory, planning, inhibition, mental flexibility, initiating action, and monitoring action (Ridderinkhof et al. 2004).

MDD patients evince less dlPFC activity than healthy controls when exposed to negative stimuli (e.g., sad faces), during working memory tasks, and when asked to deliberately modify their thoughts to attenuate or enhance an emotional reaction to a specific stimulus (Siegle et al. 2007; Johnstone et al. 2007; Joormann et al. 2011). dlPFC hypoactivity in depression might reflect an inability to control thought content (e.g., avoid focusing on aversive cues). dlPFC dysfunction would thus prevent patients from successfully employing a wide range of coping skills ranging from distraction to reframing (interpreting a situation in a new way in order to change related affect) (Delgado et al. 2008a, b; Schiller and Delgado 2010; Ochsner et al. 2002, 2004; Ochsner and Gross 2005; Kanske et al. 2010).

3.3 Representation of Context

The hippocampus is essential for contextual learning and encoding of episodic memories. MDD patients who have endured multiple episodes have significantly smaller hippocampi (Neumeister et al. 2005). MDD diagnosis is also associated with abnormal hippocampal recruitment during autobiographical memory recall (Young 2011). Specifically, depressed patients exhibit less activity in the left hippocampus and right parahippocampal gryus during memory recall while healthy volunteers display the opposite pattern. Such aberrations may underlie autobiographical memory impairment in depression. MDD patients tend to over-generalize their autobiographical memories, recalling fewer specific memories and fewer positive memories (Dalgleish et al. 2007; van Vreeswijk and de Wilde 2004; Sumner et al. 2010). Deficits in hippocampal function might impair the capacity to use historical and contextual information and to implement adaptive emotional responses.

3.4 Valuation and Choice

In response to rewards and stimuli that predict them, MDD patients evince hypoactivity in the striatum and orbitofrontal cortex (OFC) (Epstein et al. 2006; Pizzagalli et al. 2009; Smoski et al. 2009; Forbes et al. 2009; Robinson et al. 2011) regions that are also typically smaller in size in depressed versus healthy populations (Koolschijn et al. 2009; Kempton et al. 2011; Arnone et al. 2012). The medial portion of the OFC is a subdivision of the ventromedial prefrontal cortex(vmPFC), as are the subgenual and perigenual anterior cingulate cortices. In these cingulate vmPFC regions, individuals with MDD display smaller volumes but enhanced basal activity (see for review Drevets et al. 2008; Pizzagalli 2011) MDD patients who exhibit the greatest degree of elevation in baseline sub/perigenual activity are more likely than their peers to respond to a variety of antidepressant therapies (Mayberg 2003; Pizzagalli 2011). The striatum is involved in value prediction; the OFC and vmPFC, in value representation. Their dysfunction may relate to the anhedonia commonly experienced by MDD patients.

4 What Can Fear and Reward Learning Tell Us About Depression?

Having established that depression neither lives in one brain region nor results from a deficit in a single type of information processing, how should the field proceed in its efforts to understand MDD? We believe that observing and characterizing in patients the precise patterns of neural activity during associative fear and reward learning might offer new insights into the neurobiology of the illness. In contrast to depression, fear and reward learning (Box 1) are emotional phenomena uniquely suited to research, and are highly conserved processes that are easy to elicit and assess in both human and non-human animals. What is more, the brain regions that evince dysfunction in MDD overlap significantly with those involved in fear and reward learning. By focusing our inquiry on the neural substrates of simpler emotional experiences mediated by the same neural structures, we may be able to identify and reliably reproduce experimental evidence of subtle abnormalities in affective information processing that contribute to depression. Our hypothesis is buttressed by recent fMRI studies of reward learning documenting only subtle (if any) differences in the behavior of MDD patients versus control subjects, but significant differences in corresponding changes in brain activity (Pizzagalli et al. 2009; Smoski et al. 2009; Robinson et al. 2011; Kumar et al. 2008). In the subsequent sections we describe how, by conducting additional studies of reward learning, and beginning to study fear learning in depression, we might learn more about the role of three key brain regions in MDD: the striatum, the amygdala, and the vmPFC.

BOX 1: Associative Learning
Basic fear and reward learning are complementary forms of associative learning, or classical conditioning (LeDoux 2000). During conditioning, the learner encodes a relationship between a sensory cue, such as a sound or a visual stimulus, and a biologically significant outcome, such as pain or food. Knowing this relationship improves the ability to predict, based on the sensory cue, the upcoming occurrence of the biologically significant outcome. After successful conditioning, the sensory cue alone elicits a behavior similar to the response originally triggered by the biologically significant outcome (e.g., preparing to escape at the sound of an alarm).

4.1 The Striatum

Sutton and Barto developed the temporal-difference learning model to describe reward learning (Sutton 1988; Sutton and Barto 1988) In their formulation, learning is a self-perpetuating cycle in which the learner uses information about past rewards to make predications about future rewards. If the reward is greater or lesser than expected, then the learner has made a prediction error, which serves to update the next prediction.

Dopamine cells in the ventral tegmental area and substantia nigra encode prediction errors. Their phasic firing rate increases robustly following the delivery of an unexpected reward (positive prediction error). After reward omission (a negative prediction error), firing is suppressed (McClure et al. 2003; O'Doherty et al. 2003). The striatum, which receives direct input from midbrain dopamine neurons (Haber and Knutson 2010), exhibits the same pattern of change (Delgado et al. 2008a, b; McClure et al. 2003; O'Doherty et al. 2003; Sutton 1988; Sutton and Barto 1988; Schultz et al. 1997; Kishida et al. 2011; Pagnoni et al. 2002).

MDD patients evince weak prediction error signaling in the striatum during reward learning (Kumar et al. 2008; Gradin et al. 2011), which is inversely correlated with anhedonia severity (Gradin et al. 2011). These findings suggest that deficits in prediction error signaling during reward learning might cause or contribute to anhedonia in depression. If weak striatal prediction error signaling during reward learning is a genuine biomarker of depressive anhedonia, it could become an important proxy for anhedonia in animal models of MDD. Monitoring prediction error signal, in addition, could constitute a novel method of tracking illness severity. Before exploring such practical applications, it is imperative that we confirm the relationship between anhedonia and prediction error signal strength by conducting new studies, or revisiting the data from completed studies. We must also determine if signal strength changes in patients as their capacity to experience pleasure fluctuates across episodes or in response to treatment.

Because the striatum also monitors prediction errors during aversive learning (Delgado et al. 2008; Kishida et al. 2011; Li et al. 2011) it will also be important to study fear conditioning in depression. Robinson et al. reported that depression patients show normal neural responses to punishment, tentatively suggesting that they will also display normal striatal prediction error signaling during fear conditioning. If this is the case we can conclude that MDD is characterized by a valence-specific, rather than generalized, deficit in associative learning.

4.2 The Amygdala

Amygdala reactivity to sensory cues is especially robust in two conditions: early in fear conditioning before the learner understands its relationship to the primary threat; and after trials in which it is not followed by the primary threat. Though this pattern is very similar to that observed in the striatum, computational analyses reveal that amygdala and striatal responses to a given sensory cue during fear learning represent different types of information (Li et al. 2011; Kishida et al. 2011; Spoormaker et al. 2011; Kumar et al. 2008; Gradin et al. 2011). As previously described, striatal activity increases when the learner receives an unexpected reward or punishment, and decreases when an expected reward or punishment is omitted. Amygdala activity, in contrast, always increases in response to prediction error (Kishida et al. 2011; Kumar et al. 2008; Gradin et al. 2011; Harmer 1849) Amygdala and striatal responses to prediction errors also differ in their temporal pattern and magnitude.

The Pearce-Hall learning model more accurately predicts amygdala signaling during associative learning than the temporal difference model (Pearce and Hall 1980; Roesch et al. 2010; Li et al. 2011). According to Pearce and Hall (1980), surprise and uncertainty divert attention to novel or changing stimuli, resulting in reallocation of perceptual and cognitive resources thus accelerating learning. Consistent with this formulation, studies in animals (Roesch et al. 2012) and humans (Li et al. 2011) found that amygdala activity peaks early in learning, when the meaning of stimuli is unclear and they are not yet good predictors of reward or threat. As learning progresses and the rewards or punishment become more and more predictable, amygdala responses to the predictive stimuli wane, and attention supposedly returns to other processes. The amygdala thus allocates attention to stimuli that are uncertain in order to enhance learning.

In depression, the amygdala is hyperactive in response to aversive stimuli, which might result in a pervasive negative information processing bias (Calu et al. 2010; Roesch et al. 2010). It is possible that the amygdalae of depressed individuals do not accurately track the surprise associated with aversive cues, but persistently respond to negative stimuli as if uncertainty is high. Here we would expect to observe approximately equivalent increases in amygdala BOLD signal in response to both the primary threat and sensory cue throughout fear conditioning (a lack of habituation). MDD patients, as a result, would endure protracted periods of arousal whenever exposed to negative stimuli (whether the valence is innate or learned). Over time, the allostatic load conferred by excessive activation of the body's stress response systems might trigger depressive symptoms.

Alternatively, it is possible that the amygdala accurately tracks the surprise associated with aversive cues, but that the magnitude of the negative surprise signal is excessive. Here we would expect MDD patients to display excessive increases in amygdala activity when novel stimuli are presented early in conditioning, but a normal pattern of signal attenuation over time. They would therefore be predisposed to form durable negative associations after few, or even one (potentially stochastic) pairing of the primary threat with a novel sensory cue. The persistence of such a state in which an individual is constantly "overreacting", and quickly learning to expect aversive experiences, might engender feelings of hopelessness.

Lastly, it is possible that amygdala's sensitivity to surprise induced by threatening cues is different from the one induced by rewarding cues. In this case, perhaps a combination of hypersensitivity to negative and hyposensitivity to positive stimuli is required for moderate or severe depression. Alternatively, the strength of amygdala responses to surprising aversive stimuli might correlate specifically with severity of depressed mood.

4.3 The vmPFC

Making sense of vmPFC dysfunction in MDD is especially challenging because of its complex cytoarchitecture and because of its complex role in many different mental processes (see Roy et al. 2012 for review). Drevets and colleagues (Drevets et al. 2008; Price and Drevets 2010; Murray et al. 2011) proposed that hyperactivity in a subregion of the vmPFC, the perigenual ACC, is the cause for the autonomic dysregulation observed in depressed patients. Pizzagalli, in contrast, emphasizes vmPFC's role during self-referential processing (Pizzagalli 2011) and postulates that the abnormal vmPFC signaling observed in MDD might correspond to excessive maladaptive rumination. Mayberg's Limbic-Cortical Dysregulation model (Mayberg 2003; Hamani et al. 2011) highlights the contribution of the perigenual and subgenual ACC, in dysregulation of executive function and autonomic and circadian systems, respectively. Studies of reward and fear learning describe specific possible roles for the vmPFC in value representation and fear inhibition. Below we discuss how these studies might provide insights about the link between vmPFC function and MDD.

4.3.1 Reward Learning and Value

As previously mentioned in Sect. 3.4, the vmPFC is engaged during value representation. In order to understand the role of the vmPFC in this process, it is necessary to revisit the temporal-difference learning model as discussed in Sect. 4.1 (The Striatum). There we described how this model conceptualizes reward learning as a self-perpetuating cycle with two components: generation of outcome predictions, and comparison of expected with actual outcomes. We described how midbrain dopaminergic projections to the striatum enable outcome comparison by encoding prediction errors. Studies of reward learning suggest that the vmPFC participates in generation of outcome predictions, representing and updating the value of the rewarding or aversive stimuli (Roy et al. 2012 for review; Maier et al. 2006; Schiller and Delgado 2010; Hare et al. 2009).

In this context, vmPFC dysfunction would not only impair associative learning, but also result in maladaptive decision-making. If MDD patients were unable to update the value of positive stimuli, they would struggle to recover from negative experiences. Experiencing rejection when applying for a job, for instance, might cause anyone to feel inadequate. Healthy individuals, however, would regain their confidence after being offered a position by another employer. An MDD patient, in contrast, might fail to recover even after receiving another job offer because of inability to generate updated predictions. Future studies of reward learning in

MDD should explore the pattern of vmPFC activity while manipulating the value of conditioned stimuli. It will also be necessary to study vmPFC function during fear conditioning to understand the effect of valence on value encodings. Such investigations will permit careful exploration of vmPFC value representation and updating in depression. If meaningful data emerges, they will also demonstrate how probing simple cognitive processes might resolve unique components of prefrontal dysfunction in MDD.

4.3.2 Fear Learning and Inhibition

One way to diminish fear is through extinction learning, when a fear conditioned stimulus is presented repeatedly without the negative outcome. This results in a new safety association, which the vmPFC could then retrieve in the appropriate context (Milad and Quirk 2002; Quirk et al. 2006; Delgado et al. 2008a, b; Seymour et al. 2004) Animal studies show that the vmPFC projection onto the amygdala can control its output and prevents the expression of conditioned fear during extinction (Milad and Quirk 2002; Quirk et al. 2003; Quirk and Beer 2006). Viewed in this light, the vmPFC hyperactivity documented in MDD might be understood as compensatory. It might reflect vmPFC's persistent effort to regulate excessive amygdala activation and the constant generation of negative emotions. It could also reflect inadequate structural or functional connectivity between the vmPFC and amygdala. Interrogating extinction learning in MDD patients might help narrow down the possible interpretations of vmPFC's role in depression.

5 Conclusions and Future Directions

In this chapter we propose that studies of associative learning in depressed patients could offer novel insight into MDD's pathophysiology. We argue that weak striatal prediction error signaling during reward learning could be a biomarker of depressive anhedonia; we draw attention to amygdala's role in allocating attentional resources to uncertain stimuli as a possible explanation for the bias toward negative information in depression; finally, we describe how the vmPFC's role in generation and representation of reward value and vmPFC-amygdala interactions in fear extinction might explain MDD patients' inability to modify negative learning and update future predictions based on novel information.

By considering what studies of associative learning in healthy volunteers have taught us about the roles of the striatum, amygdala, and vmPFC, we highlight the potential utility of conducting similar studies in depression. Understanding the role of these brain regions in associative learning provides an excellent reference as for what goes awry in depression. Examination across domains is an efficient way to distill the basic function of a brain region, with which we can begin to predict the manifestation of this function across disorders.

We believe that collaboration between scientists who investigate associative learning and clinical researchers will advance scientific knowledge of depression's neural substrates. Given that currently available treatments fail one-third of patients, it will be important to take advantage of this untapped resource.

References

American Psychiatric Association (2000) Diagnostic and statistical manual of mental disorders (DSM-IV-TR). Fourth, Text Revision ed. Washington, DC: American Psychological Association

Amsterdam JD, Settle RG, Doty RL, Abelman E, Winokur A (1987) Taste and smell perception in depression. Biol Psychiatry 22(12):1481–1485

Arnone D, McIntosh AM, Ebmeier KP, Munafò MR, Anderson IM (2012) Magnetic resonance imaging studies in unipolar depression: systematic review and meta-regression analyses. Eur Neuropsychopharmacol 22(1):1–16

Arnulf I, Quintin P, Alvarez J, Vigil L, Touitou Y, Lèbre A, et al. (2002) Mid-morning tryptophan depletion delays REM sleep onset in healthy subjects. Neuropsychopharmacology 27(5):843–851

Beck AT (1963) Thinking and depression: I. idiosyncratic content and cognitive distortions. Arch Gen Psychiatry 9(4):324–333

Berton O, McClung CA, Dileone RJ, Krishnan V, Renthal W, Russo SJ et al (2006) Essential role of BDNF in the mesolimbic dopamine pathway in social defeat stress. Science 311(5762):864–868

Blackburn I (1986) Cognitive style in depression. Br J Clin Psychol 25(4):241–251

Bunney WE Jr, Davis JM (1965) Norepinephrine in depressive reactions: a review. Arch Gen Psychiatry 13(6):483–494

Calu DJ, Roesch MR, Haney RZ, Holland PC, Schoenbaum G (2010) Neural correlates of variations in event processing during learning in central nucleus of amygdala. Neuron 68(5):991–1001

Connor TJ, Kelly JP, Leonard BE (1997) Forced swim test-induced neurochemical, endocrine, and immune changes in the rat. Pharmacol Biochem Behav 58(4):961–967

Dalgleish T, Williams JM, Golden AM, Perkins N, Barrett LF, Barnard PJ et al (2007) Reduced specificity of autobiographical memory and depression: the role of executive control. J Exp Psychol Gen 136(1):23–42

David MC (1983) On the induction of depressed mood in the laboratory: evaluation and comparison of the velten and musical procedures. Adv Behav Res Therapy 5(1):27–49

Delgado MR, Li J, Schiller D, Phelps EA (2008a) The role of the striatum in aversive learning and aversive prediction errors. Philos Trans R Soc Lond B Biol Sci 363(1511):3787–3800

Delgado MR, Nearing KI, Ledoux JE, Phelps EA (2008b) Neural circuitry underlying the regulation of conditioned fear and its relation to extinction. Neuron 59(5):829–838

Depression [Internet].: World Health Organization; 2012; cited Janurary 18, 2012. Available from: http://www.who.int/mental_health/management/depression/definition/en/

Drevets WC, Savitz J, Trimble M (2008) The subgenual anterior cingulate cortex in mood disorders. CNS Spectr 13(8):663–681

Dulawa SC, Hen R (2005) Recent advances in animal models of chronic antidepressant effects: the novelty-induced hypophagia test. Neurosci Biobehav Rev 29(4–5):771–783

Duman RS, Heninger GR, Nestler EJ (1997) A molecular and cellular theory of depression. Arch Gen Psychiatry 54(7):597–606

Duman RS, Monteggia LM (2006) A neurotrophic model for stress-related mood disorders. Biol Psychiatry 59(12):1116–1127

Emmett VJ (1968) A laboratory task for induction of mood states. Behav Res Ther 6(4):473–482

Epstein J, Pan H, Kocsis JH, Yang Y, Butler T, Chusid J et al (2006) Lack of ventral striatal response to positive stimuli in depressed versus normal subjects. Am J Psychiatry 163(10): 1784–1790sw

Forbes EE, Hariri AR, Martin SL, Silk JS, Moyles DL, Fisher PM et al (2009) Altered striatal activation predicting real-world positive affect in adolescent major depressive disorder. Am J Psychiatry 166(1):64–73

Fu CH, Williams SC, Cleare AJ, Brammer MJ, Walsh ND, Kim J et al (2004) Attenuation of the neural response to sad faces in major depression by antidepressant treatment: a prospective, event-related functional magnetic resonance imaging study. Arch Gen Psychiatry 61(9): 877–889

Gradin VB, Kumar P, Waiter G, Ahearn T, Stickle C, Milders M et al (2011) Expected value and prediction error abnormalities in depression and schizophrenia. Brain. 134(Pt 6):1751–1764

Greenberg PE, Kessler RC, Birnbaum HG, Leong SA, Lowe SW, Berglund PA, et al. (2003) The economic burden of depression in the united states: How did it change between 1990 and 2000? J Clin Psychiatry 64(12):1465–1475

Haber SN, Knutson B (2010) The reward circuit: linking primate anatomy and human imaging. Neuropsychopharmacology. 35(1):4–26

Hamani C, Mayberg H, Stone S, Laxton A, Haber S, Lozano AM (2011) The subcallosal cingulate gyrus in the context of major depression. Biol Psychiatry 69(4):301–308

Hamilton JP, Etkin A, Furman DJ, Lemus MG, Johnson RF, Gotlib IH (2012) Functional neuroimaging of major depressive disorder: a meta-analysis and new integration of baseline activation and neural response data. Am J Psychiatry 169(7):693–703

Hare TA, Camerer CF, Rangel A (2009) Self-control in decision-making involves modulation of the vmPFC valuation system. Science 324(5927):646–648

Harmer CJ (1849) Antidepressant drug action: a neuropsychological perspective. Depress Anxiety. 27(3):231–233

Hippocrates (1849) Aphorisms. In: Proceedings of the Genuine works of Hippocrates (F. Adams Trans). London: Sydenham Society. [Internet] Retrieved from the Internet Cassics Archive (D.C. Stevenson ed.). Web Atomics [Cited June 1, 2012]

Hollon S (1965) Specificity of depressotypic cognitions in clinical depression. J Abnorm Psychol 95(1):52–59

Johnstone T, van Reekum CM, Urry HL, Kalin NH, Davidson RJ (2007) Failure to regulate: counterproductive recruitment of top-down prefrontal-subcortical circuitry in major depression. J Neurosci 27(33):8877–8884

Joormann J, Cooney RE, Henry ML, Gotlib IH (2011) Neural correlates of automatic mood regulation in girls at high risk for depression. J Abnorm Psychol 121(1):61–72

Kanda K, Omori S, Yamamoto C, Miyamoto N, Kawano S, Murata Y et al (1993) Urinary excretion of stress hormones of rats in tail-suspension. Environ Med. 37(1):39–41

Kanske P, Heissler J, Schonfelder S, Bongers A, Wessa M (2010) How to regulate emotion? neural networks for reappraisal and distraction. Cereb Cortex 21(6):379–388

Katz RJ, Baldrighi G (1982) A further parametric study of imipramine in an animal model of depression. Pharmacol Biochem Behav 16(6):969–972

Katz RJ, Roth KA, Carroll BJ (1981) Acute and chronic stress effects on open field activity in the rat: implications for a model of depression. Neurosci Biobehav Rev 5(2):247–251

Kempton MJ, Salvador Z, Munafo MR, Geddes JR, Simmons A, Frangou S et al (2011) Structural neuroimaging studies in major depressive disorder. meta-analysis and comparison with bipolar disorder. Arch Gen Psychiatry 68(7):675–690

Kishida KT, Sandberg SG, Lohrenz T, Comair YG, Saez I, Phillips PE et al (2011) Sub-second dopamine detection in human striatum. PLoS ONE 6(8):e23291

Koolschijn PC, van Haren NE, Lensvelt-Mulders GJ (2009) Hulshoff Pol HE, Kahn RS. Brain volume abnormalities in major depressive disorder: a meta-analysis of magnetic resonance imaging studies. Hum Brain Mapp 30(11):3719–3735
Kumar P, Waiter G, Ahearn T, Milders M, Reid I, Steele JD (2008) Abnormal temporal difference reward-learning signals in major depression. Brain. 131(Pt 8):2084–2093
Lázaro-Muñoz G, LeDoux JE, Cain CK (2010) Sidman instrumental avoidance initially depends on lateral and basal amygdala and is constrained by central amygdala-mediated pavlovian processes. Biol Psychiatry 67(12):1120–1127
LeDoux JE (2000) Emotion circuits in the brain. Annu Rev Neurosci 23:155–184
Lehmann HE, Cahn CH, DeVerteuil RL (1958) The treatment of depressive conditions with imipramine (G 22355). Can Psychiatr Assoc J 3(4):155–164
Li J, Schiller D, Schoenbaum G, Phelps EA, Daw ND (2011) Differential roles of human striatum and amygdala in associative learning. Nat Neurosci 14(10):1250–1252
Maier SF, Amat J, Baratta MV, Paul E, Watkins LR (2006) Behavioral control, the medial prefrontal cortex, and resilience. Dialogues Clin Neurosci. 8(4):397–406
Maier SF (1984) Learned helplessness and animal models of depression. Prog Neuro-Psychopharmacol Biol Psychiatry 8(3):435–446
Maren S (2001) Neurobiology of pavlovian fear conditioning. Annu Rev Neurosci 24:897–931
Mayberg HS, Brannan SK, Mahurin RK, Jerabek PA, Brickman JS, Tekell JL et al (1997) Cingulate function in depression: a potential predictor of treatment response. NeuroReport 8(4):1057–1061
Mayberg HS (2003) Modulating dysfunctional limbic-cortical circuits in depression: towards development of brain-based algorithms for diagnosis and optimised treatment. Br Med Bull 65:193–207
McClure SM, Berns GS, Montague PR (2003) Temporal prediction errors in a passive learning task activate human striatum. Neuron 38(2):339–346
Meaney MJ (2001) Maternal care, gene expression, and the transmission of individual differences in stress reactivity across generations. Annu Rev Neurosci 24:1161–1192
Milad MR, Quirk GJ (2002) Neurons in medial prefrontal cortex signal memory for fear extinction. Nature 420(6911):70–74
Murray EA, Wise SP, Drevets WC (2011) Localization of dysfunction in major depressive disorder: prefrontal cortex and amygdala. Biol Psychiatry 69(12):e43–e54
Neumeister A, Wood S, Bonne O, Nugent AC, Luckenbaugh DA, Young T, et al. (2005) Reduced hippocampal volume in unmedicated, remitted patients with major depression versus control subjects. Biol Psychiatry 57(8):935–937
Nibuya M, Morinobu S, Duman RS (1995) Regulation of BDNF and trkB mRNA in rat brain by chronic electroconvulsive seizure and antidepressant drug treatments. J Neurosci 15(11): 7539–7547
Ochsner KN, Bunge SA, Gross JJ, Gabrieli JD (2002) Rethinking feelings: an FMRI study of the cognitive regulation of emotion. J Cogn Neurosci. 14(8):1215–1229
Ochsner KN, Gross JJ (2005) The cognitive control of emotion. Trends Cogn Sci (Regul Ed) 9(5):242–249
Ochsner KN, Ray RD, Cooper JC, Robertson ER, Chopra S, Gabrieli JD et al (2004) For better or for worse: neural systems supporting the cognitive down- and up-regulation of negative emotion. Neuroimage. 23(2):483–499
O'Doherty JP, Dayan P, Friston K, Critchley H, Dolan RJ (2003) Temporal difference models and reward-related learning in the human brain. Neuron 38(2):329–337
O'Doherty JP (2011) Contributions of the ventromedial prefrontal cortex to goal-directed action selection. Ann N Y Acad Sci 1239:118–129
Pagnoni G, Zink CF, Montague PR, Berns GS (2002) Activity in human ventral striatum locked to errors of reward prediction. Nat Neurosci 5(2):97–98
Papez JW (1937) A proposed mechanism of emotion. Arch Neurol Psychiatry 38(4):725–743
Pearce JM, Hall G (1980) A model for pavlovian learning: variations in the effectiveness of conditioned but not of unconditioned stimuli. Psychol Rev 87(6):532–552

Phelps EA, Delgado MR, Nearing KI, LeDoux JE (2004) Extinction learning in humans: role of the amygdala and vmPFC. Neuron 43(6):897–905
Pizzagalli DA, Holmes AJ, Dillon DG, Goetz EL, Birk JL, Bogdan R et al (2009) Reduced caudate and nucleus accumbens response to rewards in unmedicated individuals with major depressive disorder. Am J Psychiatry 166(6):702–710
Pizzagalli DA (2011) Frontocingulate dysfunction in depression: toward biomarkers of treatment response. Neuropsychopharmacology. 36(1):183–206
Plotsky PM, Meaney MJ (1993) Early, postnatal experience alters hypothalamic corticotropin-releasing factor (CRF) mRNA, median eminence CRF content and stress-induced release in adult rats. Mol Brain Res 18(3):195–200
Porsolt RD (1978) Behavioural despair in rats: a new model sensitive to antidepressant treatments. Eur J Pharmacol 47(4):379–391
Price JL, Drevets WC (2010) Neurocircuitry of mood disorders. Neuropsychopharmacology. 35(1):192–216
Pringle A, Browning M, Cowen PJ, Harmer CJ. (2011) A cognitive neuropsychological model of antidepressant drug action. Prog Neuro-Psychopharmacol Biol Psychiatry 35(7):1586–1592
Quirk GJ, Beer JS (2006) Prefrontal involvement in the regulation of emotion: convergence of rat and human studies. Curr Opin Neurobiol 16(6):723–727
Quirk GJ, Garcia R, González-Lima F. (2006) Prefrontal mechanisms in extinction of conditioned fear. Biol Psychiatry 60(4):337–343
Quirk GJ, Likhtik E, Pelletier JG, Pare D (2003) Stimulation of medial prefrontal cortex decreases the responsiveness of central amygdala output neurons. J Neurosci 23(25): 8800–8807
Rangel A, Hare T (2010) Neural computations associated with goal-directed choice. Curr Opin Neurobiol 20(2):262–270
Reilly JG, McTavish SF, Young AH (1997) Rapid depletion of plasma tryptophan: a review of studies and experimental methodology. J Psychopharmacol 11(4):381–392
Ridderinkhof KR, van den Wildenberg WP, Segalowitz SJ, Carter CS (2004) Neurocognitive mechanisms of cognitive control: the role of prefrontal cortex in action selection, response inhibition, performance monitoring, and reward-based learning. Brain Cognition 56(2):129–140. doi:10.1016/j.bandc.2004.09.016
Riskind JH, Rholes WS, Eggers J (1982) The velten mood induction procedure: effects on mood and memory. J Consult Clin Psychol 50(1):146–147
Robinson OJ, Cools R, Carlisi CO, Sahakian BJ, Drevets WC (2011) Ventral striatum response during reward and punishment reversal learning in unmedicated major depressive disorder. Am J Psychiatry 169(2):152−159
Roesch MR, Calu DJ, Esber GR, Schoenbaum G (2010) Neural correlates of variations in event processing during learning in basolateral amygdala. J Neurosci 30(7):2464–2471
Roesch MR, Esber GR, Li J, Daw ND, Schoenbaum G (2012) Surprise! neural correlates of pearce-hall and rescorla-wagner coexist within the brain. Eur J Neurosci 35(7):1190–1200
Roy M, Shohamy D, Wager TD (2012) Ventromedial prefrontal-subcortical systems and the generation of affective meaning. Trends Cogn Sci (Regul Ed) 16(3):147–156
Ruedi-Bettschen D, Feldon J, Pryce CR (2004) The impaired coping induced by early deprivation is reversed by chronic fluoxetine treatment in adult fischer rats. Behav Pharmacol 15(5–6): 413–421
Ruhe HG, Mason NS, Schene AH (2007) Mood is indirectly related to serotonin, norepinephrine and dopamine levels in humans: a meta-analysis of monoamine depletion studies. Mol Psychiatry 12(4):331–359
Rush AJ, Trivedi MH, Wisniewski SR, Nierenberg AA, Stewart JW, Warden D et al (2006) Acute and longer-term outcomes in depressed outpatients requiring one or several treatment steps: A STAR*D report. Am J Psychiatry 163(11):1905–1917
Sanacora G, Treccani G, Popoli M (2012) Towards a glutamate hypothesis of depression: an emerging frontier of neuropsychopharmacology for mood disorders. Neuropharmacology 62(1):63–77

Schildkraut JJ (1967) The catecholamine hypothesis of affective disorders. A review of supporting evidence. Int J Psychiatry 4(3):203–217
Schiller D, Delgado MR (2010) Overlapping neural systems mediating extinction, reversal and regulation of fear. Trends Cogn Sci (Regul Ed) 14(6):268–276
Schiller D, Levy I, Niv Y, LeDoux JE, Phelps EA (2008) From fear to safety and back: reversal of fear in the human brain. J Neurosci 28(45):11517–11525
Schoenbaum G, Takahashi Y, Liu T-, Mcdannald MA. Does the orbitofrontal cortex signal value? [Internet]; 2011 [cited 1 June 2012]
Schultz W, Dayan P, Montague PR (1997) A neural substrate of prediction and reward. Science 275(5306):1593–1599
Seligman ME (1972) Learned helplessness. Annu Rev Med 23:407–412
Seymour B, O'Doherty JP, Dayan P, Koltzenburg M, Jones AK, Dolan RJ et al (2004) Temporal difference models describe higher-order learning in humans. Nature 429(6992):664–667
Sheline YI, Barch DM, Donnelly JM, Ollinger JM, Snyder AZ, Mintun MA (2001) Increased amygdala response to masked emotional faces in depressed subjects resolves with antidepressant treatment: an fMRI study. Biol Psychiatry 50(9):651–658
Siegle GJ, Steinhauer SR, Thase ME, Stenger VA, Carter CS (2002) Can't shake that feeling: event-related fMRI assessment of sustained amygdala activity in response to emotional information in depressed individuals. Biol Psychiatry 51(9):693–707
Siegle GJ, Thompson W, Carter CS, Steinhauer SR, Thase ME (2007) Increased amygdala and decreased dorsolateral prefrontal BOLD responses in unipolar depression: related and independent features. Biol Psychiatry 61(2):198–209
Smith MA, Makino S, Kvetnansky R, Post RM (1995) Stress and glucocorticoids affect the expression of brain-derived neurotrophic factor and neurotrophin-3 mRNAs in the hippocampus. J Neurosci 15(3 Pt 1):1768–1777
Smoski MJ, Felder J, Bizzell J, Green SR, Ernst M, Lynch TR, et al. (2009) fMRI of alterations in reward selection, anticipation, and feedback in major depressive disorder. J Affect Disord. 118(1–3):69–78
Spoormaker VI, Andrade KC, Schröter MS, Sturm A, Goya-Maldonado R, Sämann PG, et al. (2011) The neural correlates of negative prediction error signaling in human fear conditioning. Neuroimage 54(3):2250–2256
Steru L, Chermat R, Thierry B, Simon P (1985) The tail suspension test: a new method for screening antidepressants in mice. Psychopharmacology (Berl) 85(3):367–370
Sumner JA, Griffith JW, Mineka S. (2010) Overgeneral autobiographical memory as a predictor of the course of depression: a meta-analysis. Behav Res Ther 48(7):614–625
Suslow T, Konrad C, Kugel H, Rumstadt D, Zwitserlood P, Schöning S, et al. (2010) Automatic mood-congruent amygdala responses to masked facial expressions in major depression. Biol Psychiatry 67(2):155–160
Sutton RRS (1988) Learning to predict by the methods of temporal differences Mach Learning 3(1):9–44
Sutton RS, Barto AG (1988) Reinforcement learning: an introduction. MIT Press, Cambridge
Trullas R, Skolnick P (1990) Functional antagonists at the NMDA receptor complex exhibit antidepressant actions. Eur J Pharmacol 185(1):1–10
van Vreeswijk MF, de Wilde EJ (2004) Autobiographical memory specificity, psychopathology, depressed mood and the use of the autobiographical memory test: a meta-analysis. Behav Res Ther 42(6):731–743
Victor TA, Furey ML, Fromm SJ, Ohman A, Drevets WC (2010) Relationship between amygdala responses to masked faces and mood state and treatment in major depressive disorder. Arch Gen Psychiatry 67(11):1128–1138
Voderholzer U, Hornyak M, Thiel B, Huwig-Poppe C, Kiemen A, König A, et al. (1998) Impact of experimentally induced serotonin deficiency by tryptophan depletion on sleep EEC in healthy subjects. Neuropsychopharmacology 18(2):112–124

Willner P, Towell A, Sampson D, Sophokleous S, Muscat R (1987) Reduction of sucrose preference by chronic unpredictable mild stress, and its restoration by a tricyclic antidepressant. Psychopharmacology 93(3):358–364
Willner P (1997) Validity, reliability and utility of the chronic mild stress model of depression: a 10-year review and evaluation. Psychopharmacology 134(4):319–329
Yan HC, Cao X, Das M, Zhu XH, Gao TM (2010) Behavioral animal models of depression. Neurosci Bull 26(4):327–337
Young KD (2011) Functional anatomy of autobiographical memory recall deficits in depression. Psychol Med. 42(02):1–13
Zimmermann RC, McDougle CJ, Schumacher M, Olcese J, Mason JW, Heninger GR et al (1993) Effects of acute tryptophan depletion on nocturnal melatonin secretion in humans. J Clin Endocrinol Metab 76(5):1160–1164

The Use of Cognitive Bias Modification and Imagery in the Understanding and Treatment of Depression

Michael Browning, Simon E. Blackwell and Emily A. Holmes

Abstract Cognitive models of depression form the natural link between neurobiological and social accounts of the illness. Interest in the role of cognition in depression has recently been stimulated by the advent of simple, computer-based "cognitive bias modification" (CBM) techniques which are able to experimentally alter cognitive habits in clinical and non-clinical populations. In this chapter, we review recent work which has used CBM techniques to address questions of aetiology and treatment in depression with a particular focus on the interface with neurobiological and social processes relevant to the illness. We find that there are early signs that CBM may be a useful tool in exploring the aetiology of depression, particularly in regard to the neural and genetic factors which influence susceptibility to the illness and response to treatment. There is also early evidence suggesting that CBM has promise in the treatment and prevention of depression. This work suggests that the beneficial effects of CBM are mediated by the interaction between cognitive functioning and environmental and social information. In summary, by providing a method for altering habitual cognitive function in experimental and clinical settings CBM techniques have begun to further the understanding of and the treatment for depression.

Keywords Depression · Cognition · Aetiology · Treatment

Contents

M. Browning (✉) · S. E. Blackwell · E. A. Holmes
Department of Psychiatry, University of Oxford, OX3 7JX, Oxford, UK
e-mail: michael.browning@ndcn.ox.ac.uk

Curr Topics Behav Neurosci (2013) 14: 243–260
DOI: 10.1007/7854_2012_212

Published Online: 8 May 2012

Depression is associated with perturbations in a broad range of systems which vary in scale from the molecular, via the cognitive, to the social (Fig. 1). Models of the illness commonly simplify this complex presentation by focussing on one of these domains at a time. However, the cost of such focussed models is that they struggle to account for the full range of symptomatology found in the illness. For example, a purely social account of depression cannot explain interindividual differences in patients' responses to a particular environment, while it is often difficult to translate molecular findings into the subjective symptoms that patients report. Cognitive models, which link characteristics of intraindividual information processing to depressive symptoms, occupy a middle ground in the conceptualisation of the illness. While cognitive models are, in themselves, no more complete a description of depression than neurobiological or social accounts, they provide a natural link by which molecular mechanisms may interact with environmental and social information, and thus cognitive models may be used to bridge the disparate accounts of depression. In this chapter, we review recent clinically focussed experimental work that has investigated cognitive abnormalities in depression and how they may be remedied. In particular, we describe work that has investigated the impact and methodologies of "cognitive bias modification" (CBM) interventions, which present individuals with structured information in order to alter cognitive habits. Historically, these bias modification interventions have most commonly been deployed to investigate aetiology and treatment mechanisms in anxiety although, where possible, we focus on more recent studies which examine their relevance to depression. Finally, we discuss how this work has been used to examine the interaction between cognitive processes and neurobiological mechanisms on the one hand and environmental factors on the other (Fig. 1).

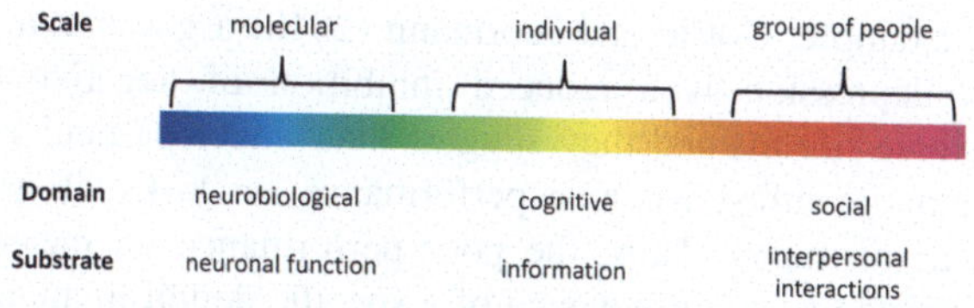

Fig. 1 Representation of some of the different scales by which the mechanisms underlying depression may be considered. Although the illness is associated with perturbations across the full range of scales it can be challenging to account for this range within a single explanatory model. Cognitive models form the natural link between the neurobiological and social domains

1 The Types of Information Processing Abnormality Associated With Depression

Over the last 40 years, a large number of cognitive models of the vulnerability to develop depression have been advanced (Beck 1967; Ingram 1984; Teasdale 1988; Williams et al. 1997; Mathews and MacLeod 2005). A common thread across these models is that susceptibility to develop the illness is explained, at least in part, by individual differences in information processing habits and abilities. As a result, much research has focussed on the identification and characterisation of the styles of information processing that are associated with either an increased or decreased risk of developing depression. Broadly, two types of information processing abnormalities have been described: frank deficits in function (cognitive deficits) and relative alterations in processing habits (cognitive biases) (see; Gotlib and Joormann 2010 for a recent review of cognitive correlates of depression).

1.1 Cognitive Deficits

Current symptoms of depression are associated with reduced performance on a range of measures of cognitive function (Austin et al. 2001; Clark et al. 2009) including in particular, gross deficits in measures of explicit memory (Gorwood et al. 2008) and executive function (Rogers et al. 2004). While it is reasonable to suppose that these global measures of cognitive dysfunction provide hints about some of the cognitive systems involved in depression, there does not seem to be a straightforward link between them and the symptoms of the illness. For example, why would a difficulty in explicitly remembering information produce symptoms of low mood? Further, it appears to be possible to ameliorate some of the deficits by altering parameters of the assessment task (for example, by reducing the opportunity to ruminate) (Gotlib and Joormann 2010). A likely cause for this explanatory gap is that measures of gross cognitive function do not have the required acuity to discriminate the more subtle aspects of processing alteration in depression and it is these more nuanced alterations that lead to the production of symptoms. For example, Gotlib and Joormann (2010) argue that a key cognitive abnormality in depression is a reduced inhibition of negative cognitions in response to adverse life events; they suggest that, under particular experimental regimes, this may manifest as poor performance on tasks of gross cognitive function (such as memory). Thus, the poor performance on measures of recall memory is interpreted as a consequence of a specific deficit in attentional control rather than as a marker of primary abnormality in memory function. By this view, symptoms result from the interaction of a deficit in attentional control and a tendency for depressed patients to preferentially process affectively negative information. Such alterations in processing habits are known as cognitive biases and are discussed in the following section.

1.2 Cognitive Biases

Cognitive biases describe changes in information processing preference that are not necessarily the product of a deficit in function. The most commonly described bias in depression is a tendency to show enhanced processing of affectively negative, relative to positive, information. Such negative cognitive biases have been described using measures of memory (Bradley et al. 1995; Gilboa-Schechtman et al. 2002), attention (Gotlib et al. 2004) and interpretation (Bouhuys et al. 1999; Rude et al. 2002). Interestingly, negative cognitive biases are also found in euthymic individuals who are at high risk of developing depression, suggesting that they may be associated with the risk of developing the illness rather than simply being markers of current depressive state. For example, the negative attentional bias described in depressed patients (Gotlib et al. 2004) is also found in previously depressed, currently euthymic patients (Joormann and Gotlib 2007) and in never depressed adolescents who are at high risk of developing the disorder because of a positive family history (Joormann et al. 2007).

Cognitive biases for affective information, which lead patients to appraise their environment in a negative manner, provide a straightforward mechanism by which the mood symptoms of the illness may be generated (Mathews and MacLeod 2005); however, other forms of information processing bias, which do not seem to directly depend on affective information, have also been described. Perhaps the most well-studied example is a tendency for patients to report less specific autobiographical memories. Thus, when presented with a cue word, depressed patients produce less specific and more overgeneral autobiographical memories (Williams et al. 2007), with this effect being apparent whether the word cues positive or negative memories. Generally, as discussed above, these apparently non-affective biases are believed to reflect deficits in attentional control and to lead to depressive symptoms by reducing an individual's efficacy at responding to challenging external events and regulating the subsequent emotional response (Williams et al. 2007).

2 Manipulating Information Processing Style Using Bias Modification Interventions

The work reviewed above demonstrates some of the information processing abnormalities that have been associated with depression. However, cognitive models of the illness view many of these abnormalities (both affective biases and specific deficits in attentional control) as being causally related to depressive symptoms, and not simply markers of depressive risk. This leads to a further, clinically interesting prediction; an intervention that acts to reverse the cognitive abnormalities should reduce the symptoms of depression. This prediction has guided the development of a number of therapeutic approaches to depression, most notably Cognitive Behavioural Therapy (CBT; Beck et al. 1979). However, by their nature, therapies such as CBT are complex and involve a diverse range of therapeutic strategies. It has therefore proven difficult to ascertain the extent to which the clinical benefits seen following therapy may be attributed to specific processes, such as alteration of cognitive bias, rather than more general factors such as therapeutic alliance (Driessen and Hollon 2010). An alternative approach to studying the role of information processing abnormalities in relation to depression is to develop interventions that target, as specifically as possible, a process of interest and measure the effect of the intervention on subsequent symptoms. Recently, a range of CBM procedures that have been designed to do just this have been described (MacLeod et al. 2009). The logic of CBM tasks is that information processing habits are malleable and tend to adapt to the particular processing demands placed on an individual (Fig. 2). The tasks are therefore designed so that they are more efficiently completed if participants adopt a particular processing style. They can therefore be used to assess whether cognitive biases precede symptoms (negative CBM, see section 3 below) and if modifying particular processing styles can improve symptoms (positive CBM, see section 4 below). For example, during attentional bias modification (Fig. 2) participants may be presented with pairs of faces, showing different emotions, on a computer screen. Depression is associated with a tendency to direct attention towards the negative emotion (Gotlib et al. 2004). During the bias modification task participants are asked to respond to a probe (e.g. either one or two small dots), which appears following the faces, in the location of one of the faces.

Positive bias modification is achieved by presenting many trials in which the probe always (or nearly always) replaces the more positive face. During the course of the CBM participants learn that the most efficient approach to the task is to direct attention towards the positive faces as these predict the location of the probes. In other words, the participants learn to deploy a positive attentional bias. A range of approaches, based on a similar logic, have been described for affective attentional (MacLeod et al. 2002; Dandeneau and Baldwin 2004), interpretative (Grey and Mathews 2000; Lang et al. 2012) and memory biases (Joormann et al. 2009), metacognitive appraisals (Lang et al. 2009; Woud et al. (2011)) as well as non-affective processing habits (for example, bias modification aimed at

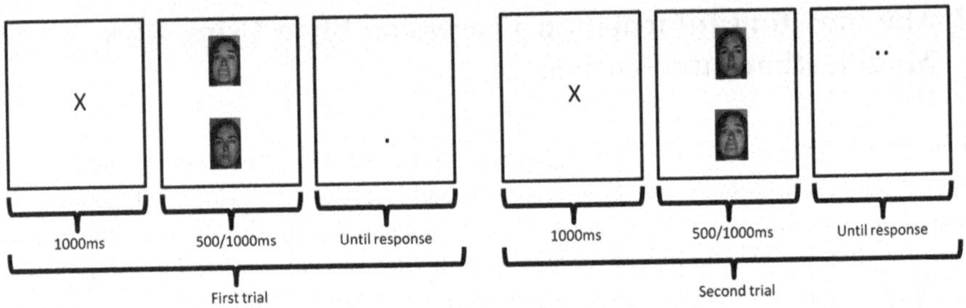

Fig. 2 Two example trials from an attentional bias training task. On each trial two face stimuli are presented with differing expressions. The faces are followed by a probe (*one or two dots*) to which the participants have to respond by button press to indicate the number of dots. During positive training (*shown*) the probe appears behind the more positive of the two faces. Over the course of many trials participants learn to direct their attention towards the positive faces, in other words they learn to deploy a positive attentional bias. Positive CBM may be compared to neutral CBM which is identical in every respect other than that the probe is equally likely to appear behind either stimulus, or to negative CBM when the probe appears behind the negative face. A similar logic underlies other forms of CBM tasks, such as those that alter interpretative or memory biases—the tasks are constructed so that their completion is facilitated if the cognitive habit to be trained is adopted

improving the deficits in attentional control processes discussed above; Siegle et al. 2007; Watkins et al. 2011). In the following sections, we consider recent work that has used CBM techniques to investigate the aetiology and treatment of the emotional disorders, where applicable highlighting the work that has focussed on depression.

3 Cognitive Bias Modification and the Aetiology of Depression

Simple changes to CBM tasks can be used to alter the *direction* of bias modification while keeping other aspects of the task constant (see Fig. 2). For example, if the attentional bias modification method described above is changed such that the probes are placed behind the negative face, then its effect reverses so that it now induces the negative bias which is thought to be causally relevant in depression. This feature has been exploited to test questions of causality, with most work having a theoretical base in the anxiety disorders. Thus, an early study demonstrated that CBM could be used to produce negative interpretative biases similar to those found in anxiety (Grey and Mathews 2000). A second, influential study randomised nonclinical participants to receive attentional bias modification either towards or away from threatening stimuli (MacLeod et al. 2002). The authors were able to show that participants developed the expected attentional bias in response to the CBM regimes. Crucially, it was also demonstrated that CBM influenced participants' reactivity to external stress (stress was induced by asking participants to try and solve insoluble anagrams under time pressure), with those who had been trained to have a negative bias reporting a greater increase in negative symptoms in response to the stressor than those who completed the relatively positive CBM.

The authors suggested that their results provided experimental proof that negative attentional biases were causally related to emotional vulnerability. Interestingly, the assessment of "negative symptoms" included separate scales that measured depression and anxiety with the CBM producing an equivalent effect on both scales. While the specific assessment scales used in this study (visual analogue scales) are unlikely to have been particularly effective in discriminating anxiety from depression and the findings of smaller studies have been inconsistent (Eldara et al. 2008; Krebs et al. 2010), these results do provide some evidence that inducing a negative attentional bias increases vulnerability to symptoms of both depression and anxiety.

4 Cognitive Bias Modification and the Treatment of Depression

CBM procedures raise a clear clinical opportunity; if used to reduce the presumptive causative biases in clinical populations they should produce an improvement in symptoms. Again, the main focus of research has been on patients with anxiety, with meta-analyses suggesting a significant reduction in the symptoms of anxiety following positive affective bias modification (Hakamata et al. 2010; Hallion and Ruscio 2011). Relative to this, the use of CBM strategies in depressed patients is in its infancy. Currently, four published studies report the effects of CBM on depressed patients. The first two of these report a case series (Blackwell and Holmes 2010) and controlled study (Lang et al. 2012) of a form of interpretative bias modification (see below for a description of this form of CBM). Both studies reported a reduction in depressive symptoms following positive CBM, with the Lang et al. (2012) study reporting that this was significantly larger than the effect of a control CBM intervention. The third study reports a relatively well powered clinical trial that compared a non-affective "concreteness training" (CBM designed to reduce over-general thinking) with either a treatment as usual or relaxation training group (Watkins et al. 2011). Concreteness training was associated with a reduction in symptoms relative to treatment as usual, but did not produce additional benefit relative to the relaxation training. Lastly, Baert et al. (2010) reported that a novel attentional bias CBM regime was ineffective in reducing symptoms in a cohort of depressed patients, although it should be noted that in this study CBM had no impact on attentional bias, suggesting that the novel form of CBM simply did not work as intended.

An alternative approach by which CBM regimes may usefully be deployed is as a "cognitive vaccine" to prevent the onset, rather than treating the symptoms of depression (Holmes et al. 2009). This approach is predicated on the findings reported above; cognitive biases are associated with the risk of becoming

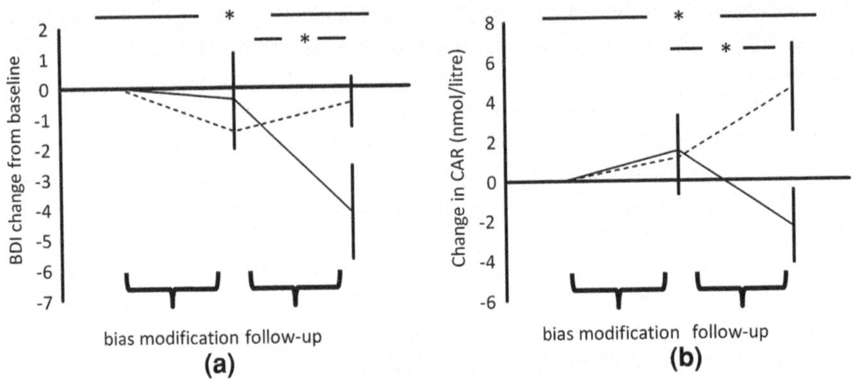

Fig. 3 The effect of attentional bias modification on **a** residual depressive symptoms (measured using the Beck Depression Inventory (BDI)) and **b** the Cortisol Awakening Response (CAR; measured as the difference between waking salivary cortisol and the peak achieved in the following hour) in previously depressed patients. The positive CBM group is represented by the *solid line* and the neutral CBM (control) group by the *dashed line*. Both outcomes show a significant effect of CBM type, specifically during the follow-up period, with positive CBM reducing the risk marker relative to neutral CBM. Results are presented as change from pretreatment baseline; error bars represent standard error of the mean. * = $p < 0.05$

depressed rather than only with current symptoms of depression. Ideally, such a strategy would be used in primary prevention, that is to prevent the onset of illness in an at-risk population. Although no study has directly tested the utility of CBM in this setting, there is evidence that bias modification techniques are efficacious in treating anxiety in adolescent patients (Eldar et al. 2012) and may be used to alter bias in non-clinical adolescent populations (Lothmann et al. 2012). There is thus emerging evidence that CBM satisfies the initial requirements of a primary preventative intervention; it is able to alter putative cognitive risk factors in young, at-risk populations. It may also be possible to use CBM techniques in the secondary prevention of depression that is, to prevent recurrence in previously depressed, currently euthymic patients. In a recently completed study of previously depressed patients, Browning et al. (in press) reported the effect of attentional bias modification on two measures of relapse risk; the extent of residual depressive symptoms (Kennedy and Paykel 2004) and the waking cortisol response (Cowen 2010). Patients were trained using either a positive or control attentional bias modification task over the course of 2 weeks and were then followed up for a further month. As can be seen from Fig. 3, positive bias modification reduced both measures of risk of relapse. Interestingly, the beneficial effects of CBM displayed a time lag, becoming apparent during the follow-up period (see section on environmental interaction below for further discussion).

In summary, there is some evidence that CBM techniques may be useful in reducing both the symptoms and the risk of recurrence in depression; however, it is unclear from the relatively few studies competed how reliable such an effect is and, more practically, what the most effective methods of CBM are. In the following section, we consider work which has investigated one method of improving CBM efficacy—the use of patients' mental imagery.

5 Can Mental Imagery be Used to Amplify the Effects of CBM?

5.1 The Nature of Mental Imagery

Mental imagery occurs "when perceptual information is accessed from memory, giving rise to the experience of 'seeing with the mind's eye', 'hearing with the mind's ear' and so on" (Kosslyn et al. 2001). When participants are provided with information about an emotionally salient event they report a more potent emotional reaction when they actively imagine rather than verbally process the information (Holmes and Mathews 2005, 2010), possibly due to mental imagery utilising many of the neural mechanisms involved in direct perception (Holmes and Mathews 2010). Using mental imagery to process the valenced information presented during bias modification may therefore be one way to enhance the impact of these procedures (Holmes et al. 2009). Manipulations of mental imagery are particularly suited to interpretative bias modification paradigms which present more complex stimuli for longer durations than those typically used during attentional bias modification. In the following section, we describe the general structure of interpretative bias modification tasks (Mathews and Mackintosh 2000) and how they have been adapted (Holmes and Mathews 2005) to investigate whether mental imagery can be used to amplify the impact of CBM.

5.2 Imagery Enhanced Interpretative Bias Modification in Non-Clinical Groups

The most common variant of interpretative bias modification involves participants listening to (or reading) brief descriptions of everyday situations, which are initially ambiguous with regard to their potential outcome, but are then resolved positively or negatively. For example,

> You ask a friend to look over a chapter you have written. They come back with some comments, which are all very *positive/critical* (resolution in italics).

By constraining the resolution of every scenario either positively or negatively, a learning contingency is established whereby participants come to automatically expect the corresponding outcome for novel ambiguous situations, that is participants learn to deploy a positive or negative interpretative bias. A series of experimental studies in nonclinical populations have demonstrated that encouraging participants to imagine, as opposed to verbally process these scenarios, increases the ability of a single session of interpretative bias modification to alter both subjectively reported mood and experimentally measured bias (Holmes and Mathews 2005; Holmes et al. 2006, 2008a, 2009). This work indicates that, at least in non-clinical populations, mental imagery may be used to amplify the effects of interpretative bias modification.

5.3 Imagery Enhanced Interpretative Bias Modification in Depressed Patients

Two of the studies described in the previous section on CBM in depressed patients, used imagery enhanced interpretative bias modification (Blackwell and Holmes 2010; Lang et al. 2012). The initial case series by Blackwell and Holmes (2010) utilised auditory scenario based CBM, before which participants received instruction in generating vivid mental imagery. The Lang et al. (2012) study supplemented the auditory scenarios with two additional forms of interpretative bias modification ("picture" (Holmes et al. 2008c) and "appraisal" (Lang et al. 2009) CBM) and also encouraged participants to utilise mental imagery. In the active condition of this study all task stimuli were resolved positively, whereas in the control condition half were resolved positively and half were resolved negatively. The results of both studies indicated a beneficial effect of positive bias modification on the symptoms of depression and are summarised in Fig. 4.

5.4 Does Practice in Using Mental Imagery Have a Direct Effect on Depressive Symptoms?

The initial rationale for combining mental imagery and interpretation CBM was that this would enhance emotional processing of the valenced stimuli used during CBM. In other words, the interpretative bias modification procedure was believed to be the primarily effective intervention in altering depressive symptoms and recruiting mental imagery was a method of enhancing this effect. This model is consistent with the preclinical experimental studies reviewed above which suggest that the use of imagery enhances the emotional impact of both positive and negative interpretative bias modification. However, recent work has suggested that major depression is associated with a marked deficit in generating positive imagery of the future, with no difference being found for negative imagery (Morina et al. 2011). A selective lack of prospective positive (but not negative) imagery is also found in dysphoric mood (Holmes et al. 2008b). This raises the possibility that difficulties in generating vivid positive imagery may be a feature, in itself, of depressive cognition and that practising producing such imagery may lead to a benefit over and above that produced by the interpretative bias modification. While

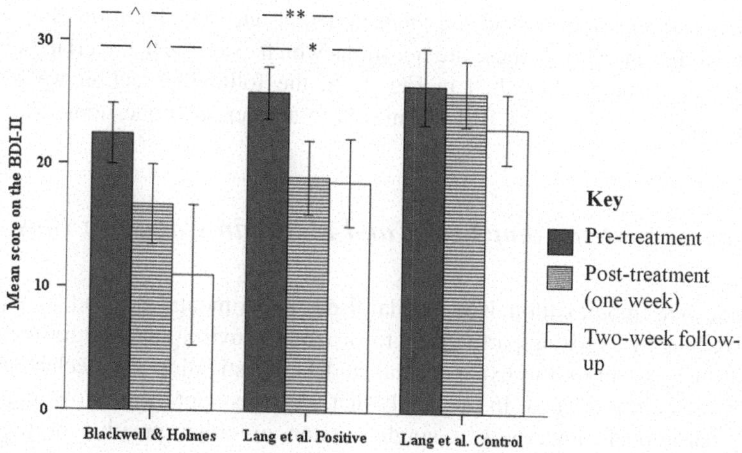

Fig. 4 Change on the Beck Depression Inventory—II in Blackwell and Holmes (2010) and Lang et al. (2012). In Blackwell and Holmes (2010), seven currently depressed participants completed daily sessions of a positive imagery and interpretation bias modification computer programme for 1 week, and provided follow-up data 2 weeks post-treatment. In Lang et al. (2012), 13 currently depressed participants completed a positive imagery and interpretation bias modification programme, and 13 participants completed a control version of the programme. The same 1-week CBM and 2-week follow-up schedule as in Blackwell and Holmes (2010) was used. The decrease in score on the BDI-II was significantly greater from pre to post-treatment in the positive imagery group compared to the control group, and the relative improvement from pretreatment to follow-up was at trend level. ^ $p < .1$, * $p < .05$, ** $p < .01$. Error bars represent standard error of the mean

the studies performed in clinical populations to date have included both bias modification and imagery enhancement components, the study designs do not allow this hypothesis to be directly tested.

In summary, there is good preclinical evidence that encouraging participants to utilise mental imagery during interpretative bias modification tasks enhances the cognitive and emotional impact of the CBM. The precise effect of encouraging the use of positive mental imagery during CBM in a clinical setting has yet to be delineated although early studies have provided encouraging results.

6 What can Cognitive Bias Modification Tell us About Neurobiological and Social Domains of Depression and Its Treatment?

The central role of biased processing in cognitive models of depression raises two key questions—what are the mechanisms by which individuals develop negative biases and how may treatments reverse these? These are the sort of questions which may be usefully tackled by considering the neurobiological systems which learn and maintain biases, and the environmental and social information which teach them. In other words these are questions which examine the overlap between the different domains described in Fig. 1. In the following section we discuss studies which have utilised CBM techniques to answer such questions.

6.1 Neural Systems that Learn and Maintain Cognitive Bias

Cognitive bias modification is a standardised, experimental method of altering cognitive bias by exposing participants to structured environmental information. It is therefore a useful tool in experimental studies investigating the mechanisms by which biases are produced. In one study that took this approach (Browning et al. 2010b) participants undertook a single session of either positive or negative attentional bias modification and then completed a task which required that they deploy their attention either in keeping with or in violation of the rules of the CBM they had received. Functional magnetic resonance imaging (fMRI) data was collected during this later task. A network of frontal regions including the lateral prefrontal cortices and rostral anterior cingulate cortex, which are known to be involved in the control of attention to emotional information (Vuilleumier 2005; Bishop 2007), was found to be specifically active when the rules of CBM were violated.

A similar pattern of activation was found in a later study which examined how participants learned to alter their attention in response to the frequency with which they were presented with negative information (Browning and Harmer 2012). These results suggest that the frontal circuitry identified attempts to maintain accurate predictions about the likelihood of encountering affectively relevant (in this case negative) stimuli and uses these predictions to direct the deployment of attention. In other words, attentional biases reflect the sort of information that an individual predicts he or she will encounter and attentional bias modification works by altering these predictions. The importance of the frontal control regions in flexibly altering attentional bias in these studies of CBM stands in contrast to the effects produced by antidepressant medication. Antidepressants have been shown to alter cognitive biases (Murphy et al. 2009); however, their most consistent neural effect is a reduction in activity within stimulus appraisal systems such as the amygdala (Harmer et al. 2006). It therefore seems that although both CBM and antidepressant medication are able to alter cognitive habits, their effects are mediated by actions on distinct neural systems. This raises a clinically interesting question about how the two forms of intervention will interact when administered concurrently (Browning et al. 2010a). Specifically, will the effects of the two types of treatment combine in an additive manner or will they interfere with each other? This was examined in a behavioural study which measured how treatment with positive attentional bias modification, the antidepressant medication citalopram or both interventions together influenced performance on a range of affective bias tasks (Browning et al. 2011). The results (Fig. 5) indicated that while both positive

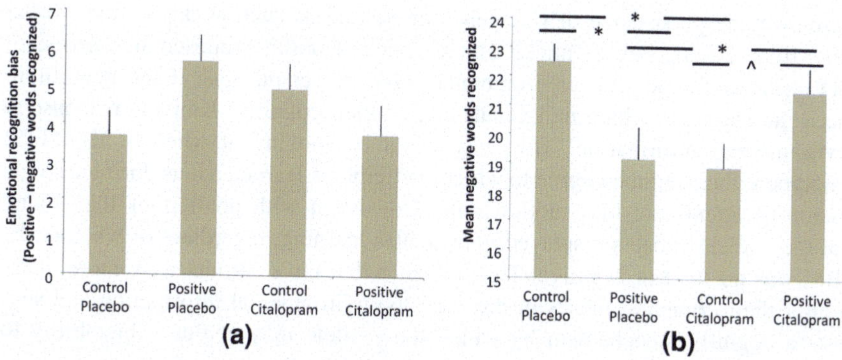

Fig. 5 The effect of positive attentional bias modification, the antidepressant citalopram or both together on memory for positive versus negative words in non-clinical participants. A control group (Control Placebo) which received no active intervention was compared with groups which received positive CBM and placebo tablets (Positive Placebo), neutral CBM and citalopram (Control Citalopram) or both interventions together (Positive Citalopram). Graphs show **a** positive bias in memory (number of positive relative to negative words recognised) and **b** the specific effect for memory of negative words. As can be seen, the individually administered interventions increase positive memory bias by reducing memory of negative words, whereas the combined intervention has no effect. * $= p < 0.05$, ^ $= p < 0.1$ for post hoc comparisons. Error bars represent standard errors of the mean

CBM and antidepressant medication, when administered individually, produced the expected increase in positive biases, when the two interventions were combined this effect was abolished. In other words, there was an interference effect whereby combining the interventions actually reduced their effectiveness. Overall, this work is consistent with neurocognitive models of attentional control (Vuilleumier 2005) which indicate that there are at least two neural systems that are able to produce alterations in cognitive biases. From a pragmatic point of view it suggests that manipulations of the two systems using interventions that individually reduce symptoms of depression can, at times, produce interference effects which would be expected to result in reduced efficacy.

6.2 Genetic Influences on Cognitive Bias Learning

A particularly active controversy in recent years has concerned the role of polymorphisms in the promoter region of the serotonin transporter gene (5HTTLPR). An influential paper suggested that the "s" allele of the gene was associated with risk of depression, but only in those who were exposed to adverse environments (Caspi et al. 2003). In other words, there was a gene × environment interaction. While this claim remains controversial (Munafo et al. 2009; Risch et al. 2009; Karg et al. 2011) balanced interactions (in which a genetic variant reduces fitness in some environments while increasing it in others) have been suggested as a reason for the persistence of risk genes in conditions such as depression (Keller and Miller 2006). Fox et al. (2011) reasoned that the proposed interaction of 5HTTLPR genotype with environmental adversity could arise if the gene influenced the degree to which individuals altered their cognitive habits in response to environmental information. This was tested by looking at whether 5HTTLPR genotype influenced the degree to which participants learned a bias during a single session of attentional bias modification. Consistent with predictions they found that the s allele carriers displayed greater bias learning, regardless of whether the CBM was positive or negative. This indicates that the s carriers were more likely to alter their cognitive biases in response to environmental information and suggests a cognitive mechanism by which the s allele may confer vulnerability to adverse environments.

6.3 Change in Mood Requires an Interaction Between Environment and Bias

As has been discussed above, CBM provides an experimental method for assessing the degree to which individuals alter their cognitive priorities in response to the environmental information they encounter. However, CBM studies also indicate that an individual's bias is not simply a function of their immediate environment, and that the effects of CBM may persist for weeks or months (Fig. 3; and Schmidt et al. 2009). In fact, some degree of bias persistence is an essential prerequisite of most of the cognitive models of depression, as without it there would be no mechanism for individual differences in illness vulnerability. One consistent finding of CBM studies (Fig. 3; MacLeod et al. 2002; Hallion and Ruscio 2011) is that the induction of an altered cognitive bias does not seem, on its own, to produce the full impact on symptoms of anxiety and depression. Rather, biases appear to confer a vulnerability or resilience to the effects of environmental events and thus the maximal change in symptoms following CBM only occurs when subjects have had sufficient exposure to environmental and (presumably) social information, which is parsed through the lens of the bias. This finding suggests that cognitive biases are one of the factors that determine an individual's reaction to his or her social situation and, as with the Fox et al. study of 5HTTLPR discussed above, indicates that cognitive biases are one route by which individual differences in neurobiological function may interact with a patient's environment.

7 Conclusions

This chapter has reviewed the recent work which has deployed CBM techniques to address questions on aetiology and treatment in depression. Although the field is relatively underdeveloped there are early signs both that bias modification may be a useful tool in exploring the cognitive aetiology of the illness and that it may even be an effective treatment itself. More generally, cognitive functioning links the neurobiological and social domains relevant to depression, and cognitive modification techniques present a rare opportunity to explore these interactions within an experimental setting.

Acknowledgments Michael Browning was supported by a grant from the Oxfordshire Health Services Research Committee (OHSRC); Emily Holmes was supported by a Wellcome Trust Clinical Fellowship (WT088217) and the National Institute for Health Research (NIHR). Simon Blackwell was supported by a grant from the Lupina Foundation awarded to Emily A. Holmes.

References

Austin MP, Mitchell P, Goodwin GM (2001) Cognitive deficits in depression: possible implications for functional neuropathology. Br J Psychiatry 178:200–206

Baert S, De Raedt R, Schacht R, Koster EH (2010) Attentional bias training in depression: therapeutic effects depend on depression severity. J Behav Ther Exp Psychiatry 41(3): 265–274

Beck AT (1967) Depression: clinical, experimental, and theoretical aspects. Harper and Row, New York

Beck AT, Rush AJ, Shaw BF, Emery G (1979) Cognitive therapy of depression. Guilford, New York

Bishop SJ (2007) Neurocognitive mechanisms of anxiety: an integrative account. Trends Cogn Sci 11(7):307–316

Blackwell SE, Holmes EA (2010) Modifying interpretation and imagination in clinical depression: a single case series using cognitive bias modification. Appl Cogn Psychol 24(3):338–350

Bouhuys AL, Geerts E, Gordijn MC (1999) Depressed patients' perceptions of facial emotions in depressed and remitted states are associated with relapse: a longitudinal study. J Nerv Ment Dis 187(10):595–602

Bradley BP, Mogg K, Williams R (1995) Implicit and explicit memory for emotion-congruent information in clinical depression and anxiety. Behav Res Ther 33(7):755–770

Browning M, Holmes EA, Charles M, Cowen P J, Harmer CJ (in press) Using attentional bias training as a cognitive vaccine against depression. Biol Psychiatry

Browning M, Holmes EA, Harmer CJ (2010a) The modification of attentional bias to emotional information: a review of the techniques, mechanisms, and relevance to emotional disorders. Cogn Affect Behav Neurosci 10(1):8–20

Browning M, Holmes EA, Murphy SE, Goodwin GM, Harmer CJ (2010b) Lateral prefrontal cortex mediates the cognitive modification of attentional bias. Biol Psychiatry 67(10):919–925

Browning M, Grol M, Ly V, Goodwin GM, Holmes EA, Harmer CJ (2011) Using an experimental medicine model to explore combination effects of pharmacological and cognitive interventions for depression and anxiety. Neuropsychopharmacology 36(13):2689–2697

Browning M, Harmer CJ (2012) Expectancy and surprise predict neural and behavioral measures of attention to threatening stimuli. Neuroimage 59(2):1942–1948

Caspi A, Sugden K, Moffitt TE, Taylor A, Craig IW, Harrington H, McClay J, Mill J, Martin J, Braithwaite A, Poulton R (2003) Influence of life stress on depression: moderation by a polymorphism in the 5-HTT gene. Science 301(5631):386–389

Clark L, Chamberlain SR, Sahakian BJ (2009) Neurocognitive mechanisms in depression: implications for treatment. Annu Rev Neurosci 32:57–74

Cowen PJ (2010) Not fade away: the HPA axis and depression. Psychol Med 40(1):1–4
Dandeneau SD, Baldwin MW (2004) The inhibition of socially rejecting information among people with high versus low self-esteem: the role of attentional bias and the effects of bias reduction training. J Soc Clin Psychol 23(4):584–602
Driessen E, Hollon SD (2010) Cognitive behavioral therapy for mood disorders: efficacy, moderators and mediators. Psychiatr Clin North Am 33(3):537–555
Eldara S, Ricona T, Bar-Haim Y (2008) Plasticity in attention: implications for stress response in children. Behav Res Ther 46(4):450–461
Eldar S, Apter A, Lotan D, Edgar KP, Naim R, Fox NA, Pine DS, Bar-Haim Y (2012) Attention bias modification treatment for pediatric anxiety disorders: a randomized controlled trial. Am J Psychiatry 169(2):213–230
Fox E, Zougkou K, Ridgewell A, Garner K (2011) The serotonin transporter gene alters sensitivity to attention bias modification: evidence for a plasticity gene. Biol Psychiatry 70(11):1049–1054
Gilboa-Schechtman E, Erhard-Weiss D, Jeczemien P (2002) Interpersonal deficits meet cognitive biases: memory for facial expressions in depressed and anxious men and women. Psychiatry Res 113(3):279–293
Gorwood P, Corruble E, Falissard B, Goodwin GM (2008) Toxic effects of depression on brain function: impairment of delayed recall and the cumulative length of depressive disorder in a large sample of depressed outpatients. Am J Psychiatry 165(6):731–739
Gotlib IH, Kasch KL, Traill SK, Joormann J, Arnow BA, Johnson SL (2004) Coherence and specificity of information processing biases in depression and social phobia. J Abnorm Psychol 113:386–398
Gotlib IH, Joormann J (2010) Cognition and depression: current status and future directions. Annu Rev Clin Psychol 6:285–312
Grey S, Mathews A (2000) Effects of training on interpretation of emotional ambiguity. Quart J Exp Psychol 53:1143–1162
Hakamata Y, Lissek S, Bar-Haim Y, Britton JC, Fox NA, Leibenluft E, Ernst M, Pine DS (2010) Attention bias modification treatment: a meta-analysis toward the establishment of novel treatment for anxiety. Biol Psychiatry 68(11):982–990
Hallion LS, Ruscio AM (2011) A meta-analysis of the effect of cognitive bias modification on anxiety and depression. Psychol Bull 137(6):940–958
Harmer CJ, Mackay CE, Reid CB, Cowen PJ, Goodwin GM (2006) Antidepressant drug treatment modifies the neural processing of nonconscious threat cues. Biol Psychiatry 59(9): 816–820
Holmes EA, Mathews A (2005) Mental imagery and emotion: a special relationship? Emotion 5(4):489–497
Holmes EA, Mathews A, Dalgleish T, Mackintosh B (2006) Positive interpretation training: effects of mental imagery versus verbal training on positive mood. Behav Ther 37(3):237–247
Holmes EA, Coughtrey AE, Connor A (2008a) Looking at or through rose-tinted glasses? Imagery perspective and positive mood. Emotion 8(6):875–879
Holmes EA, Lang TJ, Moulds ML, Steele AM (2008b) Prospective and positive mental imagery deficits in dysphoria. Behav Res Ther 46(8):976–981
Holmes EA, Mathews A, Mackintosh B, Dalgleish T (2008c) The causal effect of mental imagery on emotion assessed using picture-word cues. Emotion 8(3):395–409
Holmes EA, Lang TJ, Shah DM (2009) Developing interpretation bias modification as a "cognitive vaccine" for depressed mood: imagining positive events makes you feel better than thinking about them verbally. J Abnorm Psychol 118(1):76–88
Holmes EA, Mathews A (2010) Mental imagery in emotion and emotional disorders. Clin Psychol Rev 30(3):349–362
Ingram RE (1984) Toward an information-processing analysis of depression. Cogn Ther Res 8(5):443–477
Joormann J, Gotlib IH (2007) Selective attention to emotional faces following recovery from depression. J Abnorm Psychol 116(1):80–85

Joormann J, Talbot L, Gotlib IH (2007) Biased processing of emotional information in girls at risk for depression. J Abnorm Psychol 116(1):135–143

Joormann J, Hertel PT, LeMoult J, Gotlib IH (2009) Training forgetting of negative material in depression. J Abnorm Psychol 118(1):34–43

Karg K, Burmeister M, Shedden K, Sen S (2011) The serotonin transporter promoter variant (5-HTTLPR), stress, and depression meta-analysis revisited: evidence of genetic moderation. Arch Gen Psychiatry 68(5):444–454

Keller MC, Miller G (2006) Resolving the paradox of common, harmful, heritable mental disorders: which evolutionary genetic models work best? Behav Brain Sci 29(4):385–404 discussion 405–452

Kennedy N, Paykel ES (2004) Residual symptoms at remission from depression: impact on long-term outcome. J Affect Disord 80(2–3):135–144

Kosslyn SM, Ganis G, Thompson WL (2001) Neural foundations of imagery. Nat Rev Neurosci 2(9):635–642

Krebs G, Hirsch CR, Mathews A (2010) The effect of attention modification with explicit vs. minimal instructions on worry. Behav Res Ther 48(3):251–256

Lang TJ, Moulds ML, Holmes EA (2009) Reducing depressive intrusions via a computerized cognitive bias modification of appraisals task: developing a cognitive vaccine. Behav Res Ther 47(2):139–145

Lang TJ, Blackwell SE, Harmer CJ, Davison P, Holmes EA (2012) Cognitive bias modification using mental imagery for depression: developing a novel computerized intervention to change negative thinking styles. Eur J Pers 26:145–157

Lothmann C, Holmes EA, Chan SW, Lau JY (2012) Cognitive bias modification training in adolescents: effects on interpretation biases and mood. J Child Psychol Psychiatry 52(1): 24–32

MacLeod CM, Rutherford E, Campbell L, Ebsworthy G, Holker L (2002) Selective attention and emotional vulnerability: assessing the causal basis of their association through the experimental manipulation of attentional bias. J Abnorm Psychol 111(1):107–123

MacLeod CM, Koster EH, Fox E (2009) Whither cognitive bias modification research? Commentary on the special section articles. J Abnorm Psychol 118(1):89–99

Mathews A, Mackintosh B (2000) Induced emotional interpretation bias and anxiety. J Abnorm Psychol 109(4):602–615

Mathews A, MacLeod C (2005) Cognitive vulnerability to emotional disorders. Annu Rev Clin Psychol 1:167–195

Morina N, Deeprose C, Pusowski C, Schmid M, Holmes EA (2011) Prospective mental imagery in patients with major depressive disorder or anxiety disorders. J Anxiety Disord 25(8):1032–1037

Munafo MR, Durrant C, Lewis G, Flint J (2009) Gene X environment interactions at the serotonin transporter locus. Biol Psychiatry 65(3):211–219

Murphy SE, Yiend J, Lester KJ, Cowen PJ, Harmer CJ (2009) Short-term serotonergic but not noradrenergic antidepressant administration reduces attentional vigilance to threat in healthy volunteers. Int J Neuropsychopharmacol 12(2):169–179

Risch N, Herrell R, Lehner T, Liang KY, Eaves L, Hoh J, Griem A, Kovacs M, Ott J, Merikangas KR (2009) Interaction between the serotonin transporter gene (5-HTTLPR), stressful life events, and risk of depression: a meta-analysis. JAMA 301(23):2462–2471

Rogers MA, Kasai K, Koji M, Fukuda R, Iwanami A, Nakagome K, Fukuda M, Kato N (2004) Executive and prefrontal dysfunction in unipolar depression: a review of neuropsychological and imaging evidence. Neurosci Res 50(1):1–11

Rude SS, Wenzlaff RM, Gibbs B, Vane J, Whitney T (2002) Negative processing biases predict subsequent depressive symptoms. Cogn Emot 16:423–440

Schmidt NB, Richey JA, Buckner JD, Timpano KR (2009) Attention training for generalized social anxiety disorder. J Abnorm Psychol 118(1):5–14

Siegle GJ, Ghinassi F, Thase ME (2007) Neurobehavioral therapies in the 21st century: summary of an emerging field and an extended example of cognitive control training for depression. Cogn Ther Res 31:235–262

Teasdale JD (1988) Cognitive vulnerability to persistent depression. Cogn Emot 2(3):247–274

Vuilleumier P (2005) How brains beware: neural mechanisms of emotional attention. Trends Cogn Sci 9(12):585–594

Watkins ER, Taylor RS, Byng R, Baeyens C, Read R, Pearson K, Watson L (2011) Guided self-help concreteness training as an intervention for major depression in primary care: a Phase II randomized controlled trial. Psychol Med [epub ahead of print]

Williams JMG, Watts FN, MacLeod CM, Mathews A (1997) Cognitive psychology and emotional disorders. Wiley, New York

Williams JM, Barnhofer T, Crane C, Herman D, Raes F, Watkins E, Dalgleish T (2007) Autobiographical memory specificity and emotional disorder. Psychol Bull 133(1):122–148

Woud ML, Holmes EA, Postma P, Dalgleish T, Mackintosh B (2011) Ameliorating intrusive memories of distressing experiences using computerized reappraisal training. Emotion [epub ahead of print]

Part IV
Treatment

Pharmacological Treatment of Unipolar Depression

Ian M. Anderson

Abstract Antidepressants were first developed serendipitously 60 years ago and gave rise to the monoamine hypothesis of depression and antidepressant action which has persisted in various forms ever since. Although we have made huge strides in our understanding of the pharmacology of antidepressants, and in the neuroscience of depression, our current antidepressants have changed little since the original drugs. In this chapter I first review some controversies in the use of antidepressant drugs including whether they actually work, and then go on to describe the current state of our clinical use of antidepressants, looking both at the principles and practice of treatment and reviewing the evidence for efficacy, tolerability and safety in acute and sequenced treatments. I finally briefly consider future directions and the aspiration of developing more effective antidepressants.

Keywords Depression · Antidepressants · Treatment · Efficacy · Tolerability · Safety

Contents

I. M. Anderson (✉)
Neuroscience and Psychiatry Unit, Manchester Academic Health Science Centre, University of Manchester, Room G809, Stopford Building, Oxford Road, Manchester, M13 9PT, UK
e-mail: ian.anderson@manchester.ac.uk

Curr Topics Behav Neurosci (2013) 14: 263–289
DOI: 10.1007/7854_2012_208

Published Online: 28 April 2012

1 Introduction

Although psychoactive drugs have been used since antiquity, clinical psychopharmacology as we currently understand it is has only developed over the last 60 years. Drugs that appeared to specifically target depression were discovered by chance and astute observation, on a background of scientific advances in the understanding of biological processes and in the synthesis of new chemical compounds. The discovery that treatment with the antituberculous drug, iproniazid, and the chlorpromazine derivative, imipramine, led to improvement in 65–70 % of patients with depression led to a search for an understanding of the possible mechanisms. Iproniazid's ability to antagonise what became known as monoamine oxidase was discovered early on, with the discovery that imipramine could reverse the effects of reserpine shortly after (Ban 2001). This set the stage for the development of the monoamine hypothesis of depression in its various versions which has held sway over subsequent decades and guided, or probably restrained, drug development (Ban 2001). Subsequent developments in the production of new drugs has depended on the growth of interrelated industrial, business and marketing processes leading to the current reliance on major pharmaceutical companies for advances in the field (Preskorn 2010). This has led to considerable criticism in recent years, with increasing distrust of the marketing activities of 'big pharma' and a questioning of the transparency, validity and interpretation of clinical drug trials used for registration purposes. It has been argued that this resulted in a 'me-too' emphasis in drug development leading to failure to develop novel agents that escape the pharmacological shadow of the first drugs (Preskorn 2011).

Some aspects of clinical trials in depression such as internal validity, consistency, statistical analysis and reporting have improved since the first studies in the 1950s (Brunoni et al. 2010). However, the primary endpoints have not altered since the introduction of the Hamilton Depression Rating Scale (HDRS), and we lack useful markers of disease or treatment response which is a serious challenge to the hope of being better able to match treatment to individual circumstances or illness profile (Simon and Perlis 2010). A number of factors are likely to contribute to this including the limitations of our assessment tools, the large non-specific (placebo) component to treatment response and, more fundamentally, the current classification systems of depression which offer only heterogeneous symptom clusters (Gelenberg et al. 2008) and lack a foundation in neuropsychobiology. In addition, the signal to noise ratio in studies has decreased over time with a growth of placebo response rates and a greater proportion of 'failed' trials (Gelenberg et al. 2008). This has been fertile ground for attacks on the usefulness of antidepressants for the majority of cases of depression, especially for those presenting in primary care where most people are treated (Kirsch et al. 2008; Middleton and Moncrieff 2011). Paradoxically this is occurring at a time when the prescription of antidepressants is growing fast (Pratt et al. 2011) and underlines the need to have better tools to target treatment.

In this chapter I will discuss some of the current controversies in the use of antidepressants to treat depression and then outline current approaches to treatment finishing with briefly considering future developments.

2 Some Current Controversies

2.1 Do Antidepressants Work in Treating Depression?

The last decade has seen a challenge to the assertion that antidepressants are useful in the treatment of depression. There are essentially three main versions of the challenge that are amenable to empirical examination (I will leave aside here more emotive philosophical or moral arguments about using antidepressants to treat mental 'distress' and concerns about their widespread use in society). The arguments are (1) that antidepressants do not work better than placebo once publication bias has been taken into account, (2) if they do work it is due to an augmented placebo effect because of failed blinding/side-effects and (3) even if there is a specific effect it is too small to be clinically useful and/or the risks and disadvantages outweigh the small benefits.

There is not space here to do the arguments full justice but (1) is easily countered by systematic reviews that include unpublished studies and show clear benefit for antidepressants over placebo, although the effect size is reduced to about 0.3 (Kirsch et al. 2008; Turner et al. 2008). Argument (2) is difficult to address directly but implausible for a number of reasons. First, a meta-analysis of randomised controlled trials (RCTs) of tricylic antidepressants against 'active' placebo (anticholinergic drugs), in order to match the adverse effects of the treatment arms and prevent unblinding, still showed a benefit for antidepressants

(Moncrieff et al. 2004), while studies comparing tricyclic antidepressants against the better tolerated selective serotonin reuptake inhibitors (SSRIs) show equal efficacy (Anderson 2001). Conversely differences in efficacy have been demonstrated between second-generation antidepressants with similar side-effect profiles (see Sect. 4.2). Second it is difficult to explain any differential effect of depression severity on response to placebo and antidepressants using this argument. More severely ill patients respond less well to placebo but the response to antidepressants is maintained across the severity range (e.g. Kirsch et al. 2008); if the response to antidepressants is simply an enhanced placebo response the drug-placebo difference should be as good, or even larger, in milder depression (Fountoulakis and Moller 2011). Finally the robust relapse prevention data (Geddes et al. 2003) are difficult to explain away as a placebo effect given that patients are currently well (however see Sect. 2.2).

Perhaps the most challenging argument is (3). Meta-analyses at the level of studies (Kirsch et al. 2008) and individual patient data (Fournier et al. 2010) using continuous measures of improvement suggest rather small, and it is argued clinically questionable, differences between drug and placebo at mild and moderate severity depression, i.e. for the vast majority of patients, only reaching an effect size of 0.5, or an HDRS difference of 3, at HDRS scores of 25–28 (Kirsch et al. 2008; Fournier et al. 2010). There are a number of complications in addressing this, including debate about the accuracy of the figures (Fountoulakis and Moller 2011), their generalisability (they are based on a relatively small number of studies/patients with considerable measurement error) and how meaningful it is to apply, a strict threshold of clinical importance (Anderson and Haddad 2011). The method of analysis is also important. A large European meta-analysis of 56 published and unpublished RCTs of SSRIs and serotonin and noradrenaline reuptake inhibitors (SNRIs) in regulatory submissions analysed outcomes by a responder analysis (i.e. the proportion achieving a 50 % or greater reduction in HDRS) (Melander et al. 2008). Their results appear to challenge the previous analyses as there was only a non-significant marginal effect of severity on the drug-placebo difference in response across the range from mild to severe depression (15–20 %) and few would argue, that a number needed to treat (NNT) between 5 and 7 lacks clinical significance. However, this has been countered by the argument that response rates derived from continuous measures provide only an illusion of efficacy (Kirsch and Moncrieff 2007). Response defined arbitrarily as a threshold of improvement in a continuous measure means that the difference between responder and non-responder numbers reflect those who have moved over the threshold rather than a true difference in those improving and those not improving. The majority of people will cluster around the mean improvement and apparent large differences in response proportions may reflect only a small difference in mean HDRS reductions in the two treatment arms. In addition, in milder severity depression, the difference in mean HDRS reductions will be smaller than in more severe depression so that a 15 % difference in response rates may only reflect a difference of 1 or 2 HDRS points or even less.

This argument is however based on a normal distribution in the change in HDRS scores. If improvement is instead more bimodal, with a 'responder group' that have a substantial improvement and a 'non-responder group' tending to change rather little, the argument weakens considerably. A recent reanalysis of data comparing escitalopram and placebo, using a mixture model to identify subgroups, found that a bimodal distribution fitted the data much better than a unimodal modal (explaining 60 % of the variance compared with 6 %) (Thase et al. 2011). The Montgomery–Åsberg Depression Rating Scale (MÅDRS) dropped from a mean of 30 to 10 for those who benefited and only from 30 to 25 for those who did not. The difference in proportion of benefiters and nonbenefiters was 14 % (NNT 7) for the mild to moderately ill patient group (mean initial MÅDRS 26) and 23 % (NNT 5) for more severely ill patients (mean MÅDRS 33). It is important also to take into account the relapse prevention data where there is an extremently robust effect of antidepressants for those with a high risk of recurrence (Geddes et al. 2003) although interpretation of these data have been challenged (see Sect. 2.2).

What do we take away from this debate? Given the increase in the use of antidepressants it must be right to subject the claims of their efficacy and usefulness to rigorous inspection. We are blessed with abundant data but cursed by potential confounds, endless analysis, reanalysis and subanalysis of the data not to mention vested interests. In my view the current evidence does support a clinically useful benefit from antidepressants for many depressed patients but there remains some uncertainty at the mild to moderate end of the spectrum in acute treatment. The threshold where benefit outweighs disadvantages is not fixed and has to be judged against individual patient needs together with the appropriateness and availability of alternative or conjunctive treatments (Anderson and Haddad 2011).

2.2 Sensitisation to Relapse

The robust efficacy of antidepressants in preventing relapse in patients with highly recurrent depression (Geddes et al. 2003) has been a cornerstone of the argument for antidepressant efficacy and usefulness. There have previously been challenges to the interpretation of these relapse prevention RCTs on the basis of the potential confound from discontinuation symptoms either being mistaken for, or even precipitating, depressive relapse (Moncrieff 2006). Recently there has been an even stronger version of this argument proposing that antidepressants bring about oppositional tolerance in long-term use in which homeostatic mechanisms counter the pharmacological effect of antidepressants. This means that when they are stopped there is an 'overshoot' pushing monaminergic neurotransmission in the opposite direction resulting in the triggering of a depressive episode (Andrews et al. 2011). In support of their thesis the authors compared the relapse rate in the placebo arm in standard relapse prevention studies following remission on an antidepressant (antidepressant-placebo) with placebo continuation studies in which patients had remitted on placebo acute treatment (placebo-placebo). They found a 42–56 % relapse rate depending on drug class in the antidepressant-placebo studies compared with only 25 % on placebo-placebo. They report that this remained after correcting for illness characteristics and differential drop-outs (substantially more in the placebo-placebo group). Furthermore, the relapse rate in the antidepressant-placebo group was predicted by the degree of monoamine perturbation (specifically serotonin and noradrenaline) caused by individual antidepressants in the medial prefrontal cortex of rodents using microdialysis (Andrews et al. 2011). Separate evidence indicates that the shorter the length of time without antidepressants in remitted patients the higher the risk of relapse with tryptophan depletion (Ruhe et al. 2007) consistent with the oppositional tolerance hypothesis.

This is an intriguing hypothesis with fundamental implications for our treatment of depression. However, in spite of attempting to correct for patient characteristics, the crucial factor of the underlying risk of relapse is poorly reported in studies, so that it remains highly likely that the population entered into the continuation studies were significantly different to the relapse-prevention study patients who are specifically chosen for a high relapse risk. Uncertainties in extrapolation from animal studies, selection of brain area and study selection for quantifying monoamine perturbation also warrants caution in the pharmacological correlation seen. If the hypothesis is true then one might predict that the longer the period of antidepressant treatment before randomisation to placebo the greater would be the risk of relapse on placebo; in fact the opposite occurs (Geddes et al. 2003). Treatment with antidepressants in healthy volunteers might be also expected to bring about oppositional tolerance and cause vulnerability to mood lowering by tryptophan depletion but this does not happen, at least with treatment over 6 weeks (Barr et al. 1997) which should be long enough to result in adaptive changes. A naturalistic follow-up study of young patients randomised to antidepressant, cognitive behaviour therapy, the combination or placebo did not find that original treatment allocation predicted relapse over a 5-year follow-up (Curry et al. 2011).

At present, therefore, the evidence is not compelling that this type of sensitisation to relapse occurs with antidepressants but, given the importance of the question, a properly designed prospective study to investigate it is warranted.

2.3 Suicide

Over the last two decades concerns have been raised about whether SSRIs (and other antidepressants) might actually cause an increased risk of suicidal acts and completed suicide during treatment. This has been particularly pertinent with regard to their use in adolescents where their benefit in treating depression has been questioned. Resolving the issue has been complicated by a number of factors: the background raised risk of suicidal acts in depression, detecting an increased risk in the context of a decrease in suicidality by effective treatment, the definition of suicidality in children and adolescents where impulsive and self-harm acts are relatively high and the inadequacy of data from studies which have not been designed to address this question. To a large extent the controversy has been replaced by a degree of consensus (Moller et al. 2012). It is now known that suicidal acts are most common before treatment starts (and this applies to psychological as well as drug treatment) and decrease through the early treatment period, so there is indeed a raised risk of suicidal behaviour in the early period of treatment. This is however predominantly due to the nature of the illness itself (Simon and Savarino 2007) with the risks of self-harm greatest in younger age-groups, especially adolescents (Simon and Savarino 2007). On top of this background risk, antidepressant treatment generally (not just SSRIs), compared with placebo, appears to slightly increase the risk of suicidal ideation or suicidal behaviour in adolescents and younger adults up to age of 25 years (number needed to harm about 100 for adolescents, 200 for younger adults). However, in adults older than 25 years antidepressants have a neutral or protective effect which increases with age (number needed to benefit about 160 for those aged 65 and older) (Hammad et al. 2006; Stone et al. 2009). The reasons for this are not understood with suggestions including a link between suicidality and nonresponse to treatment, a higher rate undiagnosed bipolar disorder in younger subjects or an age-related effect of changes in neurotransmitter, especially serotonin, function on impulsive behaviours, activation/drive-enhancement or side-effects such as akathisia (Moller et al. 2012). It is important to recognise, however, that that these data cannot resolve the question as to whether antidepressants affect the risk of dying as the rate of completed suicide in studies is so low (Stone et al. 2009).

The implications of these findings are that all patients should be carefully assessed and monitored for suicide risk before and during treatment, and that particular care should be taken with younger patients who should be monitored more frequently (National Institute for Health and Clinical Excellence 2009a). The risk–benefit balance when prescribing antidepressant treatment in younger patients needs to take into account that there might be a slightly higher risk of suicidal behaviour with antidepressants early in treatment.

2.4 Emotional Blunting

Reports of a blunting of both positive and negative emotions with SSRIs go back 20 years (Opbroek et al. 2002) but there has been relatively little exploration of the phenomenon. Two small studies, one observational (Opbroek et al. 2002) and one qualitative (Price et al. 2009), have concluded that at least some patients do experience emotional blunting or detachment which they attribute to the medication and can distinguish from the effects of depression. The consequences varied between individuals but some found it sufficiently problematic to make them consider stopping, or actually stop, treatment (Price et al. 2009). Some neuroimaging results have provided support for a general reduction of neural responses to both pleasant and unpleasant stimuli with SSRIs e.g. McCabe et al. (2010).

This is a potentially previously unrecognized side-effect of SSRIs, and other serotonergic antidepressants (Price et al. 2009) that may also have implications for our understanding of how antidepressants alter emotional processing. However evidence to date is based on small and largely self-selected samples so further investigation is needed to determine if it is a true effect of treatment and, if so, what proportion of patients is affected.

3 Principles and Practice of Treatment

The obvious aim of treatment is to return the sufferer from depression to being 'well' again. However, the patient may have different expectations and emphasis compared with the doctor, and goals may need to be tempered by practical considerations and unwanted effects of treatment. Optimising treatment on an individual basis involves careful assessment, application of the best evidence and sensitivity to the wishes, concerns and beliefs of the patient. This necessarily involves a mutual enterprise based on negotiation which recognises the different roles of doctor and patient and the rights of the patient to make informed choices. I consider here some factors that need to be considered in treating patients with antidepressants but do not aim to be comprehensive and will not discuss in any detail the assessment of risk and the need to adapt drug treatment in the face of factors such as age, medical and psychiatric comorbidity, pregnancy and breast-feeding. These are all important but outside the scope of this chapter.

3.1 Goals of Treatment

The emphasis that clinicians tend to have on symptoms and their improvement does not capture what patients desire from treatment (Zimmerman et al. 2006) which includes an improvement in their overall well-being or quality of life (QoL). In addition to the presence of depressive symptoms this is impacted by the burdens of treatment (such as side effects) as well as wider aspects of functioning, optimism and engagement in social and occupational activities. In addition, much of the focus has been on short-term outcomes whereas depression often has a chronic course with persistent symptoms and relapses. This requires consideration of longer term treatment in which the acceptable balance of benefit to acceptability is likely to be very different; readily illustrated by the different short- and long-term impact of two common side effects of antidepressants, sexual impairment and weight gain.

Although there are many measures that attempt to capture QoL, none have received sufficient general acceptance to become primary, or even required, outcomes in clinical trials. In practice symptom, function and QoL measures tend to improve together and persisting depressive symptoms are associated with ongoing functional and QoL impairments as well as increased risk of relapse (Keller 2003). This has led to an increasing emphasis on absence, or near absence, of symptoms ('remission') rather than significant improvement ('response') as the goal of treatment (Keller 2003). Remission is usually measured by a minimum number and severity of depressive symptoms (e.g. 7 or less on the observer-rated 17-item HDRS) and was adopted as the primary outcome in the large Sequenced Treatment Alternatives to Relieve Depression (STAR*D) (Rush et al. 2006a). This higher bar, however, emphasises how disappointing the outcomes of treatment are, with typically only about a third of depressed patients in the STAR*D trial achieving remission with initial antidepressant treatment compared with about 50 % of patients meeting criteria for a response (Rush et al. 2006a).

3.2 When to Treat with Antidepressants

Treatment with antidepressants, as with all drug treatments, needs to be based on a balance of benefits and risks. The risks need to be judged on an individual basis, for example medical comorbidity or pregnancy can raise threshold for antidepressant drug treatment on the basis of safety. The balance of evidence, discussed above, supporting an increasing active drug-placebo difference in outcome with increasing severity of depression (Angst and Stabl 1992; Kirsch et al. 2008), has led recent guidelines (Anderson et al. 2008; National Institute for Health and Clinical Excellence 2009a) to restrict the primary recommendation for antidepressants to depression with moderate of greater severity while acknowledging that a trial of antidepressants is warranted for milder degrees of depression that have failed to respond to non-drug treatments. The British Association for Psychopharmacology guidelines suggest a dimensional approach with a DSM-IV moderate major depressive episode and a duration of 3 months as a guide to where clinically important benefit starts to be seen (Anderson et al. 2008). Another way of putting this is to consider using antidepressants in situations where spontaneous improvement, or response to nonspecific measures, are unlikely. One of the questions underlying the increase in prescribing of antidepressants is to what degree this represents 'inappropriate' prescribing (i.e in those who would get better spontaneously or with simple psychosocial interventions) as opposed to an increasing awareness and appropriate treatment of significant depression. The truth is likely to be a combination of both. The poor response to antidepressants alone, however, emphasizes that drugs always need to be combined with psychosocial approaches including specific psychotherapies, particularly if there is insufficient response or high risk of relapse.

3.3 How to Treat with Antidepressants

Paradoxically the discussions about personalising medicine come at a time when treatment has never been so impersonal and continuity of care so fragmented. The nonspecific or placebo component to treatment response highlights the importance of how, as well what, treatment is given. Expectations (patient and clinician) play a considerable role in outcome. For example when patients recruited into RCTs know they are going to receive an antidepressant in non-placebo-controlled comparative trials the remission rate is on average 50 % higher than for patients on active treatment in placebo-controlled trials (when they know they have a 50:50 chance of receiving placebo) (Sinyor et al. 2010). Patient preference not surprisingly influences whether a treatment is accepted (King et al. 2005) but the evidence that patient preference when receiving active treatment influences outcome is less established. However, recent studies in depression have support a better outcome when the treatment type (drug or psychological treatment) matches the one preferred (Lin et al. 2005; Kocsis et al. 2009; Mergl et al. 2011). Interestingly this might be a stronger effect for psychological treatment than for antidepressant drugs (Mergl et al. 2011). As well as assessing, and matching where possible, expectations to treat the process is also important. More frequent follow-up assessments for patients receiving placebo in RCTs increases the degree of improvement and may account for as much as 40 % of the placebo response (Posternak and Zimmerman 2007). Undergoing standardised assessments with implementation of time-critical treatment changes according to an algorithm improves response for those receiving active treatment (Trivedi et al. 2004). Patient engagement in treatment with feedback of progress from assessments has been shown to improve symptomatic outcome, at least with psychological treatment (Newnham et al. 2010).

The implications from research are therefore that simply providing a prescription for an antidepressant is not enough. Matching or influencing patient preference, and the process of management including assessments, monitoring and probably feedback/engagement are at least as important as the pharmacological agent if treatment benefit is to be maximised.

3.4 Duration of Treatment

Preventing relapse and recurrence is a high priority and the findings from STAR*D (Rush et al. 2006a) show just how high this risk can be. The effectiveness of antidepressants in preventing relapse is well established (Geddes et al. 2003) (but again see Sect. 2.2) with a consistent large benefit in reducing relapse to about a third relative to the risk of relapse on placebo. This is an especially important effect in those with higher risk of relapse. When considering the duration of treatment with antidepressants therefore it is important to take a dimensional view and match the potential benefit to an individual's risk of relapse. This involves taking into account the history of number, frequency and severity of previous relapses, other risk factors such as psychosocial stress and residual symptoms, protective factors such as change in circumstances or receiving specific psychotherapies targeting relapse and, often forgotten, the consequences of relapse. In effect we need to get rid of the idea of fixed durations of treatment and to use the evidence to individualise treatment according to need. In practice this means a minimum duration of 6–9 months treatment in those with minimal risk factors extending to indefinite treatment in highly recurrent depression (Anderson et al. 2008). This one of the most important areas where we can personalise treatment at present.

4 Choice of Antidepressant

4.1 Current Antidepressants

As stated in the introduction, the drugs currently marketed as antidepressants show the legacy of their first discovery and all act primarily on monoamine systems. The latest drug to be marketed, agomelatine, is unusual in that one of its main actions is as a melatonin agonist, although it is also a 5-HT_{2C} antagonist, a property it shares with other antidepressant drugs such as mirtazapine and trazodone (Hickie and Rogers 2011). Melatonin, although a monoamine, is not one previously strongly implicated in antidepressant action although its role in circadian rhythms and sleep means it has links to theories relating disordered chronobiology with depression (Hickie and Rogers 2011). Theories of the mechanism of action of antidepressants have attempted to move beyond the monoamines, ranging from changes in neuroplasticity and immune function to neuropeptides and neurohumeral effects (Tanti and Belzung 2010). To date this has not radically changed the landscape of drugs available for treatment and the link between pharmacology and the alleviation of depression remains tantalizingly obscure. The link between pharmacology and adverse or unwanted effects is rather better understood.

The classification, or nomenclature, of antidepressants is another problem that does not help in their understanding or clinical use. It has grown in a random way with inconsistencies as to whether they are named according to structure (e.g. tricyclic antidepressant), history (e.g. second generation), mode of action (e.g. SSRI), mode of action with a marketing spin (e.g. noradrenaline and specific serotonin antagonist or NaSSA) and those without a clear nomenclature as yet (e.g. agomelatine). The World Health Organisation (WHO) system is based on monoamine pharmacology and is limited because most recent antidepressants, and putative antidepressants, are simply listed under 'other antidepressants' (Nutt 2009). Nutt (2009) has suggested that a more meaningful approaches to nomenclature might be to include neurotransmitter targets and site of action (reuptake sites, receptors, enzymes) combined with a more generic classification as to whether they act at single or multiple sites. Unfortunately our lack of understanding of how pharmacology is translated into efficacy means that all current attempts at classification are going to either have to squeeze atypical or novel drugs into our current monoamine scheme (e.g. lithium as a functional inhibitor of 5-HT_2-mediated function, S-adenosylmethionine as a modulator of monoamine neurotransmission), or continue the ramshackle construction depending on empirical evidence (such as adding actions at melatonin, glutamate or glucocorticoid receptors).

Although the basic pharmacology of marketed antidepressants has not changed radically over the last 60 years the pattern of usage has altered. The search for 'cleaner' antidepressants with fewer side-effects and greater safety in overdose resulted in the marketing of a series of single action drugs, the SSRIs, in the 1980s. Over the next decade these became the standard first-line treatment, especially once they became available as generic drugs and the acquisition costs reduced. Since then the focus of clinical use has shifted to the current dominance of newer antidepressants over the original tricyclic antidepressants and irreversible monoamine oxidase inhibitors (MAOIs).

4.2 *Efficacy of Antidepressants*

Arguably more ink has been used (or wasted) on the topic of relative efficacy and/or tolerability of antidepressants than on any other aspect of their use. This has of course been driven strongly by marketing needs, but also the genuine recognition that, given the limited efficacy and tolerability of individual antidepressants, it is important to try and choose the best drug for a patient. In addition, differences between drugs have a potential heuristic value in the search for understanding mechanisms and improving treatments. Clinical experience makes it difficult to believe that antidepressant drugs do not have important differences at an individual patient level, but this has proved difficult to show at a group level in RCTs, and even more difficult to demonstrate differences that are clinically important.

In terms of efficacy there is limited evidence from meta-analyses of RCTs that there might be small advantages (typically 5–10 % difference in response rates, effect sizes of 0.15–0.2) for amitriptyline over SSRIs, for venlafaxine (at doses of 150 mg or more) over SSRIs and for escitalopram 20 mg over other SSRIs including citalopram (Anderson 2001; Smith et al. 2002; Kennedy et al. 2009). The advantage to venlafaxine, an SNRI, over SSRIs may reflect a more general small efficacy benefit for drugs increasing both serotonin and noradrenaline function (SNRIs, mirtazapine, moclobemide) over a single action at the serotonin reuptake site (Papakostas et al. 2007b). A network, or multiple-treatments, meta-analysis of 12 newer antidepressants, which was able to include direct and indirect drug comparisons, was largely consistent with this proposal with mirtazapine, escitalopram, venlafaxine and sertraline being ranked as the top four most effective drugs (in that order), and significantly so compared with fluoxetine, paroxetine, fluvoxamine and reboxetine; the last being less effective than all other antidepressants (Cipriani et al. 2009). A more recent meta-analysis using both direct and indirect comparisons found fewer differences between drugs (Gartlehner et al. 2011). In head-to-head comparisons, sertraline and venlafaxine were more effective than fluoxetine, and escitalopram more effective than citalopram, but the differences were modest (about 5 % difference in response rate). Indirect comparisons only found significant differences in favour of escitalopram over both fluoxetine and duloxetine. The question raised over the efficacy of reboxetine has also led to further attempts to clarify this with the predictable result that one meta-analysis including unpublished data found it did not separate significantly from placebo and was less effective than SSRIs in leading to remission (Eyding et al. 2010) but a previous meta-analysis of essentially the same data against SSRIs concluded they did not differ in efficacy (Papakostas et al. 2008b). These differing results highlight how much outcomes of meta-analyses are dependent on their methodology and suggest that any differences between drugs are not robust. In addition if antidepressants do differ in efficacy it is hard to find a straightforward pharmacological property to explain this: in particular why sertraline and escitalopram might be more effective than other SSRIs. For the former an additional effect in inhibiting dopamine reuptake may be a factor (Carrasco and Sandner 2005), and for the highly selective drug, escitalopram, its unique binding properties at the serotonin transporter could be relevant (Zhong et al. 2012), but the truth is that we have yet to really understand the basis for differences in antidepressant efficacy.

The limited efficacy of antidepressants, and the rationale that multiple pharmacological actions may be more effective than a single action, has lead to the question as to whether combining treatment from the start may be more effective than using a single antidepressant. Two recent RCTs attempting to answer this have reached opposite conclusions. Blier et al. (2010) reported that mirtazapine combined with fluoxetine, venlafaxine or bupropion was greatly superior to fluoxetine monotherapy with equal tolerability and an impressive doubling of remission rates from 25 % to over 50 %. Limitations of this study were the lack of a mirtazapine monotherapy arm, low numbers assigned to each treatment arm and surprisingly few patients remitting on fluoxetine alone. A larger RCT comparing escitalopram monotherapy with escitalopram plus bupropion and venlafaxine plus mirtazapine found no difference in efficacy between treatments but more adverse events with the venlafaxine-mirtazapine combination (Rush et al. 2011). There is therefore a lack of evidence to recommend combination antidepressant treatment first-line at present.

Although overall efficacy might not differ substantially between antidepressants it remains possible that there may be differences between antidepressants in other aspects of efficacy. However, in a comprehensive meta-analysis of newer antidepressants (Gartlehner et al. 2011) it was not possible to find any good evidence for clinical predictors of differential response to specific antidepressants. There is some evidence that mirtazapine may act more quickly than SSRIs although it is not clear that this generalizes to other antidepressants such as venlafaxine and tricyclic antidepressants (Anderson 2001; Gartlehner et al. 2011) and among the SSRIs fluoxetine may have a slower onset of action (Anderson 2001); if these are correct they may reflect pharmacodynamic and pharmacokinetic properties respectively given mirtazapine's direct receptor action and fluoxetine's long half-life.

4.3 Tolerability of Antidepressants

Whereas the efficacy differences between antidepressants are subtle, the differences in side-effects and tolerability are clearer. In RCTs more patients stop treatment due to adverse-effects on tricyclic antidepressants than on SSRIs, although the absolute difference is only about 3–4 % (Anderson 2001). Among the more recent antidepressants in a network meta-analysis treatment acceptability, as measured by patients dropping out of treatment for all reasons, was greatest for escitalopram, citalopram, sertraline and bupropion and worst for reboxetine, duloxetine and milnacipran (an SNRI). In direct comparisons, escitalopram and sertraline led to significantly fewer discontinuations than did duloxetine, fluvoxamine, paroxetine, reboxetine and venlafaxine (Cipriani et al. 2009).

However, equally as important as the rather crude measure of tolerability measured by dropouts, is the side-effect profile of different drugs (see Anderson et al. (2008) for a summary table). SSRIs compared with TCAs cause more gastrointestinal (nausea, diarrhoea, anorexia) and stimulant (agitation, anxiety and insomnia) side effects, whereas TCAs cause more anticholinergic side effects (dry mouth, constipation, blurred vision), sweating and dizziness (Anderson 2001). In a systematic review of newer antidepressants, bupropion caused significantly less sexual dysfunction than SSRIs; venlafaxine more nausea and vomiting than SSRIs; sertraline more diarrhoea than other SSRIs, venlafaxine, bupropion, mirtazapine and nefazodone; traxodone more sedation than SSRIs, venlafaxine and bupropion; and mirtazapine more weight-gain than SSRIs and venlafaxine (Gartlehner et al. 2011). Reboxetine causes more anticholinergic-like side-effects and insomnia than SSRIs (Papakostas et al. 2008b). Treatment-emergent sexual dysfunction is an often under-reported adverse effect that has important implications for patients. A meta-analysis found that SSRIs, venlafaxine, the TCA imipramine and the MAOI phenelzine caused greater sexual dysfunction than placebo (ranging from 26 to 80 %) with the worst being sertraline and venlafaxine and the least escitalopram and fluvoxamine (Serretti and Chiesa 2009). Antidepressants with a similar rate of sexual dysfunction to placebo included agomelatine, mirtazapine and moclobemide (Serretti and Chiesa 2009). The discontinuation syndrome on stopping newer antidepressants seems most common with paroxetine and venlafaxine and least with fluoxetine (Gartlehner et al. 2011).

4.4 Safety of Antidepressants

Depression is associated with an increased risk of eventual suicide ranging from a fourfold increase over the general population risk (0.5 %) in outpatient populations, to a 16-fold increased risk in patients hospitalized with suicidal attempts or ideation (Bostwick and Pankratz 2000). The considerable risk of death from overdose of tricyclic and MAOI antidepressants was one driver for the development of newer drugs that would be safer if taken in overdose. The fatal toxicity index (FTI) has been used as measure of lethality in overdose, and describes the mortality rate per 1 million prescriptions of a drug. As a crude measure it is liable to a number of potential confounds, including the risk of overdose in the population being treated with a specific drug which might inflate or mask the true toxicity; this is likely to have contributed to the high FTI for venlafaxine (Rubino et al. 2007). A recent study used the case fatality index (CTI), which is the rate ratio for poisoning mortality relative to self-poisoning (deaths/100,000 self-poisonings), to control for this confound. Tricyclic antidepressants were the most toxic (rate ratio 8.6–23.3) and greater than venlafaxine (rate ratio 2.5) and mirtazapine (rate ratio 1.9) which in turn were higher than SSRIs (rate ratio 0.5) (Hawton et al. 2010). Of the SSRIs citalopram was the most toxic in overdose (rate ratio 1.1) compared with other SSRIs (rate ratios 0–0.4) (Hawton et al. 2010). The finding with citalopram is consistent with recent data showing that it affects cardiac conduction with increases in the QT interval at higher treatment doses (Food and Drug Administration 2011) and which has led to dose restrictions for citalopram and escitalopram.

Antidepressants vary in their pharmacokinetic properties and a particular issue is their ability to interact with other medication through inhibition or competition in catabolic pathways in the liver, particularly involving cytochrome P450 (CYP450) enzymes. This is important in those on treatment for physical health problems and especially for the elderly population who are increasingly on multiple other treatments. A recent guideline from the National Institute for Health and Clinical Excellence specifically considers the treatment of depression in the medically ill providing useful information about drug interactions and a recommendation to reserve antidepressants for more severe degrees of depression because of concerns about the risk–benefit balance (National Institute for Health and Clinical Excellence 2009b). Antidepressants with little or no interactions at the level of CYP450 enzymes include sertraline, citalopram/escitalopram, mirtazapine and venlafaxine. The National Institute for Health and Clinical Excellence (2009b) also draws attention to the risk of increased gastrointestinal bleeding with SSRIs and SNRIs, particularly when combined with aspirin or non-steroidal anti-inflammatory drugs.

4.5 Choosing an Antidepressant

Relative efficacy is therefore not a main criterion for choosing between currently available antidepressants, at least for first-line use, although it seems reasonable to choose an antidepressant with established efficacy. The different tolerability and side-effect profiles of current antidepressants do provide information that is useful in attempting to match treatment to patient (although it is difficult to predict to what degree an individual will be affected). Particular care is needed to consider adverse effects that have long-term impact such as weight gain and sexual dysfunction.

Table 1 Choosing a specific antidepressant

Previous history of treatment response, tolerability and adverse effects to a particular antidepressant drug/drug class
Comorbid psychiatric disorder that may indicate a particular treatment (e.g. obsessive compulsive disorder and SSRIs)
Likely side effect profile
Low lethality in overdose if history or likelihood of overdose
Concurrent medical illness or condition that may make the antidepressant more noxious or less well-tolerated
Concurrent medication that may interact with the antidepressant drug
Strong family history of differential antidepressant response

Adapted from Anderson et al. (2008)

The factors described above have led most guidelines to recommend that SSRIs are a general first-line choice when prescribing an antidepressant, and other factors being equal, sertraline or escitalopram may be reasonable choices (but note recent cardiac concerns with escitalopram). However, it is important to take a rounded view in choosing a drug and Table 1 lists some considerations to consider when prescribing an antidepressant.

5 Treatment Non-Response and Sequencing

Inadequate response to initial treatment is unfortunately a common problem and the STAR*D study found that only about half responded and a third remitted at the first stage (Rush et al. 2006a). In this section, I will concentrate on reviewing pharmacological treatment options but it must be emphasized that they should be undertaken after an assessment of possible reasons for inadequate response including re-evaluation of the diagnosis and treatment goals, attempting to identify factors that may be impairing response to treatment (including nonadherence) and considering the whole range of further treatment options, especially psychological treatment.

It is common to hear the term ‘treatment resistant’ depression used as a shorthand in patients with poor response to treatment, but it is unsatisfactory as it lacks a clear definition and does not help to identify the degree of, or reasons for, lack of improvement on treatment (Anderson et al. 2008; National Institute for Health and Clinical Excellence 2009a). An emphasis on considering ‘next-step’ treatment given the person’s current situation and treatment history is preferable, and has been adopted in some recent guidelines (Anderson et al. 2008; National Institute for Health and Clinical Excellence 2009a).

The STAR*D study was an ambitious attempt to investigate treatment sequencing. It highlights many important points but I will only mention a few here. First, that poor treatment tolerability is a common reason for changing treatment, not just inadequate response. Second, patients have treatment preferences about next-step treatment and it was not possible to examine the relative efficacy of switching and augmenting treatment because there was little willingness to be randomised between these types of treatment option (although there was within treatment type). This seemed to be related to the adverse effects and degree of improvement on the first treatment. Third, the response and remission rates declined with each treatment step so that by the fourth step only about 15 % of patients were achieving remission (Rush 2007). Amidst the plethora of publications coming from the STAR*D study the primary next-step outcomes are reported in 6 studies (Trivedi et al. 2006; Fava et al. 2006; McGrath et al. 2006; Nierenberg et al. 2006; Rush et al. 2006b; Thase et al. 2007). No treatments were demonstrated to be more effective than any other on the primary outcome of remission although bupropion compared with buspirone augmentation, and the mirtazapine-venlafaxine combination compared with the MAOI tranylcypromine, led to greater reductions in continuous measures of depression, and cognitive behaviour therapy was slower in its action than medication. There were tolerability differences with later stage treatments tending to lead to more discontinuations and lithium and buspirone augmentation were poorly tolerated, probably due to the dosing regime with the former. This disappointing outcome has led to soul-searching but some of the problems may lie with the details and complexity of the study design, and the size of effect that could be detected (Anderson 2009).

The treatment choices faced by clinicians and patients when current treatment has not resulted in sufficient improvement are (1) to persist longer with current treatment, (2) to increase the dose or intensity, (3) to switch treatment and (4) to combine treatments. For antidepressants there has been a debate about how long to treat with antidepressants before changing treatment given the recent recognition that the greatest degree of improvement in depressive symptoms occurs in the first week and 60 % of the improvement seen at 6 weeks occurs by 2 weeks (Posternak and Zimmerman 2005). In spite of the conclusion by the authors of one meta-analysis (Szegedi et al. 2009) that treatment could be changed as early as 2–3 weeks into treatment for nonimprovers (less that 20 % improvement in rating scale scores) other 8-week studies found negative predictive values (proportion of non-improvers at a specific timepoint failing to respond by end of study, NPV) of 55–64 % at 2 weeks, 80–82 % at 4 weeks and 90–93 % at 6 weeks (Nierenberg et al. 1995; Nierenberg et al. 2000; Baldwin et al. 2009). Caution is required in extrapolating from these data to clinical practice; the STAR*D study found that eventual responders following about 12 weeks periods of treatment first responded on average after 5.5 weeks at step 1 increasing to 8.3 weeks at step 4 (with similar figures for remitters) (Rush et al. 2006a). Of eventual remitters over a third did so after 8 weeks showing that even an 8-week trial may not be long enough consistent with a separate 12-week open study which found that the NPV for nonremission at 12 weeks based on nonimprovement at 4 weeks was only 49 % (Quitkin et al. 2003). The decision about when to change treatment therefore cannot be based on a fixed timepoint but needs to be based on whether there is any improvement, or trajectory of improvement, and on the eventual chance of response/remission.

This will depend on many factors especially the number of failed treatments and the type of next step being considered. It is reasonable to start thinking about changing treatment at 3–4 weeks, especially dose increase, when there is no improvement (National Institute for Health and Clinical Excellence 2009a), but for many patients longer trials are warranted.

5.1 Dose Increase

In spite of it being common clinical practice evidence from RCTs is that increasing the dose of SSRIs is not more helpful than continuing the same dose in patients who have not responded (Adli et al. 2005), with a lack of evidence for other antidepressants although a trend was seen with high dose venlafaxine (Thase et al. 2006). There is indirect evidence from dose response data for trying a higher dose of tricyclic antidepressants, venlafaxine (>150 vs. 75 mg), escitalopram (20 vs. 10 mg) and MAOIs (Rudolph et al. 1998; Adli et al. 2005; Anderson et al. 2008) and a recent meta-analysis suggests there may be small advantage at higher than starting doses for other SSRIs at the cost of poorer tolerability (Papakostas et al. 2010). Therefore, in spite of the limited evidence increasing the dose, provided side-effects and safety allow, may be a reasonable step especially as there is wide interindividual variability in plasma concentration of antidepressants and associated uncertainty about what is an effective dose for an individual patient. Increasing the dose may also keep a patient in treatment to allow adequate time to respond.

In patients who have failed to respond to previous treatments high-dose antidepressants are sometimes considered. This is usually an off-label use and needs to be discussed with patients. Clinical experience and limited data suggests that some patients do benefit particularly when treated with high-dose TCAs, venlafaxine or MAOIs (Adli et al. 2005) but caution need to be exercised and monitoring for adverse events.

5.2 Switching Antidepressants

Open and controlled studies of antidepressant switching show widely varying response rates (25–70 %) (Anderson 2003). It is important to remember that RCTs of antidepressant switching show up to 50 % response rate by simply continuing on the same antidepressant at the same dose (Shelton et al. 2005; Corya et al. 2006). There may be marginal benefit from switching between antidepressant class than to a second drug of the same class (Papakostas et al. 2008a), largely accounted for by studies switching from an SSRI to venlafaxine (Ruhe et al. 2006). Older studies have also suggested that switching from a reuptake inhibitor to an MAOI may be effective (Anderson 2003) but the STAR*D study did not find tranylcypromine had greater efficacy than a venlafaxine-mirtazapine combination (McGrath et al. 2006).

How to switch antidepressants has been little studied and there is a need to balance the needs of safety (arising from the potential drug interactions) and tolerability against those of avoiding discontinuation reactions and timeliness. Immediate switching appears reasonable with drugs of similar pharmacology and shorter half-lives and STAR*D did not report problems switching between class. If two drugs can reasonably be combined then immediate switching (between modest doses) of the two drugs appears safe and well-tolerated. However, potentially toxic interactions do need to be considered and these are compounded if the initial drug has long-lasting effects (e.g. fluoxetine to a tricyclic antidepressant, MAOI to a serotonergic drug).

5.3 *Combining or Augmenting Antidepressants*

There has been considerable interest in combining drug treatments since the first antidepressants were developed and two of the major clinically applicable advances in pharmacological treatment have been lithium augmentation and, more recently, atypical antipsychotic augmentation. The rationale for combining antidepressants has been discussed above when considering antidepressant efficacy. However, although commonly done, evidence for it being an effective strategy is largely lacking. In practice the safety of antidepressant combinations is a key factor in determining their choice. Table 2 summarises common combination/augmentation strategies.

Some of the strongest evidence is for lithium augmentation of monoamine reuptake inhibitors, mostly tricyclic antidepressants (Bauer et al. 2010). However, the studies are small and there are few recent trials so this evidence is less secure than it once seemed particularly as lithium augmentation was poorly tolerated and not particularly effective in the STAR*D study (Nierenberg et al. 2006). However, as in bipolar disorder, lithium augmentation does appear to decrease the risk of suicide (Guzzetta et al. 2007) making it a potentially valuable treatment. In clinical practice its use appears to have declined and we do not know whether it has efficacy added to newer antidepressants such as SSRIs or non-reuptake inhibiting antidepressants such as mirtazapine. The big development in the last decade has been the evidence that atypical antipsychotics are effect as augmenting agents for SSRIs (Papakostas et al. 2007a; Anderson et al. 2008; National Institute for Health and Clinical Excellence 2009a). Pooled data and some individual studies have provided evidence for quetiapine, olanzapine, aripiprazole and risperidone, although further data are needed as well as longer term studies. The best evidence currently is for quetiapine at a lower dose than used as an antimanic or antipsychotic agent (McIntyre et al. 2009). Atypical antipsychotics are reasonably well-tolerated but there is an increased side-effect burden, including weight gain and sedation (National Institute for Health and Clinical Excellence 2009a).

Table 2 Common antidepressant augmentation/combination strategies

Combination/augmentation	Efficacy	Safety/tolerability
Lithium + TCA	++	+
Lithium + SSRI/SNRI/mirtazapine	?	+
T3 + TCA/SSRI	+	+
Pindolol + SSRI	−	+
Buspirone + SSRI	−	+
Atypical antipsychotic + SSRI	++	+
SSRI[a] + mirtazapine	+	+
+bupropion (amfebutamone)	?	+
+mianserin	(−)	+
+reboxetine	?	+
+trazodone	?	+
+TCAs[a]	?	±
+agomelatine	?	+
Venlafaxine + mirtazapine	?	+
TCA + MAOI	?	±

Key: ++ good evidence from meta-analysis of randomised controlled trials (RCTs); + some evidence from RCTs for efficacy, acceptable tolerability although cautions may be required; − meta-analysis/RCT evidence of lack of efficacy, (−) conflicting evidence; ? unknown; ± safety depends on combinations used

SNRI: serotonin and noradrenaline reuptake inhibitor, *SSRI* selective serotonin reuptake inhibitor, *T3* tri-iodothyronine, *TCA* tricyclic antidepressant

[a] SSRIs that inhibit hepatic cytochrome P450 enzymes, especially fluoxetine, paroxetine and fluvoxamine, may elevate plasma concentrations of TCAs and antidepressants

Adapted from Friedman and Anderson (2010)

Evidence for other strategies is weaker. Results with triiodothyronine (T3) augmentation suggest efficacy but it has not been extensively studied and longer term experience with continuing treatment is lacking (Anderson et al. 2008; National Institute for Health and Clinical Excellence 2009a); it was better tolerated and nonsignificantly better than lithium in the STAR*D study (Nierenberg et al. 2006). In contrast pindolol augmentation has been shown to be an ineffective strategy for improving outcome, although it does speed improvement (Anderson et al. 2008; Portella et al. 2011), and buspirone augmentation is also ineffective (Anderson et al. 2008) and it was less well-tolerated that bupropion in the STAR*D study (Trivedi et al. 2006).

The evidence-base for combining antidepressants is even weaker, including for the popular strategy of mirtazapine augmentation of SSRIs/venlafaxine (Anderson et al. 2008; National Institute for Health and Clinical Excellence 2009a). A variety of other augmentation strategies have been used. Tryptophan augmentation of MAOIs has some support (Anderson 2003) and clinically it has been used as an adjunct with lithium-MAOI and lithium-TCA combinations with the rationale that it increases serotonin availability. There has been interest in adding treatments that might be considered 'complementary' to antidepressants such as the omega-3 fatty acid eicosapentanoic acid (Anderson et al. 2008) but the evidence is equivocal.

The disappointing efficacy of strategies for treating depression poorly responsive to antidepressants has lead to a wide variety of proof-of-concept studies based on plausible pharmacological rationales but a full review is beyond the scope of this chapter.

Augmentation/combination strategies are increasingly seen as important next-step treatments and are probably being used earlier in treatment algorithms; caution is needed to make sure they are used safely. Many combinations involve off-label use and discussion with the patient is required and informed consent obtained. Augmentation strategies are particularly useful when current treatment has produced some benefit and stopping the current antidepressant risks losing this, and when dose increase and switching strategies have failed.

6 Conclusions and Future Directions

In this chapter, I have briefly reviewed the use of antidepressants in the treatment of unipolar depression and tried to put it in the historical context of drug development as well as considering some of the current controversies about the efficacy and adverse effects of antidepressants. There is no doubt that there are major limitations in our current ability to treat the syndrome of major depression. The efficacy of antidepressants is limited and has not changed in the years since their discovery, although we have a much better knowledge of their pharmacology and adverse effects, providing a 'palette' of drugs with different side-effects that at least allows some personalisation of treatment in that aspect. The relative failure to improve outcomes so far underlines how much of a challenge there is in developing better antidepressants although this remains an aspiration for researchers, clinicians and sufferers alike.

What are the challenges and where are advances likely to lie? The thrust so far has been to produce new compounds that might have advantages over current treatments which has been done by working largely in an empirical way based on known pharmacology and animal models. It remains to be seen whether new pharmacological approaches such as glutamatergic drugs will offer rapid onset of action and greater efficacy (e.g. Zarate et al. (2006) or whether our current lack of understanding of how neurobiology maps onto mood, psychological processes and symptom clusters means that progress in improving outcomes in such a broadly defined groups a major depression is a vain hope. Although there is current enthusiasm for stratified or personalised treatment we are still a long way from being able to identify predictors of response (Simon and Perlis 2010) that will allow us to target treatment more effectively for individuals. The present interest in affective processing and its neurobiology, together with increasing sophistication in functional brain imaging techniques, holds some promise for greater understanding of abnormalities that occur in depression and the effects of pharmacological manipulation. This may provide a means for moving beyond animal models which arguably have anchored us to the pharmacology of the past and to oversimple behaviours that do not sufficiently model or capture what occurs in human depression.

Perhaps, unfortunately, we are destined to be defeated by combination of the complexities of the human brain and the neural networks underlying human thought and experience, the multiple causes of depression and the variations in its meaning for individuals. This may simply mean that crude manipulations of neurotransmitter pathways by systemic administration will never be able to take us farther than the modest 'improvement-enabling' effect of current drug treatments. The challenge in that case could be to combine what we already have with smarter targeted psychological or brain region-specific physical interventions rather than to continue to seek to develop a 'better' antidepressant drug as currently conceived.

The definition of depression itself is generally accepted to be too broad and probably encompasses heterogenous conditions, or at least facets, at a more basic level (Holtzheimer and Mayberg 2011). This confusion helps little in furthering our understanding of the neurobiology of depression, its social and psychological context and causes, nor in the development of better treatments. It has been argued on this basis that we need to target treatments at more narrowly defined symptoms or symptom sets, or have a different concept of depression (Holtzheimer and Mayberg 2011). It is of interest that a recent antidepressant treatment study found that symptoms dimensions, but not standard rating scales, could identify drug-specific effects of a serotonergic compared with a noradrenergic antidepressant (Uher et al. 2009) which, if replicated, argues that we need to alter our outcome measures in treatment trials if we want to truly personalise treatment and develop new targeted agents.

Finally, although there have been attempts to identify genetic predictors of treatment none have been sufficiently robust to be useful at an individual patient level. We may simply be misguided in thinking that a single gene, or a combination of a few genes, will ever yield a large enough effect to be clinically useful. Alternatively we may need to be more sophisticated in looking for gene × environment interactions (or gene × gene interactions) to explain variance in outcomes (e.g. Keers et al. 2011) and hence to be able to harness knowledge about genes as predictors of outcomes.

References

Adli M, Baethge C, Heinz A et al (2005) Is dose escalation of antidepressants a rational strategy after a medium-dose treatment has failed? A systematic review. Eur Arch Psychiatry Clin Neurosci 255:387–400

Anderson IM (2001) Meta-analytical studies on new antidepressants. Br Med Bull 57:161–178

Anderson IM (2003) Drug treatment of depression: reflections on the evidence. Adv Psychiatr Treat 9:11–20

Anderson I (2009) Commentary on STAR*D: a summary and UK perspective. J Psychopharmacol 23:613–614

Anderson IM, Haddad PM (2011) Prescribing antidepressants for depression: time to be dimensional and inclusive. Br J Gen Pract 61:50–52

Anderson IM, Ferrier IN, Baldwin RC et al (2008) Evidence-based guidelines for treating depressive disorders with antidepressants: a revision of the 2000 British Association for Psychopharmacology guidelines. J Psychopharmacol 22:343–396

Andrews PW, Kornstein SG, Halberstadt LJ et al (2011) Blue again: perturbational effects of antidepressants suggest monoaminergic homeostasis in major depression. Front Psychol 2:159 Epub 7 Jul 2011

Angst J, Stabl M (1992) Efficacy of moclobemide in different patient groups: a meta-analysis of studies. Psychopharmacology 106(Suppl):S109–S113

Baldwin DS, Stein DJ, Dolberg OT et al (2009) How long should a trial of escitalopram treatment be in patients with major depressive disorder, generalised anxiety disorder or social anxiety disorder? an exploration of the randomised controlled trial database. Hum Psychopharmacol 24:269–275

Ban TA (2001) Pharmacotherapy of mental illness—a historical analysis. Prog Neuropsychopharmacol Biol Psychiatry 25:709–727

Barr LC, Heninger GR, Goodman W et al (1997) Effects of fluoxetine administration on mood response to tryptophan depletion in healthy subjects. Biol Psychiatry 41:949–954

Bauer M, Adli M, Bschor T et al (2010) Lithium's emerging role in the treatment of refractory major depressive episodes: augmentation of antidepressants. Neuropsychobiology 62:36–42

Blier P, Ward HE, Tremblay P et al (2010) Combination of antidepressant medications from treatment initiation for major depressive disorder: a double-blind randomized study. Am J Psychiatry 167:281–288

Bostwick JM, Pankratz VS (2000) Affective disorders and suicide risk: a reexamination. Am J Psychiatry 157:1925–1932

Brunoni AR, Tadini L, Fregni F (2010) Changes in clinical trials methodology over time: a systematic review of six decades of research in psychopharmacology. PLoS One 5:9479

Carrasco JL, Sandner C (2005) Clinical effects of pharmacological variations in selective serotonin reuptake inhibitors: an overview. Int J Clin Pract 59:1428–1434

Cipriani A, Furukawa TA, Salanti G et al (2009) Comparative efficacy and acceptability of 12 new-generation antidepressants: a multiple-treatments meta-analysis. Lancet 373:746–758

Corya SA, Williamson D, Sanger TM et al (2006) A randomized, double-blind comparison of olanzapine/fluoxetine combination, olanzapine, fluoxetine, and venlafaxine in treatment-resistant depression. Depress Anxiety 23:364–372

Curry J, Silva S, Rohde P et al (2011) Recovery and recurrence following treatment for adolescent major depression. Arch Gen Psychiatry 68:263–269

Eyding D, Lelgemann M, Grouven U et al (2010) Reboxetine for acute treatment of major depression: systematic review and meta-analysis of published and unpublished placebo and selective serotonin reuptake inhibitor controlled trials. BMJ 341:c4737. doi:10.1136/bmj.c4737

Fava M, Rush AJ, Wisniewski SR et al (2006) A comparison of mirtazapine and nortriptyline following two consecutive failed medication treatments for depressed outpatients: a STAR*D report. Am J Psychiatry 163:1161–1172

Food and Drug Administration (2011) FDA Drug Safety Communication: abnormal heart rhythms associated with high doses of Celexa (citalopram hydrobromide). Safety Announcement [08-24-2011]. http://www.fda.gov/Drugs/DrugSafety/ucm269086.htm

Fountoulakis KN, Moller HJ (2011) Efficacy of antidepressants: a re-analysis and re-interpretation of the Kirsch data. Int J Neuropsychopharmacol 14:405–412

Fournier JC, DeRubeis RJ, Hollon SD et al (2010) Antidepressant drug effects and depression severity: a patient-level meta-analysis. JAMA 303:47–53

Friedman ES, Anderson IM (2010) Managing depression in clinical practice. Springer-Verlag, London

Gartlehner G, Hansen RA, Morgan LC et al (2011) Comparative benefits and harms of second-generation antidepressants for treating major depressive disorder: an updated meta-analysis. Ann Intern Med 155:772–785

Geddes JR, Carney SM, Davies C et al (2003) Relapse prevention with antidepressant drug treatment in depressive disorders: a systematic review. Lancet 361:653–661

Gelenberg AJ, Thase ME, Meyer RE et al (2008) The history and current state of antidepressant clinical trial design: a call to action for proof-of-concept studies. J Clin Psychiatry 69:1513–1528

Guzzetta F, Tondo L, Centorrino F et al (2007) Lithium treatment reduces suicide risk in recurrent major depressive disorder. J Clin Psychiatry 68:380–383

Hammad TA, Laughren T, Racoosin J (2006) Suicidality in pediatric patients treated with antidepressant drugs. Arch Gen Psychiatry 63:332–339

Hawton K, Bergen H, Simkin S et al (2010) Toxicity of antidepressants: rates of suicide relative to prescribing and non-fatal overdose. Br J Psychiatry 196:354–358

Hickie IB, Rogers NL (2011) Novel melatonin-based therapies: potential advances in the treatment of major depression. Lancet 378:621–631

Holtzheimer PE, Mayberg HS (2011) Stuck in a rut: rethinking depression and its treatment. Trends Neurosci 34:1–9

Keers R, Uher R, Huezo-Diaz P et al (2011) Interaction between serotonin transporter gene variants and life events predicts response to antidepressants in the GENDEP project. Pharmacogenomics J 11:138–145

Keller MB (2003) Past, present, and future directions for defining optimal treatment outcome in depression: remission and beyond. JAMA 289:3152–3160

Kennedy SH, Andersen HF, Thase ME (2009) Escitalopram in the treatment of major depressive disorder: a meta-analysis. Curr Med Res Opin 25:161–175

King M, Nazareth I, Lampe F et al (2005) Impact of participant and physician intervention preferences on randomized trials: a systematic review. JAMA 293:1089–1099

Kirsch I, Moncrieff J (2007) Clinical trials and the response rate illusion. Contemp Clin Trials 28:348–351

Kirsch I, Deacon BJ, Huedo-Medina TB et al (2008) Initial severity and antidepressant benefits: a meta-analysis of data submitted to the food and drug administration. PLoS Med 5:e45

Kocsis JH, Leon AC, Markowitz JC et al (2009) Patient preference as a moderator of outcome for chronic forms of major depressive disorder treated with nefazodone, cognitive behavioral analysis system of psychotherapy, or their combination. J Clin Psychiatry 70:354–361

Lin P, Campbell DG, Chaney EF et al (2005) The influence of patient preference on depression treatment in primary care. Ann Behav Med 30:164–173

McCabe C, Mishor Z, Cowen PJ et al (2010) Diminished neural processing of aversive and rewarding stimuli during selective serotonin reuptake inhibitor treatment. Biol Psychiatry 67:439–445

McGrath PJ, Stewart JW, Fava M et al (2006) Tranylcypromine versus venlafaxine plus mirtazapine following three failed antidepressant medication trials for depression: a STAR*D report. Am J Psychiatry 163:1531–1541

McIntyre RS, Muzina DJ, Adams A et al (2009) Quetiapine XR efficacy and tolerability as monotherapy and as adjunctive treatment to conventional antidepressants in the acute and maintenance treatment of major depressive disorder: a review of registration trials. Expert Opin Pharmacother 10:3061–3075

Melander H, Salmonson T, Abadie E et al (2008) A regulatory Apologia—a review of placebo-controlled studies in regulatory submissions of new-generation antidepressants. Eur Neuropsychopharmacol 18:623–627

Mergl R, Henkel V, Allgaier AK et al (2011) Are treatment preferences relevant in response to serotonergic antidepressants and cognitive-behavioral therapy in depressed primary care patients? results from a randomized controlled trial including a patients' choice arm. Psychother Psychosom 80:39–47

Middleton H, Moncrieff J (2011) 'They won't do any harm and might do some good': time to think again on the use of antidepressants? Br J Gen Pract 61:47–49

Moller HJ, Bitter I, Bobes J et al (2012) Position statement of the European Psychiatric Association (EPA) on the value of antidepressants in the treatment of unipolar depression. Eur Psychiatry 27:114–128

Moncrieff J (2006) Why is it so difficult to stop psychiatric drug treatment? it may be nothing to do with the original problem. Med Hypotheses 67:517–523
Moncrieff J, Wessely S, Hardy R (2004) Active placebos versus antidepressants for depression. Cochrane Database Syst Rev CD003012
National Institute for Health and Clinical Excellence (2009a) Clinical guideline 90. Depression in adults (update): full guideline. http://guidance.nice.org.uk/CG90/Guidance
National Institute for Health and Clinical Excellence (2009b) Clinical guideline 91. Depression in adults with a chronic health problem: full guideline. http://guidance.nice.org.uk/CG91/Guidance
Newnham EA, Hooke GR, Page AC (2010) Progress monitoring and feedback in psychiatric care reduces depressive symptoms. J Affect Disord 127:139–146
Nierenberg AA, McLean NE, Alpert JE et al (1995) Early nonresponse to fluoxetine as a predictor of poor 8-week outcome. Am J Psychiatry 152:1500–1503
Nierenberg AA, Farabaugh AH, Alpert JE et al (2000) Timing of onset of antidepressant response with fluoxetine treatment. Am J Psychiatry 157:1423–1428
Nierenberg AA, Fava M, Trivedi MH et al (2006) A comparison of lithium and T(3) augmentation following two failed medication treatments for depression: a STAR*D report. Am J Psychiatry 163:1519–1530
Nutt DJ (2009) Beyond psychoanaleptics—can we improve antidepressant drug nomenclature? J Psychopharmacol 23:343–345
Opbroek A, Delgado PL, Laukes C et al (2002) Emotional blunting associated with SSRI-induced sexual dysfunction: do SSRIs inhibit emotional responses? Int J Neuropsychopharmacol 5:147–151
Papakostas GI, Shelton RC, Smith J et al (2007a) Augmentation of antidepressants with atypical antipsychotic medications for treatment-resistant major depressive disorder: a meta-analysis. J Clin Psychiatry 68:826–831
Papakostas GI, Thase ME, Fava M et al (2007b) Are antidepressant drugs that combine serotonergic and noradrenergic mechanisms of action more effective than the selective serotonin reuptake inhibitors in treating major depressive disorder? A meta-analysis of studies of newer agents. Biol Psychiatry 62:1217–1227
Papakostas GI, Fava M, Thase ME (2008a) Treatment of SSRI-resistant depression: a meta-analysis comparing within- versus across-class switches. Biol Psychiatry 63:699–704
Papakostas GI, Nelson JC, Kasper S et al (2008b) A meta-analysis of clinical trials comparing reboxetine, a norepinephrine reuptake inhibitor, with selective serotonin reuptake inhibitors for the treatment of major depressive disorder. Eur Neuropsychopharmacol 18:122–127
Papakostas GI, Charles D, Fava M (2010) Are typical starting doses of the selective serotonin reuptake inhibitors sub-optimal? A meta-analysis of randomized, double-blind, placebo-controlled, dose-finding studies in major depressive disorder. World J Biol Psychiatry 11: 300–307
Portella MJ, de Diego-Adelino J, Ballesteros J et al (2011) Can we really accelerate and enhance the selective serotonin reuptake inhibitor antidepressant effect? a randomized clinical trial and a meta-analysis of pindolol in nonresistant depression. J Clin Psychiatry 72:962–969
Posternak MA, Zimmerman M (2005) Is there a delay in the antidepressant effect? a meta-analysis. J Clin Psychiatry 66:148–158
Posternak MA, Zimmerman M (2007) Therapeutic effect of follow-up assessments on antidepressant and placebo response rates in antidepressant efficacy trials: meta-analysis. Br J Psychiatry 190(287–92):287–292
Pratt L, Brody D, Gu Q (2011) Antidepressant use in persons aged 12 and over: United States, 2005–2008. NCHS data brief, no 76. National Center for Health Statistics, Hyattsville. http://www.cdc.gov/nchs/data/databriefs/db76.htm
Preskorn SH (2010) CNS drug development: part I: the early period of CNS drugs. J Psychiatr Pract 16:334–339
Preskorn SH (2011) CNS drug development: part III: future directions. J Psychiatr Pract 17: 49–52

Price J, Cole V, Goodwin GM (2009) Emotional side-effects of selective serotonin reuptake inhibitors: qualitative study. Br J Psychiatry 195:211–217

Quitkin FM, Petkova E, McGrath PJ et al (2003) When should a trial of fluoxetine for major depression be declared failed? Am J Psychiatry 160:734–740

Rubino A, Roskell N, Tennis P et al (2007) Risk of suicide during treatment with venlafaxine, citalopram, fluoxetine, and dothiepin: retrospective cohort study. BMJ 334:242

Rudolph RL, Fabre LF, Feighner JP et al (1998) A randomized, placebo-controlled, dose-response trial of venlafaxine hydrochloride in the treatment of major depression. J Clin Psychiatry 59:116–122

Ruhe HG, Huyser J, Swinkels JA et al (2006) Switching antidepressants after a first selective serotonin reuptake inhibitor in major depressive disorder: a systematic review. J Clin Psychiatry 67:1836–1855

Ruhe HG, Mason NS, Schene AH (2007) Mood is indirectly related to serotonin, norepinephrine and dopamine levels in humans: a meta-analysis of monoamine depletion studies. Mol Psychiatry 12:331–359

Rush AJ (2007) STAR*D: what have we learned? Am J Psychiatry 164:201–204

Rush AJ, Trivedi MH, Wisniewski SR et al (2006a) Acute and longer-term outcomes in depressed outpatients requiring one or several treatment steps: a STAR*D report. Am J Psychiatry 163:1905–1917

Rush AJ, Trivedi MH, Wisniewski SR et al (2006b) Bupropion-SR, sertraline, or venlafaxine-XR after failure of SSRIs for depression. N Engl J Med 354:1231–1242

Rush AJ, Trivedi MH, Stewart JW et al (2011) Combining medications to enhance depression outcomes (CO-MED): acute and long-term outcomes of a single-blind randomized study. Am J Psychiatry 168:689–701

Serretti A, Chiesa A (2009) Treatment-emergent sexual dysfunction related to antidepressants: a meta-analysis. J Clin Psychopharmacol 29:259–266

Shelton RC, Williamson DJ, Corya SA et al (2005) Olanzapine/fluoxetine combination for treatment-resistant depression: a controlled study of SSRI and nortriptyline resistance. J Clin Psychiatry 66:1289–1297

Simon GE, Perlis RH (2010) Personalized medicine for depression: can we match patients with treatments? Am J Psychiatry 167:1445–1455

Simon GE, Savarino J (2007) Suicide attempts among patients starting depression treatment with medications or psychotherapy. Am J Psychiatry 164:1029–1034

Sinyor M, Levitt AJ, Cheung AH et al (2010) Does inclusion of a placebo arm influence response to active antidepressant treatment in randomized controlled trials? results from pooled and meta-analyses. J Clin Psychiatry 71:270–279

Smith D, Dempster C, Glanville J et al (2002) Efficacy and tolerability of venlafaxine compared with selective serotonin reuptake inhibitors and other antidepressants: a meta-analysis. Br J Psychiatry 180(396–404):396–404

Stone M, Laughren T, Jones ML et al (2009) Risk of suicidality in clinical trials of antidepressants in adults: analysis of proprietary data submitted to US food and drug administration. BMJ 339:b2880. doi:10.1136/bmj.b2880

Szegedi A, Jansen WT, van Willigenburg AP et al. (2009) Early improvement in the first 2 weeks as a predictor of treatment outcome in patients with major depressive disorder: a meta-analysis including 6562 patients. J Clin Psychiatry 70:344–353

Tanti A, Belzung C (2010) Open questions in current models of antidepressant action. Br J Pharmacol 159:1187–1200

Thase ME, Shelton RC, Khan A (2006) Treatment with venlafaxine extended release after SSRI nonresponse or intolerance: a randomized comparison of standard- and higher-dosing strategies. J Clin Psychopharmacol 26:250–258

Thase ME, Friedman ES, Biggs MM et al (2007) Cognitive therapy versus medication in augmentation and switch strategies as second-step treatments: a STAR*D report. Am J Psychiatry 164:739–752

Thase ME, Larsen KG, Kennedy SH (2011) Assessing the 'true' effect of active antidepressant therapy v. placebo in major depressive disorder: use of a mixture model. Br J Psychiatry 199:501–507

Trivedi MH, Rush AJ, Crismon ML et al (2004) Clinical results for patients with major depressive disorder in the Texas Medication Algorithm Project. Arch Gen Psychiatry 61: 669–680

Trivedi MH, Fava M, Wisniewski SR et al (2006) Medication augmentation after the failure of SSRIs for depression. N Engl J Med 354:1243–1252

Turner EH, Matthews AM, Linardatos E et al (2008) Selective publication of antidepressant trials and its influence on apparent efficacy. N Engl J Med 358:252–260

Uher R, Maier W, Hauser J et al (2009) Differential efficacy of escitalopram and nortriptyline on dimensional measures of depression. Br J Psychiatry 194:252–259

Zarate CA Jr, Singh JB, Carlson PJ et al (2006) A randomized trial of an N-methyl-D-aspartate antagonist in treatment-resistant major depression. Arch Gen Psychiatry 63:856–864

Zhong H, Haddjeri N, Sanchez C (2012) Escitalopram, an antidepressant with an allosteric effect at the serotonin transporter—a review of current understanding of its mechanism of action. Psychopharmacology 219:1–13

Zimmerman M, McGlinchey JB, Posternak MA et al (2006) How should remission from depression be defined? the depressed patient's perspective. Am J Psychiatry 163:148–150

New Approaches in the Treatment of Bipolar Depression

Kate E. A. Saunders and G. M. Goodwin

Abstract Patients with bipolar disorder more frequently suffer from depression than from mania. Indeed depressive symptomatology represents the majority of the illness burden experienced by bipolar patients and is also associated with elevated suicide risk. The treatment of bipolar depression differs from that of recurrent major depression in that the efficacy of conventional antidepressant drugs is not well-established. Some evidence-based treatments for bipolar depression are available and include the anticonvulsant drug, lamotrigine, and the atypical antipsychotic, quetiapine. Lithium also has some benefits in the prevention of depression and its use is associated with a significant reduction in suicidal behaviour. Several other treatments are in clinical development including dopamine agonists, glutamatergic modulating drugs and antioxidants. The role of these interventions in management is uncertain. The clinical uncertainties in the pharmacological treatment of bipolar depression mean that treatment has to be individualised and empirical; there is also an important role for psychological therapies, which again need to be tailored to suit the needs of the individual patient and their family.

Keywords Bipolar depression · Lamotrigine · Quetiapine

Contents

K. E. A. Saunders · G. M. Goodwin (✉)
University Department of Psychiatry, Warneford Hospital, Oxford OX3 7JX, UK
e-mail: guy.goodwin@psych.ox.ac.uk

K. E. A. Saunders
e-mail: kate.saunders@psych.ox.ac.uk

Curr Topics Behav Neurosci (2013) 14: 291–307
DOI: 10.1007/7854_2012_209

Published Online: 19 August 2012

Bipolar disorder (BD) is a life-long disorder characterised by periods of elated and depressed mood. In Bipolar I disorder (BD-I), mood elevation is extreme and causes marked disruption and impairment of daily activities and interpersonal relations. In Bipolar II disorder (BD-II), mood elevation is not impairing, but it is distinct and often obvious to others. However, it is the depressive symptoms that are more common, more enduring, and associated with greater morbidity and mortality in BD-I and BD-II. Existing treatments for unipolar depression are often insufficient and polypharmacy is common. Patients with BD-I and BD-II often suffer quite chronic depressive symptoms and may describe themselves as well for only about half the time (Judd et al. 2003). Thus, depression consumes the lives of BD patients more markedly than manic or hypomanic episodes. Suicide risk is increased in the depressive phase (Leverich et al. 2003; Post et al. 2003) and frequent depressive episodes are associated with greater psychosocial impairment compared with mania (Calabrese et al. 2004). Subsyndromal depressive symptoms are also associated with functionally significant impairment (Altshuler et al. 2002). Despite the disease burden of bipolar depression studies of treatments are relatively scarce and much treatment lacks a clear strategy.

1 Antidepressants

From a regulatory or licensing perspective a major depressive episode is treated similarly whether the illness course is unipolar or bipolar. Most trials have excluded bipolar cases, so despite their widespread use in bipolar patients, no antidepressant has a specific licence for use in bipolar disorder and data supporting their use in bipolar depression is far from convincing. It is an open question whether one can fairly extrapolate the results of the large number of studies of antidepressants in unipolar depression to the management of bipolar depression. While a systematic review in 2004 concluded that selective serotonin re-uptake inhibitors (SSRIS) were superior to placebo in the acute treatment of bipolar disorder (Gijsman et al. 2004), patient numbers were low and treatment with antidepressants was usually in combination with a mood stabilizer.

In contrast, the STEP-BD acute treatment arm found no benefit for adjunctive antidepressants (Sachs et al. 2007) when compared to mood stabilisers alone, but participants were allowed to receive additional (effective) psychotherapy and it is difficult to accept that this trial should simply outweigh previous positive data rather than be seen as a failed study. Tricyclic antidepressants (Koszewska and Rybakowski 2009) and venlafaxine (Leverich et al. 2006) have been associated with increased switching into the manic phase and should be used with caution. Recent naturalistic data suggest that depressed patients who fail to respond to antidepressants are more likely to be diagnosed subsequently as bipolar disorder which adds to the impression that antidepressants may be a less effective choice for bipolar patients than unipolar cases (Li et al. 2012). Antidepressants are often used in bipolar patients both in acute indications and in the long term where evidence for efficacy is even more limited (Ghaemi et al. 2001). When given alone antidepressants are associated with a considerable risk of manic relapse.

In summary, the status of antidepressants in the first line treatment of bipolar patients is uncertain. Where other strategies have failed, antidepressants probably do merit a trial in individual patients. Where antidepressants and other monotherapies have failed, the next step in treatment resistant bipolar patients is not established, and there is little option but to follow strategies for treatment resistant depression in general (Anderson et al. 2008).

2 Mood Stabilizers

Lithium has long been regarded as effective in treating bipolar depression despite a limited evidence base of predominantly old trials (Bhagwagar and Goodwin 2002). The EMBOLDEN trial failed to demonstrate a significant acute antidepressant effect for those patients treated with lithium (Young et al. 2010) when compared to placebo. A delay in the onset of antidepressant action for lithium might explain the negative findings. Maintenance studies have generally suggested that lithium is more effective against the manic pole of the illness than the depressive, but in favour of an important antidepressant action is its apparent antisuicidal efficacy (a unique demonstration in psychiatry) (Cipriani et al. 2005).

Valproate has rarely been examined for efficacy in bipolar depression. Just four randomised trials in acute bipolar depression have been condcuted and their results are mixed (Bond et al. 2010). In its indication as a maintenance treatment, valproate appeared to be superior to lithium in preventing depressive relapse in those who have already responded to valproate during a manic phase (Bowden et al. 2000); however, this was in an unplanned secondary analysis of a trial that had failed to show separations from placebo on the primary outcome. Valproate appeared to be as effective (or as ineffective) as lithium in the maintenance of rapid cycling bipolar disorder (Calabrese et al. 2005). The BALANCE study allowed a fair comparison of lithium and valproate in a real world patient sample. Results from this study suggested that lithium was probably equivalent (and synergistic) in preventing admission to hospital for mania but superior to valproate in preventing depressive relapse (Geddes et al. 2010). There is no convincing reason to prioritize valproate in the treatment of bipolar depression.

Carbamazepine's main described efficacy is antimanic and there have been no published studies of its use in acute bipolar depression. Maintenance studies suggest that carbamazepine has less prophylactic efficacy than lithium, again compatible with a limited potential role in bipolar depression (Greil and Kleindienst 1999; Greil et al. 1997; Hartong et al. 2003).

3 Lamotrigine

Lamotrigine is an anticonvulsant which acts primarily on ion channels underlying the action potential. Its use in acute bipolar depression is limited by the need for a slow upward dose taper to reduce the risk of skin reactions which can be very serious. The efficacy of lamotrigine in acute bipolar depression is, partly in consequence, not established to universal satisfaction. It has been investigated against placebo by GlaxoSmithKline, who conducted five randomized, double-blind studies which individually failed to show a statistically significant result. However, a meta-analysis of individual patient data from these studies (Geddes et al. 2009) found evidence of a modest benefit for lamotrigine, which was more striking when analysis was confined to the most severe patients at baseline. An independent add on study also demonstrated an antidepressant effect in bipolar depression (Van der Loos et al. 2009). No differences in response to lamotrigine have been found between BD-I and BD-II depressed subgroups; lamotrigine's limited or absent antimanic effects may make it a possible monotherapy choice for BD-II but not for BD-I. Support for this conclusion in rapid cycling patients comes from a study that randomised 182 patients to double-blind maintenance treatment with lamotrigine versus placebo. Secondary endpoints revealed a significant difference between the treatment groups in those with rapid cycling BD-II but not BD-I (Calabrese et al. 2000). Thus, a recent review of the efficacy of lamotrigine in the various phases of bipolar disorder, summarised the absent or negative evidence to support the use of lamotrigine in the acute treatment of mania, mixed episodes or rapid cycling in BD-I (Amann et al. 2010).

The use of lamotrigine in the longer term prevention of depressive episodes in BD-I is convincing. Bipolar patients, who could be stabilised on lamotrigine monotherapy in two trials (after respectively a manic or a depressive episode), were randomized to lamotrigine, placebo or lithium for double-blind maintenance treatment. A pooled analysis of these trials confirmed that for the primary endpoint of time to intervention for a new mood episode, lamotrigine but not lithium was superior to placebo in the prevention of depressive relapse (Goodwin et al. 2004), while lithium was more effective than lamotrigine in the prevention of manic relapse. There was no evidence that lamotrigine caused relapse to the manic pole of the illness. The efficacy of long-term use of lamotrigine in BD-II is less clear. A 16-week equipoise study found that lamotrigine was associated with a higher recovery rate in BD-I and II when compared to risperidone and inositol (Nierenberg et al. 2006). In other positive studies, lamotrigine appeared to be efficacious in a retrospective study of patients with treatment resistant BD-II depression (Sharma et al. 2008) and open-label adjunctive lamotrigine was associated with sustained benefit in BD-II (Chang et al. 2010). However, there have been a number of unpublished studies which report negative findings (Benazzi 2007).

4 Antipsychotics

Olanzapine monotherapy was found to be weakly effective as an acute treatment for BD-I depression when compared with placebo in studies conducted by Eli Lilly (Tohen et al. 2003). The combination of olanzapine and fluoxetine was superior to olanzapine monotherapy for all measures of depressive symptoms with higher rates of response and remission and faster times to response and remission. The combination of olanzapine and fluoxetine has also been shown to be more effective as an acute treatment than lamotrigine (Brown et al. 2006) although it was associated with significantly more emergent adverse events, greater weight gain and some elevated metabolic markers. There have been no studies where fluoxetine alone has been used as a comparator to olanzapine. In long-term maintenance studies enriched for an acute treatment response, olanzapine monotherapy was superior to placebo (Tohen et al. 2006) but not lithium (Tohen et al. 2005) or valproate (Tohen et al. 2003) in the prevention of depressive relapse.

Quetiapine has demonstrated a striking acute treatment effect for depression in BD-I and BD-II patients. Astra-Zeneca conducted the BOLDER studies as large, multicentre, randomised, placebo controlled trials including patients with rapid cycling disorder (Weisler et al. 2008; Thase et al. 2006; Gajwani et al. 2007). Quetiapine was associated with significant and sustained reductions in the majority of core mood symptoms including suicidal ideation. No added improvement was associated with 600 mg as opposed to 300 mg of quetiapine, which is already a relatively sedative dose of the medicine. The EMBOLDEN studies respectively compared quetiapine with paroxetine or lithium monotherapy (Young et al. 2010; McElroy et al. 2010). Patients taking quetiapine showed greater improvements in depression compared to those taking paroxetine or lithium. In relapse prevention studies, either as monotherapy or when added to lithium or valproate, quetiapine significantly reduced the time to relapse of a mood event (Suppes et al. 2008; Vieta et al. 2008b). These studies had significant power to demonstrate that quetiapine was superior in preventing relapse to each individual mood state (mania, depression and mixed). A further study, described in preliminary form but not fully unpublished, compared lithium with quetiapine (SPARCLE) and demonstrated both to be superior to placebo. Quetiapine appears to have been superior to lithium in the prevention of depressive episodes, but this result will in part have reflected study design since participants were enriched for an acute response to quetiapine.

Risperidone's proper place in the treatment of bipolar depression is not established. Two open label trials lend some support to its use as an augmentation strategy. The STEP-BD study (Nierenberg et al. 2006) reported only a 5 % response when risperidone was added to a mood stabiliser but low response rates to all augmentation strategies characterised this study.

Depot risperidone as monotherapy has been shown to be effective in preventing manic episodes (Vieta et al. 2008a; Savas et al. 2006; Quiroz et al. 2010; Malempati et al. 2008; Han et al. 2007; Fagiolini et al. 2010). There is little to suggest that it has benefit in bipolar depression (Yatham et al. 2007; Pavuluri et al. 2010).

Aripiprazole monotherapy did not show a statistically significant benefit in the acute treatment of bipolar depression (Thase et al. 2008) and in maintenance therapy aripiprazole as a monotherapy or an adjunct prevented manic but not depressive episodes (Marcus et al. 2011; Keck et al. 2006a; Keck et al. 2007). This is slightly paradoxical since three studies in treatment resistant depression have shown a statistically and clinically convincing effect of aripiprazole as an adjunct to SSRIs (Pae et al. 2011).

5 Novel Treatments

In recent years a number of novel treatments have been suggested for the treatment of bipolar depression. Some have emerged from a better understanding of the neurobiology of mood disorders support by preclinical research. These treatment targets include the dopaminergic, glutamatergic and cholinergic systems, neuropeptides and intracellular metabolic pathways.

6 The Dopaminergic System

Pramipexole is a highly selective D2/D3 receptor agonist, whose primary indication is Parkinson's disease. Its affinity for the D3 receptor subtype is thought to account for its antidepressant efficacy in unipolar depression. Two small studies have demonstrated apparent efficacy in bipolar depression (Zarate et al. 2004; Goldberg et al. 2004). Goldberg et al. (2004) demonstrated a significant improvement in depressive symptoms with pramipexole as opposed to placebo in a randomized placebo controlled trial; one patient developed hypomanic symptoms while taking the treatment. In a placebo-controlled proof of concept study of pramipexole in bipolar II depression a significant reduction in depressive symptoms was reported and two of the treatment group developed hypomanic symptoms (Zarate et al. 2004). Minimum doses (below 1 mg daily) may be desirable because additional problems in practice will be the emergence of abnormal movements or punding behaviours (complex prolonged, purposeless and stereotyped behaviour) as a result of dopamine receptor stimulation.

Modafinil is a drug which is approved for use in the USA for the treatment of sleepiness associated with conditions such as obstructive sleep apnoea. Its mechanism of action is unknown but it appears to differ from other stimulants in a number of respects, most notably the absence of addictive and abuse potential. The addition of modafinil to antidepressant treatment has been found to be effective in that 70 % of subjects with unipolar depressive disorders reported their symptoms to be "much or very much improved". Frye et al. (2007) randomly assigned 85 patients with bipolar depression which was inadequately responsive to a mood stabiliser to adjunctive modafinil or placebo for 6 weeks. A significant improvement in depressive symptoms was found in the modafinil group and both response and remission rates were significantly higher. No difference in emergent mania or hypomania was observed between the groups.

7 Melatonergic System

It has been suggested that the pathophysiology of bipolar disorder may be related to disturbances in circadian rhythm and the normalisation of these rhythms has been associated with successful antidepressant treatment. For this reason modulation of the melatonin receptors in bipolar disorder is an attractive treatment approach. Agomelatine is a potent agonist of MT1 and MT2 as well as being a 5HT2c receptor antagonist which has been well-tolerated in studies of unipolar depression. There has only been one open label study of agomelatine in acute bipolar depression which showed that a 25 mg dose added to lithium or valproate was associated with an improvement in depressive symptoms in patients with BD-I disorder (Calabrese et al. 2007). To date no controlled studies for the use of agomelatine in BD have been published.

8 The Glutamatergic System

The glutamatergic system is a potential treatment target because of the various roles it plays in neurotransmission, cellular resistance and neuroplasticity (Machado-Vieira et al. 2009). Potential targets are either the glutamate receptors (NMDA, AMPA or metabotropic) or glutamate itself prior to its release into the extracellular space. Ketamine is a noncompetitive NMDA receptor antagonist which has an established safety profile in research and clinical settings. The administration of i.v. ketamine to a group of patients with bipolar depression already established on Lithium or Valproate has been found to be associated with a rapid improvement in mood when compared to placebo (Diazgranados et al. 2010). This is a preliminary finding; it was only significant for the first 3 days after administration and involved a small treatment resistant patient group with a long history of bipolar disorder. However, the results are consistent with those described in treatment resistant unipolar depression and support the potential role for the NMDA receptor complex in the treatment of bipolar depression. The current challenge is to find a way of preserving acute benefits with an oral well-tolerated drug having a related action.

Memantine is an NMDA receptor antagonist licensed for oral use in the treatment of Alzheimer's disease. Antidepressant properties have been described in animal models but there is little data in humans. In two patients with bipolar depression the addition of memantine was associated with improvements in depressive symptoms and cognitive performance (Teng and Demetrio 2006). Logically, memantine could be a continuation treatment for patients responding to ketamine acutely but further systematic studies are necessary to determine its actual utility in bipolar depression.

Riluzole acts on the glutamate system in a number of ways. It inhibits glutamate release, enhances AMPA trafficking by increasing the insertion of the GLUR1 and GLUR2 subunits into the cell membrane and enhances the activity of reuptake (Fumagalli et al. 2008). It also stimulates Brain-Derived Neurotrophic Factor which has been implicated in the mechanism of antidepressant action. Riluzole has been added to lithium in a small open label trial in patients with BD-I and BD-II disorder who were currently depressed and produced a reduction in depressive symptoms with no switch to mania or hypomania (Zarate et al. 2005). This improvement has subsequently been found to be positively associated with increased N-acetylaspartate in the anterior cingulate cortex (Brennan et al. 2010), but again remains preliminary.

The glutamatergic system appears to be a promising target in the search for effective treatments for depressive episodes in the context of bipolar disorder.

9 Cholinergic System

An imbalance between the adrenergic and cholinergic systems was once proposed as central to the pathophysiology of mood disorders (Janowsky and Davis 1978) and there is now evidence from a PET imaging study that muscarinic M2 receptor binding in the anterior cingulate cortex is reduced in individuals with bipolar as compared with unipolar disorder (Cannon et al. 2006). The specific association of BD depression with a cholinergic mechanism has not been complemented by studies of antimuscarinic drugs in bipolar disorder but bipolar participants were included in a double-blind placebo-controlled, randomised, crossover study involving i.v. scopolamine versus placebo (Furey et al. 2010). Scopolamine was found to be associated with significant albeit temporary improvements in MADRS scores in unipolar and bipolar patients (Furey et al. 2010).

10 Intracellular Metabolic Pathways: Oxidative Stress

Peripheral and central abnormalities of antioxidant pathways are associated with bipolar disorder (Steckert et al. 2010) and increased oxidative stress has been proposed to underlie some of the phenomena of bipolar disorder and its comorbidity. Both lithium and valproate are thought to act upon relevant pathways and successful treatment is associated with improvements in antioxidant function. Glutathione is a major agent for mopping up oxygen free radicals. The production of glutathione is rate-limited by its precursor cysteine. Its acetylated derivative N-acetyl-cysteine (NAC) increases blood glutathione levels and has been suggested to be neuroprotective in a number of neurodegenerative diseases (Andreassen et al. 2000; Rothstein et al. 1994). One randomised placebo-controlled study has explored the efficacy and tolerability of NAC in treating depressive symptoms in people with BDI and II (Berk et al. 2008). Participants were given 2 g of NAC daily (or matching placebo) for 24 weeks. NAC (in addition to usual medication) was associated with significant reduction of symptoms on the MADRS at 24 weeks and MADRS scores were significantly lower overall visits for the treatment group compared with the placebo group. These improvements were paralleled by improvements in quality of life and other measures of function. However, treatment did not impact upon time to next mood episode. This study raises interesting questions. Was depression the most logical endpoint for this treatment: would cognitive function be more appropriate? How long should the study period be to capture relevant endpoints?.

Altered cerebral energy metabolism may be relevant in the pathophysiology of depression because functional imaging studies have suggested hypometabolism in the prefrontal cortex in bipolar depression (Ketter et al. 2001). Creatine plays a pivotal role in cerebral metabolism as it acts as a buffer for cellular adenosine triphosphate. Increased rates of brain oxygen utilisation (Persky and Brazeau 2001) have been demonstrated following the administration of oral creatine. In an open label study eight depressed and two bipolar depressed patients who were still depressed despite treatment were given a 4 week course of creatine monohydrate. Significant improvements in mood were noted in the unipolar depressed participants but both bipolar individuals developed hypomania/mania (Roitman et al. 2007).

Cytidine, a pyrimidine which plays a role in phospholipid metabolism and membrane stability appears to exert a neuroprotective effect by increasing cerebral high energy phosphate levels and reversing excess glutamate transmission. Given that mitochondrial dysfunction has been observed in bipolar disorder and the synthesis of pyrimidines such as cytidine are energy requiring, exogenous administration of cytidine may reduce the energy requirement upon mitochondria and normalise glutamate recycling in bipolar disorder. Thirty-five patients with bipolar depression were randomly assigned to valproate plus cyctidine or placebo (Yoon et al. 2009). Cytidine was associated with earlier improvement in depressive symptoms when compared to placebo; however, response rates did not differ significantly between cytidine and placebo.

The hypothesis that overactive cell signalling mechanisms are involved in the pathophysiology of bipolar disorder has been the basis of a number of studies which have explored the effects of fatty acids in this patient group. The first placebo controlled-double-blind study was conducted in 1999 (Stoll et al. 1999) when 44 patients with bipolar I or bipolar II were given 16 weeks of omega fatty acids or placebo in addition to their usual treatment. Adjunctive treatment with fatty acids was associated with significant improvement depression scores and a striking reduction in relapse rates. Although, the study was not designed to provide definitive data on antidepressant effects of omega-3 fatty acids most of the patients receiving placebo exhibited depressive exacerbations or recurrence. A randomised double-blind placebo controlled study of ethyl-eicosapentanoate (EPA) in bipolar depression showed a greater reduction in HDRS scores in the treated group (Frangou et al. 2006) although changes in medications were permitted as symptoms worsened. In a negative 4-month trial of EPA as an adjunctive treatment for bipolar depression or rapid cycling bipolar disorder there were no significant effects (Keck et al. 2006b).

The promise of therapeutic strategies focussed on oxidative stress or metabolism remains tantalizing. The underlying physiology appears to be credible, but the strength of the clinical trial data is not currently convincing enough to change practice.

11 Psychotherapies

While there is strong evidence to support significant genetic involvement in bipolar disorder, it would be wrong to be negative about the role of psychosocial interventions since the environment clearly plays a major part in shaping the expression of the illness. Life events and other stressors appear to play a significant role in illness onset and relapse (Johnson and Roberts 1995; Johnson and Miller 1997). The STEP-BD trial examined pharmacologic and psychosocial treatments across 22 US centres in acute bipolar depression. In one subgroup, patients were randomised to one of three intensive psychosocial treatments (family-focused therapy, cognitive behavioural therapy or interpersonal social rhythm therapy) or a control treatment (just three psychotherapy sessions over 6 weeks focussing on relapse prevention) all in conjunction with medication. Over 1 year intensive psychotherapies were associated with higher rates of recovery from bipolar depression suggesting that psychotherapy has an potential role to play in recovery from an acute episode (Miklowitz et al. 2007).

Most of the literature on psychosocial interventions has regarded relapse prevention as a more logical application for talking therapies. BD is associated with high levels of subsyndromal symptoms that provide a plausible treatment target. Group psychoeducation appears to be an effective psychotherapeutic adjunct in bipolar disorder (Colom et al. 2003) showing reduced rates of relapse. Simple case management (Bauer et al. 2006) also improves outcomes over treatment 'as usual'. In the largest trial to date (Simon et al. 2005) patients were randomised to a 2-year multicomponent care management intervention or treatment as usual. Those who received the intervention had significantly lower mania scores over the 2 years and spent less time in a hypomanic state. These interventions stress consistency of activities, adherence to medication and action plans to detect early evidence of relapse.

Cognitive behavioural therapy has received wide acceptance as a successful intervention in unipolar depression and the cognitive model for maintenance of depression would not necessarily exclude bipolar cases. Nevertheless the role for CBT in relapse prevention remains uncertain. In euthymic individuals, CBT was associated with a reduction in syndromal relapse, manic symptoms and higher social functioning when compared to treatment as usual (Lam et al. 2003). However, in a larger pragmatic study of CBT (Scott et al. 2006) no benefit was demonstrated when compared with treatment as usual.

Family focussed therapy (FFT) has been explored in a number of different settings including relapse prevention and mood stabilisation following a depressive episode. A 35–40 % reduction in recurrence rates at 2 years has been reported with FFT and a 48 % increase in recovery rates over 1 year (Rea et al. 2003; Miklowitz et al. 2003). It appears to be more effective at preventing depressive as opposed to manic relapse and has also been shown to improve adherence with medication regimes. Its application is most obvious in teenage patients.

Interpersonal and social rhythm therapy emphasises the role of social and circadian rhythm deficits in preventing relapse in bipolar disorder. When given in the acute phase of the illness it is associated with longer survival time prior to recurrence in the maintenance phase when compared with treatment as usual. However, continued treatment in the maintenance phase did not affect recurrence rates (Frank et al. 2005).

The variety of content in these different psychotherapies may be more apparent than real. A review of the key ingredients of the different approaches revealed surprising convergence on a surprisingly limited number of ingredients across studies (Miklowitz et al. 2008). All have the important limitation of being time intensive and potentially expensive to deliver. The challenge is to find the steps required to enhance patient management, starting with minimalist self help and perhaps escalating to much more focused bipolar specific psychotherapy. We are a long way from achieving a coherent pathway to this end. Since bipolar disorder is associated with high levels of subsyndromal symptoms, it has been proposed that improved management of this core by behavioural change may be the key to simplicity and to treatment development (Miklowitz et al. 2012).

12 Conclusion

There is a paucity of evidence to allow us to differentiate the efficacy of different agents in bipolar depression and clinical practice often simply follows that in unipolar depression. On balance, there is much to criticise in this approach, but little confidence that we yet have a coherent alternative strategy. The use of antidepressants in the treatment of bipolar depression remains the major uncertainty in current practice. The emergence of lamotrigine and quetiapine as new antidepressant treatments in bipolar depression is promising although further comparative evaluation is desirable. Lithium can prevent depressive relapse and lowers the risk of suicide. Central to the development of more effective treatments will be a better understanding of the neurobiology and phenomenology of bipolar disorder. Novel treatments are emerging informed by preclinical research but not based on a certain understanding of pathophysiology. Guided serendipity has been the way in psychiatry. The challenge in bipolar depression will be to conduct the necessary pragmatic trials which will be sufficiently convincing to give us a coherent new approach to treatment. At present, we have a menu of reasonable choices but little guidance on what to choose for the individual patient.

References

Altshuler LL, Gitlin MJ, Mintz J, Leight KL, Frye MA (2002) Subsyndromal depression is associated with functional impairment in patients with bipolar disorder. J Clin Psychiatry 63:807–811

Amann B, Born C, Crespo JM, Pomarol-Clotet E, Mckenna P (2010) Lamotrigine: when and where does it act in affective disorders? A systematic review. J Psychopharmacol 25(10):1289–1294

Anderson IM, Ferrier IN, Baldwin RC, Cowen PJ, Howard L, Lewis G, Matthews K, Mcallister-Williams RH, Peveler RC, Scott J, Tylee A (2008) Evidence-based guidelines for treating depressive disorders with antidepressants: a revision of the 2000 British Association for Psychopharmacology guidelines. J Psychopharmacol 22:343–396

Andreassen OA, Dedeoglu A, Klivenyi P, Beal MF, Bush AI (2000) N-acetyl-L-cysteine improves survival and preserves motor performance in an animal model of familial amyotrophic lateral sclerosis. NeuroReport 11:2491–2493

Bauer MS, Mcbride L, Williford WO, Glick H, Kinosian B, Altshuler L, Beresford T, Kilbourne AM, Sajatovic M (2006) Collaborative care for bipolar disorder: part II. Impact on clinical outcome, function, and costs. Psychiatr Serv 57:937–945

Benazzi F (2007) Bipolar II disorder: epidemiology, diagnosis and management. CNS Drugs 21:727–740

Berk M, Copolov DL, Dean O, Lu K, Jeavons S, Schapkaitz I, Anderson-Hunt M, Bush AI (2008) N-acetyl cysteine for depressive symptoms in bipolar disorder—a double-blind randomized placebo-controlled trial. Biol Psychiatry 64:468–475

Bhagwagar Z, Goodwin GM (2002) Role of mood stabilizers in bipolar disorder. Expert Rev Neurother 2:239–248

Bond DJ, Lam RW, Yatham LN (2010) Divalproex sodium versus placebo in the treatment of acute bipolar depression: a systematic review and meta-analysis. J Affect Disord 124:228–234

Bowden CL, Calabrese JR, Mcelroy SL, Gyulai L, Wassef A, Petty F, Pope HG Jr, Chou JC, Keck PE Jr, Rhodes LJ, Swann AC, Hirschfeld RM, Wozniak PJ (2000) A randomized, placebo-controlled 12-month trial of divalproex and lithium in treatment of outpatients with bipolar I disorder. Divalproex maintenance study group. Arch Gen Psychiatry 57:481–489

Brennan BP, Hudson JI, Jensen JE, Mccarthy J, Roberts JL, Prescot AP, Cohen BM, Pope HG Jr, Renshaw PF, Ongur D (2010) Rapid enhancement of glutamatergic neurotransmission in bipolar depression following treatment with riluzole. Neuropsychopharmacology 35:834–846

Brown EB, Mcelroy SL, Keck PE Jr, Deldar A, Adams DH, Tohen M, Williamson DJ (2006) A 7 week, randomized, double-blind trial of olanzapine/fluoxetine combination versus lamotrigine in the treatment of bipolar I depression. J Clin Psychiatry 67:1025–1033

Calabrese JR, Guelfi JD, Perdrizet-Chevallier C (2007) Agomelatine adjunctive therapy for acute bipolar depression: preliminary open data. Bipolar Disord 9:628–635

Calabrese JR, Hirschfeld RM, Frye MA, Reed ML (2004) Impact of depressive symptoms compared with manic symptoms in bipolar disorder: results of a U.S. community-based sample. J Clin Psychiatry 65:1499–1504

Calabrese JR, Shelton MD, Rapport DJ, Youngstrom EA, Jackson K, Bilali S, Ganocy SJ, Findling RL (2005) A 20 month, double-blind, maintenance trial of lithium versus divalproex in rapid-cycling bipolar disorder. Am J Psychiatry 162:2152–2161

Calabrese JR, Suppes T, Bowden CL, Sachs GS, Swann AC, Mcelroy SL, Kusumakar V, Ascher JA, Earl NL, Greene PL, Monaghan ET (2000) A double-blind, placebo-controlled, prophylaxis study of lamotrigine in rapid-cycling bipolar disorder. Lamictal 614 Study Group. J Clin Psychiatry 61:841–850

Cannon DM, Carson RE, Nugent AC, Eckelman WC, Kiesewetter DO, Williams J, Rollis D, Drevets M, Gandhi S, Solorio G, Drevets WC (2006) Reduced muscarinic type 2 receptor binding in subjects with bipolar disorder. Arch Gen Psychiatry 63:741–747

Chang JS, Moon E, Cha B, Ha K (2010) Adjunctive lamotrigine therapy for patients with bipolar II depression partially responsive to mood stabilizers. Prog Neuropsychopharmacol Biol Psychiatry 34:1322–1326

Cipriani A, Pretty H, Hawton K, Geddes JR (2005) Lithium in the prevention of suicidal behavior and all-cause mortality in patients with mood disorders: a systematic review of randomized trials. Am J Psychiatry 162:1805–1819

Colom F, Vieta E, Martinez-Aran A, Reinares M, Goikolea JM, Benabarre A, Torrent C, Comes M, Corbella B, Parramon G, Corominas J (2003) A randomized trial on the efficacy of group psychoeducation in the prophylaxis of recurrences in bipolar patients whose disease is in remission. Arch Gen Psychiatry 60:402–407

Diazgranados N, Ibrahim L, Brutsche NE, Newberg A, Kronstein P, khalife S, Kammerer WA, Kammerer WA, Quezado Z, Luckenbaugh DA, Salvadore G, Machado-Vieira R, Manji HK, Zarate CA Jr (2010) A randomized add-on trial of an N-methyl-D-aspartate antagonist in treatment-resistant bipolar depression. Arch Gen Psychiatry 67:793–802

Fagiolini A, Casamassima F, Mostacciuolo W, Forgione R, Goracci A, Goldstein BI (2010) Risperidone long-acting injection as monotherapy and adjunctive therapy in the maintenance treatment of bipolar I disorder. Expert Opin Pharmacother 11:1727–1740

Frangou S, Lewis M, Mccrone P (2006) Efficacy of ethyl-eicosapentaenoic acid in bipolar depression: randomised double-blind placebo-controlled study. Br J Psychiatry 188:46–50

Frank E, Kupfer DJ, Thase ME, Mallinger AG, Swartz HA, Fagiolini AM, Grochocinski V, Houck P, Scott J, Thompson W, Monk T (2005) Two-year outcomes for interpersonal and social rhythm therapy in individuals with bipolar I disorder. Arch Gen Psychiatry 62:996–1004

Frye MA, Grunze H, Suppes T, Mcelroy SL, Keck PE Jr, Walden J, Leverich GS, Altshuler LL, Nakelsky S, Hwang S, Mintz J, Post RM (2007) A placebo-controlled evaluation of adjunctive modafinil in the treatment of bipolar depression. Am J Psychiatry 164:1242–1249

Fumagalli E, Funicello M, Rauen T, Gobbi M, Mennini T (2008) Riluzole enhances the activity of glutamate transporters GLAST, GLT1 and EAAC1. Eur J Pharmacol 578:171–176

Furey ML, Khanna A, Hoffman EM, Drevets WC (2010) Scopolamine produces larger antidepressant and antianxiety effects in women than in men. Neuropsychopharmacology 35:2479–2488

Gajwani P, Muzina DJ, Kemp DE, Gao K, Calabrese JR (2007) Update on quetiapine in the treatment of bipolar disorder: results from the BOLDER studies. Neuropsychiatr Dis Treat 3:847–853

Geddes JR, Calabrese JR, Goodwin GM (2009) Lamotrigine for treatment of bipolar depression: independent meta-analysis and meta-regression of individual patient data from five randomised trials. Br J Psychiatry 194:4–9

Geddes JR, Goodwin GM, Rendell J, Azorin JM, Cipriani A, Ostacher MJ, Morriss R, Alder N, Juszczak E (2010) Lithium plus valproate combination therapy versus monotherapy for relapse prevention in bipolar I disorder (BALANCE): a randomised open-label trial. Lancet 375:385–395

Ghaemi SN, Lenox MS, Baldessarini RJ (2001) Effectiveness and safety of long-term antidepressant treatment in bipolar disorder. J Clin Psychiatry 62:565–569

Gijsman HJ, Geddes JR, Rendell JM, Nolen WA, Goodwin GM (2004) Antidepressants for bipolar depression: a systematic review of randomized, controlled trials. Am J Psychiatry 161:1537–1547

Goldberg JF, Burdick KE, Endick CJ (2004) Preliminary randomized, double-blind, placebo-controlled trial of pramipexole added to mood stabilizers for treatment-resistant bipolar depression. Am J Psychiatry 161:564–566

Goodwin GM, Bowden CL, Calabrese JR, Grunze H, Kasper S, White R, Greene P, Leadbetter R (2004) A pooled analysis of 2 placebo-controlled 18-month trials of lamotrigine and lithium maintenance in bipolar I disorder. J Clin Psychiatry 65:432–441

Greil W, Kleindienst N (1999) Lithium versus carbamazepine in the maintenance treatment of bipolar II disorder and bipolar disorder not otherwise specified. Int Clin Psychopharmacol 14:283–285

Greil W, Ludwig-Mayerhofer W, Erazo N, Schochlin C, Schmidt S, Engel RR, Czernik A, Giedke H, Muller-Oerlinghausen B, Osterheider M, Rudolf GA, Sauer H, Tegeler J, Wetterling T (1997) Lithium versus carbamazepine in the maintenance treatment of bipolar disorders–a randomised study. J Affect Disord 43:151–161

Han C, Lee MS, Pae CU, Ko YH, Patkar AA, Jung IK (2007) Usefulness of long-acting injectable risperidone during 12-month maintenance therapy of bipolar disorder. Prog Neuropsychopharmacol Biol Psychiatry 31:1219–1223

Hartong EG, Moleman P, Hoogduin CA, Broekman TG, Nolen WA (2003) Prophylactic efficacy of lithium versus carbamazepine in treatment-naive bipolar patients. J Clin Psychiatry 64:144–151

Janowsky DS, Davis JM (1978) Adrenergic-cholinergic balance in affective disorders. Psychopharmacol Bull 14:58–60

Johnson SL, Miller I (1997) Negative life events and time to recovery from episodes of bipolar disorder. J Abnorm Psychol 106:449–457

Johnson SL, Roberts JE (1995) Life events and bipolar disorder: implications from biological theories. Psychol Bull 117:434–449

Judd LL, Akiskal HS, Schettler PJ, Coryell W, Endicott J, Maser JD, Solomon DA, Leon AC, Keller MB (2003) A prospective investigation of the natural history of the long-term weekly symptomatic status of bipolar II disorder. Arch Gen Psychiatry 60:261–269

Keck PE Jr, Calabrese JR, Mcintyre RS, Mcquade RD, Carson WH, Eudicone JM, Carlson BX, Marcus RN, Sanchez R (2007) Aripiprazole monotherapy for maintenance therapy in bipolar I disorder: a 100-week, double-blind study versus placebo. J Clin Psychiatry 68:1480–1491

Keck PE Jr, Calabrese JR, Mcquade RD, Carson WH, Carlson BX, Rollin LM, Marcus RN, Sanchez R (2006a) A randomized, double-blind, placebo-controlled 26-week trial of aripiprazole in recently manic patients with bipolar I disorder. J Clin Psychiatry 67:626–637

Keck PE Jr, Mintz J, Mcelroy SL, Freeman MP, Suppes T, Frye MA, Altshuler LL, Kupka R, Nolen WA, Leverich GS, Denicoff KD, Grunze H, Duan N, Post RM (2006b) Double-blind, randomized, placebo-controlled trials of ethyl-eicosapentanoate in the treatment of bipolar depression and rapid cycling bipolar disorder. Biol Psychiatry 60:1020–1022

Ketter TA, Kimbrell TA, George MS, Dunn RT, Speer AM, Benson BE, Willis MW, Danielson A, Frye MA, Herscovitch P, Post RM (2001) Effects of mood and subtype on cerebral glucose metabolism in treatment-resistant bipolar disorder. Biol Psychiatry 49:97–109

Koszewska I, Rybakowski JK (2009) Antidepressant-induced mood conversions in bipolar disorder: a retrospective study of tricyclic versus non-tricyclic antidepressant drugs. Neuropsychobiology 59:12–16

Lam DH, Watkins ER, Hayward P, Bright J, Wright K, Kerr N, Parr-Davis G, Sham P (2003) A randomized controlled study of cognitive therapy for relapse prevention for bipolar affective disorder: outcome of the first year. Arch Gen Psychiatry 60:145–152

Leverich GS, Altshuler LL, Frye MA, Suppes T, Keck PE, Mcelroy SL, Denicoff KD, Obrocea G, Nolen WA, Kupka R, Walden J, Grunze H, Perez S, Luckenbaugh DA, Post RM (2003) Factors associated with suicide attempts in 648 patients with bipolar disorder in the Stanley Foundation Bipolar Network. J Clin Psychiatry 64:506–515

Leverich GS, Altshuler LL, Frye MA, Suppes T, Mcelroy SL, Keck PE Jr, Kupka RW, Denicoff KD, Nolen WA, Grunze H, Martinez MI, Post RM (2006) Risk of switch in mood polarity to hypomania or mania in patients with bipolar depression during acute and continuation trials of venlafaxine, sertraline, and bupropion as adjuncts to mood stabilizers. Am J Psychiatry 163:232–239

Li CT, Bai YM, Huang YL, Chen YS, Chen TJ, Cheng JY, Su TP (2012) Association between antidepressant resistance in unipolar depression and subsequent bipolar disorder: cohort study. Br J Psychiatry 200:45–51

Machado-Vieira R, Manji HK, Zarate CA (2009) The role of the tripartite glutamatergic synapse in the pathophysiology and therapeutics of mood disorders. Neuroscientist 15:525–539

Malempati RN, Bond DJ, Yatham LN (2008) Depot risperidone in the outpatient management of bipolar disorder: a 2-year study of 10 patients. Int Clin Psychopharmacol 23:88–94

Marcus R, Khan A, Rollin L, Morris B, Timko K, Carson W, Sanchez R (2011) Efficacy of aripiprazole adjunctive to lithium or valproate in the long-term treatment of patients with bipolar I disorder with an inadequate response to lithium or valproate monotherapy: a multicenter, double-blind, randomized study. Bipolar Disord 13:133–144

Mcelroy SL, Weisler RH, Chang W, Olausson B, Paulsson B, Brecher M, Agambaram V, Merideth C, Nordenhem A, Young AH (2010) A double-blind, placebo-controlled study of quetiapine and paroxetine as monotherapy in adults with bipolar depression (EMBOLDEN II). J Clin Psychiatry 71:163–174

Miklowitz D, Price J, Holmes E, Rendell J, Bell S, Budge K, Christensen J, Wallace J, Simon J, Goodwin G, Geddes J (2012) Facilitated integrated mood management (FIMM) for adults with bipolar disorder. Bipolar Disord 14(2):185–197

Miklowitz DJ, George EL, Richards JA, Simoneau TL, Suddath RL (2003) A randomized study of family-focused psychoeducation and pharmacotherapy in the outpatient management of bipolar disorder. Arch Gen Psychiatry 60:904–912

Miklowitz DJ, Goodwin GM, Bauer MS, Geddes JR (2008) Common and specific elements of psychosocial treatments for bipolar disorder: a survey of clinicians participating in randomized trials. J Psychiatr Pract 14:77–85

Miklowitz DJ, Otto MW, Frank E, Reilly-Harrington NA, Wisniewski SR, Kogan JN, Nierenberg AA, Calabrese JR, Marangell LB, Gyulai L, Araga M, Gonzalez JM, Shirley ER, Thase ME, Sachs GS (2007) Psychosocial treatments for bipolar depression: a 1-year randomized trial from the Systematic Treatment Enhancement Program. Arch Gen Psychiatry 64:419–426

Nierenberg AA, Ostacher MJ, Calabrese JR, Ketter TA, Marangell LB, Miklowitz DJ, Miyahara S, Bauer MS, Thase ME, Wisniewski SR, Sachs GS (2006) Treatment-resistant bipolar depression: a STEP-BD equipoise randomized effectiveness trial of antidepressant augmentation with lamotrigine, inositol, or risperidone. Am J Psychiatry 163:210–216

Pae CU, Forbes A, Patkar AA (2011) Aripiprazole as adjunctive therapy for patients with major depressive disorder: overview and implications of clinical trial data. CNS Drugs 25:109–127

Pavuluri MN, Henry DB, Findling RL, Parnes S, Carbray JA, Mohammed T, Janicak PG, Sweeney JA (2010) Double-blind randomized trial of risperidone versus divalproex in pediatric bipolar disorder. Bipolar Disord 12:593–605

Persky AM, Brazeau GA (2001) Clinical pharmacology of the dietary supplement creatine monohydrate. Pharmacol Rev 53:161–176

Post RM, Denicoff KD, Leverich GS, Altshuler LL, Frye MA, Suppes TM, Rush AJ, Keck PE, Mcelroy SL, Luckenbaugh DA, Pollio C, Kupka R, Nolen WA (2003) Morbidity in 258 bipolar outpatients followed for 1 year with daily prospective ratings on the NIMH life chart method. J Clin Psychiatry 64:680–690 (quiz 738–739)

Quiroz JA, Yatham LN, Palumbo JM, Karcher K, Kushner S, Kusumakar V (2010) Risperidone long-acting injectable monotherapy in the maintenance treatment of bipolar I disorder. Biol Psychiatry 68:156–162

Rea MM, Tompson MC, Miklowitz DJ, Goldstein MJ, Hwang S, Mintz J (2003) Family-focused treatment versus individual treatment for bipolar disorder: results of a randomized clinical trial. J Consult Clin Psychol 71:482–492

Roitman S, Green T, Osher Y, Karni N, Levine J (2007) Creatine monohydrate in resistant depression: a preliminary study. Bipolar Disord 9:754–758

Rothstein JD, Bristol LA, Hosler B, Brown RH Jr, Kuncl RW (1994) Chronic inhibition of superoxide dismutase produces apoptotic death of spinal neurons. Proc Natl Acad Sci U S A 91:4155–4159

Sachs GS, Nierenberg AA, Calabrese JR, Marangell LB, Wisniewski SR, Gyulai L, Friedman ES, Bowden CL, Fossey MD, Ostacher MJ, Ketter TA, Patel J, Hauser P, Rapport D, Martinez JM, Allen MH, Miklowitz DJ, Otto MW, Dennehy EB, Thase ME (2007) Effectiveness of adjunctive antidepressant treatment for bipolar depression. N Engl J Med 356:1711–1722

Savas HA, Yumru M, Ozen ME (2006) Use of long-acting risperidone in the treatment of bipolar patients. J Clin Psychopharmacol 26:530–531
Scott J, Paykel E, Morriss R, Bentall R, Kinderman P, Johnson T, Abbott R, Hayhurst H (2006) Cognitive-behavioural therapy for severe and recurrent bipolar disorders: randomised controlled trial. Br J Psychiatry 188:313–320
Sharma V, Khan M, Corpse C (2008) Role of lamotrigine in the management of treatment-resistant bipolar II depression: a chart review. J Affect Disord 111:100–105
Simon GE, Ludman EJ, Unutzer J, Bauer MS, Operskalski B, Rutter C (2005) Randomized trial of a population-based care program for people with bipolar disorder. Psychol Med 35:13–24
Steckert AV, Valvassori SS, Moretti M, Dal-Pizzol F, Quevedo J (2010) Role of oxidative stress in the pathophysiology of bipolar disorder. Neurochem Res 35:1295–1301
Stoll AL, Severus WE, Freeman MP, Rueter S, Zboyan HA, Diamond E, Cress KK, Marangell LB (1999) Omega 3 fatty acids in bipolar disorder: a preliminary double-blind, placebo-controlled trial. Arch Gen Psychiatry 56:407–412
Suppes T, Hirschfeld RM, Vieta E, Raines S, Paulsson B (2008) Quetiapine for the treatment of bipolar II depression: analysis of data from two randomized, double-blind, placebo-controlled studies. World J Biol Psychiatry 9:198–211
Teng CT, Demetrio FN (2006) Memantine may acutely improve cognition and have a mood stabilizing effect in treatment-resistant bipolar disorder. Rev Bras Psiquiatr 28:252–254
Thase ME, Jonas A, Khan A, Bowden CL, Wu X, Mcquade RD, Carson WH, Marcus RN, Owen R (2008) Aripiprazole monotherapy in nonpsychotic bipolar I depression: results of 2 randomized, placebo-controlled studies. J Clin Psychopharmacol 28:13–20
Thase ME, Macfadden W, Weisler RH, Chang W, Paulsson B, Khan A, Calabrese JR (2006) Efficacy of quetiapine monotherapy in bipolar I and II depression: a double-blind, placebo-controlled study (the BOLDER II study). J Clin Psychopharmacol 26:600–609
Tohen M, Calabrese JR, Sachs GS, Banov MD, Detke HC, Risser R, Baker RW, Chou JC, Bowden CL (2006) Randomized, placebo-controlled trial of olanzapine as maintenance therapy in patients with bipolar I disorder responding to acute treatment with olanzapine. Am J Psychiatry 163:247–256
Tohen M, Greil W, Calabrese JR, Sachs GS, Yatham LN, Oerlinghausen BM, Koukopoulos A, Cassano GB, Grunze H, Licht RW, Dell'Osso L, Evans AR, Risser R, Baker RW, Crane H, Dossenbach MR, Bowden CL (2005) Olanzapine versus lithium in the maintenance treatment of bipolar disorder: a 12-month, randomized, double-blind, controlled clinical trial. Am J Psychiatry 162:1281–1290
Tohen M, Vieta E, Calabrese J, Ketter TA, Sachs G, Bowden C, Mitchell PB, Centorrino F, Risser R, Baker RW, Evans AR, Beymer K, Dube S, Tollefson GD, Breier A (2003) Efficacy of olanzapine and olanzapine-fluoxetine combination in the treatment of bipolar I depression. Arch Gen Psychiatry 60:1079–1088
van der Loos ML, Mulder PG, Hartong EG, Blom MB, Vergouwen AC, de Keyzer HJ, Notten PJ, Luteijn ML, Timmermans MA, Vieta E, Nolen WA (2009) Efficacy and safety of lamotrigine as add-on treatment to lithium in bipolar depression: a multicenter, double-blind, placebo-controlled trial. J Clin Psychiatry 70:223–231
Vieta E, Nieto E, Autet A, Rosa AR, Goikolea JM, Cruz N, Bonet P (2008a) A long-term prospective study on the outcome of bipolar patients treated with long-acting injectable risperidone. World J Biol Psychiatry 9:219–224
Vieta E, Suppes T, Eggens I, Persson I, Paulsson B, Brecher M (2008b) Efficacy and safety of quetiapine in combination with lithium or divalproex for maintenance of patients with bipolar I disorder (international trial 126). J Affect Disord 109:251–263
Weisler RH, Calabrese JR, Thase ME, Arvekvist R, Stening G, Paulsson B, Suppes T (2008) Efficacy of quetiapine monotherapy for the treatment of depressive episodes in bipolar I disorder: a post hoc analysis of combined results from 2 double-blind, randomized, placebo-controlled studies. J Clin Psychiatry 69:769–782
Yatham LN, Fallu A, Binder CE (2007) A 6-month randomized open-label comparison of continuation of oral atypical antipsychotic therapy or switch to long acting injectable risperidone in patients with bipolar disorder. Acta Psychiatr Scand Suppl 434:50–56

Yoon SJ, Lyoo IK, Haws C, Kim TS, Cohen BM, Renshaw PF (2009) Decreased glutamate/glutamine levels may mediate cytidine's efficacy in treating bipolar depression: a longitudinal proton magnetic resonance spectroscopy study. Neuropsychopharmacology 34:1810–1818

Young AH, Mcelroy SL, Bauer M, Philips N, Chang W, Olausson B, Paulsson B, Brecher M (2010) A double-blind, placebo-controlled study of quetiapine and lithium monotherapy in adults in the acute phase of bipolar depression (EMBOLDEN I). J Clin Psychiatry 71:150–162

Zarate CA Jr, Payne JL, Singh J, Quiroz JA, Luckenbaugh DA, Denicoff KD, Charney DS, Manji HK (2004) Pramipexole for bipolar II depression: a placebo-controlled proof of concept study. Biol Psychiatry 56:54–60

Zarate CA Jr, Quiroz JA, Singh JB, Denicoff KD, de Jesus G, Luckenbaugh DA, Charney DS, Manji HK (2005) An open-label trial of the glutamate-modulating agent riluzole in combination with lithium for the treatment of bipolar depression. Biol Psychiatry 57:430–432

Molecular and Cellular Mechanisms of Antidepressant Action

Trevor Sharp

Abstract A long-standing theory is that brain monoamine signalling is critically involved in the mechanisms of antidepressant drug treatment. Theories on the nature of these mechanisms commenced with ideas developed in the 1960s that the drugs act simply by increasing monoamine availability in the synapse. However, this thinking has advanced remarkably in the last decade to concepts which posit that antidepressant drug action on monoamine signalling is just the starting point for a complex sequence of neuroadaptive molecular and cellular changes that bring about the therapeutic effect. These changes include activation of one or more programmes of gene expression that leads to the strengthening of synaptic efficacy and connectivity, and even switching neural networks into a more immature developmental state. It is thought that through this increase in plasticity, key neural circuits within the limbic system are more easily remodelled by incoming emotionally relevant stimuli. This article attempts to bring together previous and current knowledge of antidepressant drug action on monoamine signalling at molecular and cellular levels, and introduces current thinking that these changes interact with neuropsychological processes ultimately to elevate mood.

Keywords Antidepressant · 5-HT · Noradrenaline · Neural plasticity · Gene expression · Neurotrophic factors · Autoreceptors

T. Sharp (✉)
University Department of Pharmacology, Mansfield Road, Oxford OX1 3QT, UK
e-mail: trevor.sharp@pharm.ox.ac.uk

Curr Topics Behav Neurosci (2013) 14: 309–325
DOI: 10.1007/7854_2012_216

Published Online: 4 August 2012

Contents

1 Introduction

The foundation for the use of drugs in the clinical management of depression was laid in the 1950s by the serendipitous discovery of the first antidepressant drugs, imipramine and isoniazid. Although the primary pharmacological targets of imipramine and isoniazid were unknown at the time of their discovery, these targets were revealed in subsequent studies; imipramine was found to block the reuptake of the monoamines serotonin (5-hydroxytryptamine; 5-HT) and noradrenaline, whereas isoniazid blocked 5-HT and noradrenaline metabolism through the inhibition of monoamine oxidase. From these findings, and the introduction of other antidepressants with similar pharmacology, it was reasonably hypothesised that the drugs were acting by increasing availability of 5-HT or noradrenaline or both, and that depression was caused by a monoamine deficiency (Coppen 1967; Schildkraut 1965).

These first antidepressants were not pharmacologically selective and were often prone to pharmacokinetic interactions with other drugs. As a consequence, this generation of antidepressants were hampered by adverse effects, poor tolerability and safety concerns, and a lack of efficacy in a significant proportion of patients. In the 1970s, a second generation of antidepressants, the 'atypical antidepressants', was developed in an attempt to improve this situation. These drugs were so-called because even though they targeted monoamine neurons in one way or another, their pharmacological profile was not 'typical' of the first antidepressants (Tyrer and Marsden 1985). For example, included in this group are mianserin and its successor mirtazapine whose main pharmacological action includes blockade of the $\alpha 2$ adrenoceptor which serves as an inhibitory autoreceptor, which when blocked increases monoamine availability. A third generation of selective monoamine reuptake inhibitors followed in the 1990s. Fluoxetine was the first selective serotonin reuptake inhibitor (SSRI) that was extensively marketed (after zimelidine was withdrawn), and was followed by other SSRIs as well as selective noradrenaline reuptake inhibitors (e.g. reboxetine) and dual 5-HT/noradrenaline reuptake inhibitors (e.g. venlafaxine).

2 The Role of Monoamines in Antidepressant Drug Action

As evident from the examples given above, the major classes of antidepressant drugs, all still in widespread clinical use, target monoamine neurons in one way or another. Some of the most direct evidence that the therapeutic effect of antidepressant drugs is actually mediated by monoamines comes from monoamine depletion studies. These studies involve testing the effect of acute lowering of monoamine availability using amino acid mixtures or enzyme inhibitors, in depressed patients in remission but taking antidepressant medication (Ruhe et al. 2007). Overall, these studies found that monoamine depletion elicits a striking return of clinical depression in a high proportion of medicated depressed patients in remission. For example, patients responding effectively to a course of SSRI treatment demonstrate a dramatic relapse within hours of depletion of the 5-HT precursor tryptophan (Delgado et al. 1990). These and similar findings with catecholamine depletion suggest that maintaining synaptic monoamine levels is critical for the maintenance of the therapeutic effect of antidepressant drugs (Ruhe et al. 2007).

The evidence derived from the monoamine depletion studies suggest that monoamines play a key role in the therapeutic effect of antidepressant drugs, and this is in keeping with evidence from biological studies of altered 5-HT and noradrenaline markers in depressed patients (Cowen and Harmer 2009; Sharp and Cowen 2011). It should be noted however that many of the important monoamine changes detected in depression persist with clinical recovery, suggesting that the changes are a marker of depression vulnerability rather than driving the depressive symptoms.

Crucially, a significant problem for the original monoamine hypothesis (Coppen 1967; Schildkraut 1965) is the paradoxical dissociation in time between the acute pharmacological effects of these drugs and their maximal therapeutic effect, which requires a course of treatment of several weeks (Gelenberg and Chesen 2000). The early onset of the pharmacological effect is strikingly evident in positron emission tomography studies, which demonstrate near maximal occupancy of the 5-HT transporter within hours following administration of a single clinical dose of an SSRI (Meyer et al. 2004). Thus, although antidepressant drugs rapidly enter the brain and block monoamine reuptake or metabolism, a course of treatment of several weeks is required for the full antidepressant response.

3 Adaptation of the Monoamine System to Antidepressants

An early hypothesis put forward in the 1980 s to explain the paradoxical lag in the onset of therapeutic effect was that the drugs induce adaptive changes in the sensitivity in postsynaptic monoamine receptors (Stahl and Palazidou 1986). For instance, a well-replicated observation was that antidepressant drugs of different types, when administered to rodents over several weeks (but not acutely), caused a decrease in the binding and function of central 5-HT_2 receptor and β-adrenoceptors (Ogren and Fuxe 1985). These data lead to the formulation of the 'neurotransmitter receptor' or 'receptor sensitisation' hypothesis (Stahl and Palazidou 1986), which posited that antidepressants act by down-regulating 5-HT_2 receptor and β-adrenoceptors. This finding did not sit easily with the idea that antidepressants produce their therapeutic effect by increasing monoamine function, but it was argued that with increased synaptic monoamine levels the net effect at the synapse was to increase monoamine transmission (Duman et al. 1997).

Whilst 5-HT_2 and β-adrenoceptor down-regulation may be important to the therapeutic effect of antidepressants, it is equally plausible that these changes contribute little to the antidepressant action. Thus, the down-regulation in receptors could simply be a homeostatic response to increased monoamine availability. Indeed, clinical data suggest that β-adrenoceptor or 5-HT_2 receptor blockade by itself is not sufficient to evoke an antidepressant effect (Artigas et al. 1996; Duman et al. 1997).

Alternative theories of antidepressant action took greater prominence in the 1990s, with particular emphasis being placed on the idea that autoreceptor desensitization played a critical role, with a particular focus on somatodendritic 5-HT_{1A} autoreceptors (Artigas et al. 1996; Blier and de Montigny 1994). This theory posited that acute inhibition of monoamine reuptake or metabolism evokes a rise in extracellular 5-HT levels that is immediately sensed by the 5-HT_{1A} autoreceptors to initiate a 'brake' on 5-HT cell firing and release (Gartside et al. 1995). Evidence derived using a range of experimental models of 5-HT_{1A} autoreceptor function in animals and humans suggested that with continued antidepressant treatment, these autoreceptors desensitize leading to a maintained increase in monoamine availability in the synapse (Blier and de Montigny 1994; Artigas et al. 1996). Although the mechanism underlying desensitization was initially unclear, recent findings indicate increased internalisation of somatodendritic 5-HT_{1A} receptors rather than a net reduction in receptor levels (Riad et al. 2004).

The 5-HT_{1A} autoreceptor desensitization theory predicted that 5-HT_{1A} receptor blockade would augment antidepressant action, and this was verified in preclinical studies (e.g. Castro et al. 2003). These findings translated into clinical trials of depression that tested the $5\text{-HT}_{1A}/\beta$-adrenoceptor ligand pindolol in combination with SSRI treatment. Two recent meta-analyses of the pindolol trials report an increase in speed of onset of clinical effect (Portella et al. 2011; Whale et al. 2010). However, a recent study in depression administering an SSRI in combination with a selective 5-HT_{1A} receptor antagonist observed no benefit compared to SSRI treatment alone (Scorza et al. 2012). Although the latter study was small and the findings await confirmation the data suggest that, contrary to predictions, 5-HT_{1A} receptor blockade may not be a useful for antidepressant augmentation strategy. It is possible that the antidepressant augmenting properties of pindolol observed in some studies relate to the apparent selectivity of pindolol for pre- versus postsynaptic 5-HT_{1A} receptors (Castro et al. 2000; Rabiner et al. 2000).

Note that a recent study using an interesting genetic mouse model of altered expression of the 5-HT_{1A} autoreceptor reported that desensitization of the 5-HT_{1A} autoreceptor is not required for an antidepressant behavioural effect (Richardson-Jones et al. 2010). Specifically, it was found that mice with high levels of 5-HT_{1A} autoreceptor expression failed to show an antidepressant response to repeated administration of an SSRI in a novelty supressed feeding (NSF) paradigm, despite these mice demonstrating evidence of 5-HT_{1A} autoreceptor desensitization. Whilst these data question the importance of 5-HT_{1A} autoreceptor desensitization in the mechanism of action of antidepressants, additional work is needed to confirm the conclusions. Thus, the method used to assess 5-HT_{1A} autoreceptor function was indirect (5-HT_{1A} receptor agonist induced hypothermia), and it is possible that in the presence of high 5-HT_{1A} autoreceptor expression, desensitization was incomplete. The experiment also did not include depression models (in many contexts, the NSF model is used as a test of anxiety).

In addition to 5-HT_{1A} autoreceptors, adaptation of other monoamine feedback systems may play a role in antidepressant action and thus have potential as antidepressant targets. Such feedback mechanisms include nerve terminal 5-HT_{1B} and $\alpha 2$-adrenoceptor autoreceptors. For instance, antagonists of the $\alpha 2$-adrenoceptor demonstrate antidepressant augmentation effects in both animal models (Yanpallewar et al. 2010) and depressed patients (Cappiello et al. 1995; Sanacora et al. 2004). The feedback control of 5-HT neurons by postsynaptic 5-HT receptors, including 5-HT_{2A} and 5-HT_{2C} receptor subtypes, may also be relevant in this context (Sharp et al. 2007; Queree et al. 2009). Interestingly, recent evidence suggests that 5-HT_2 receptor-mediated feedback may contribute to the acute adaptive response of 5-HT neurones to antidepressant treatment (Mongeau et al. 2010). Moreover, there are both preclinical and clinical data suggesting that 5-HT_2 receptor antagonists have SSRI augmenting properties (Marek et al. 2003), and this might link to blockade of 5-HT_2 receptor feedback mechanisms. It would be interesting to know whether the down-regulation of 5-HT_2 receptors by antidepressant treatment (see above), includes the population of 5-HT_2 receptors involved in feedback control of 5-HT neurons.

Encouraged by the promising findings of the above antidepressant augmentation trials, drug development programmes have identified molecules with combined monoamine uptake/autoreceptor blocking properties (e.g. Hughes et al. 2007; Millan 2009; Serres et al. 2012), although little is currently know regarding the actions of such molecules in man.

4 Adaptation of Downstream Signalling Cascades

Evidence accumulated largely through preclinical research since the end of the 1990s, suggests that changes in monoamine feedback are just part of a series of neuroadaptive events that occur during a course of antidepressant treatment and lead up to the therapeutic effect. In particular, research focused on signalling events downstream of postsynaptic monoamine receptors has identified changes in expression of a variety of genes linked to mechanisms of neural plasticity.

Initial studies demonstrated that repeated antidepressant treatment to rodents increased expression of brain-derived neurotrophic factor (BDNF) (Lindefors et al. 1995; Nibuya et al. 1995). This was extended to other trophic factors, in particular vascular endothelial growth factor (Warner-Schmidt and Duman 2008) and fibroblast growth factor 2 (Elsayed et al. 2012). In addition to trophic factors, expression of a host of other genes important for processes linked to neural plasticity have been found to increase in response to repeated antidepressant treatment including specific gene transcription factors (CREB) (Thome et al. 2000; Blendy 2006; Tardito et al. 2009) and other activity-dependent genes (Arc) (Pei et al. 2003; Larsen et al. 2007; Calabrese et al. 2011), vesicular glutamate transporters (VGluT1) (Tordera et al. 2005; Farley et al. 2012), postsynaptic density proteins (Homer, Shank) (Serres et al. 2012); receptor chaperones (P11, Arc) (Pei et al. 2003; Svenningsson et al. 2006) and other transmitter signalling molecules and enzymes (Wnt2, MAP kinase) (Duric et al. 2010; Okamoto et al. 2010).

These examples of genes altered by antidepressant administration are likely to be part of a much more complex picture because genome-wide analyses using microarray methodology indicate that antidepressant treatment is associated with altered expression of many tens if not hundreds of genes (Conti et al. 2007; Sillaber et al. 2008; Gaska et al. 2012). Although a precise and consistent molecular 'fingerprint' of antidepressant action has yet to emerge from the genome-wide analysis approach, the large number of genes that change is consistent with the activation of multiple gene programmes involved in many neuroadaptive processes ranging from synaptic formation and strengthening through to increased neuronal survival and even neurovascular support.

Interestingly, the gene changes are time-dependent in that the onset of change of some genes is quicker than others (Conti et al. 2007; Serres et al. 2012). The latter findings raise the possibility that speeding up the sequence of gene changes associated with antidepressant drug treatment may provide a marker of putative antidepressants with a faster onset of action. Interestingly, there are reports that when combined with antidepressants, 5-HT$_{1A}$ receptor and α2-adrenoceptor antagonists accelerate the onset of expression of Arc and BDNF expression (Castro et al. 2003; Yanpallewar et al. 2010).

The large diversity of genes influenced by a course of antidepressant treatment makes it challenging to attribute weight of importance to individual genes and their products, and specific downstream effects. For example, which genes amongst the many are key to the antidepressant effects and thereby make reasonable targets for drug development? Interestingly, there are preclinical data from experiments using genetic and pharmacological tools that demonstrate that specific genes activated by antidepressant treatment are necessary and sufficient for behavioural and molecular effects of the treatment, as summarised in Table 1. The effects of antidepressant treatment measured in these experiments themselves are simple, proxy markers of the complex neuropsychological and behavioural

Table 1 Genes shown to be necessary and sufficient for antidepressant effects in animal models

Gene	Manipulation	Antidepressant	Experimental read-out	Outcome	Reference
CREB	Knockout	Desipramine Fluoxetine	Behaviour (TST, FST Molecular (BDNF) Endocrine (corticosterone)	Prevented molecular but not behavioural or endocrine effects	Conti et al. 2002
BDNF	Knockout TrkB dominant -ve	Fluoxetine Imipramine Ketamine	Behaviour (FST)	Prevented behavioural effects of imipramine and fluoxetine but not ketamine	Saarelainen et al. 2003; Lindholm et al. 2012
VEGF	Drug inhibitor	Desipramine Fluoxetine	Behaviour (NSF, FST) Neurogenesis	Prevented behavioural and cellular effects	Warner-Schmidt and Duman 2008; Greene et al. 2009
FGF2	Drug inhibitor	Fluoxetine Imipramine	Behaviour (NSF, FST)	Prevented behavioural effects	Elsayed et al. 2012
5-HT_{1A} receptor	Knockout	Fluoxetine Paroxetine Desipramine Imipramine	Behaviour (TST, FST) Neurogenesis	Prevented behavioural and cellular effects of fluoxetine and paroxetine but not imipramine or desipramine	Mayorga et al. 2001; Santarelli et al. 2003
5-HT transporter	Knockout	Fluoxetine Imipramine Desipramine	Behaviour (TST)	Prevented behavioural effects of fluoxetine but not imipramine or desipramine	Holmes et al. 2002
5-HT transporter	Knockout	Paroxetine Citalopram	Electrophysiology (5-HT cell firing)	Prevented electrophysiological effects	Mannoury la Cour et al. 2001

(continued)

Table 1 (continued)

Gene	Manipulation	Antidepressant	Experimental read-out	Outcome	Reference
Dopamine-β-hydroxylase	Knockout	Desipramine Reboxetine Pargyline Fluoxetine Sertraline Paroxetine Citalopram	Behaviour (TST)	Prevented behavioural effects of all drugs bar citalopram	Cryan et al. 2004
P11	Knockout	Imipramine Fluoxetine	Behaviour (TST, NSF) Neurogenesis	Prevented behavioural and cellular effects	Svenningsson et al. 2006; Egeland et al. 2010
P11 (targeted to cortical pyramidal cells)	Conditional knockout	Fluoxetine	Behaviour (FST, TST, NSF)	Prevented behavioural effects	Schmidt et al. 2012

CREB Cyclic AMP response element binding protein; *BDNF* Brain-derived neurotrophic factor; *VEGF* Vascular endothelial growth factor; *FGF2* Fibroblast growth factor 2; *TST* Tail suspension test; *FST* Forced swim test; *NSF* Novelty suppressed feeding test therapeutic process that in itself is not well understood. However, the finding that these same markers are sensitive to knockout of the 5-HT transporter and the noradrenaline forming enzyme dopamine-β-hydroxylase (see Table 1), provides good predictive validity to the data.

The monoamine receptors triggering the downstream molecular and cellular effects of antidepressant treatment is presently uncertain, but identification of the receptor subtypes and signalling pathways involved will provide an important guide to future antidepressant drug development programmes. Monoamine receptors that are currently seen as important include 5-HT receptor and adrenoceptor subtypes positively coupled to either adenylate cyclase (5-HT_4, 5-HT_6, 5-HT_7, $\beta_{1/2}$) or Ca^{2+} dependent kinases (5-$HT_{2A/C}$, $\alpha 1$) that are likely to activate the expression of CREB, BDNF, and also MAP kinase, possibly in a linked and coordinated way. The latter are considered key nodal points in the signalling pathways that are thought to underpin the therapeutic effects of not only classical antidepressants but also the putative rapid acting NMDA antagonist ketamine (Duman and Voleti 2012; Murrough 2012). Interestingly, as would be predicted from this schema, recent data show that 5-HT_4 and 5-HT_6 receptor agonists demonstrate early onset antidepressant-like molecular and behavioural effects in preclinical models (Lucas et al. 2007; Svenningsson et al. 2007). Moreover, there are ongoing clinical trials into the antidepressant potential of psychedelic drug, psilocybin, reinforced by evidence that a 5-HT_2 receptor agonist action may evoke changes in neural plasticity to produce long-lasting improvement in mood (Vollenweider and Kometer 2010).

A recent quite different perspective is that antidepressant-induced changes in gene expression may be mediated via epigenetic mechanisms; that is the regulation of DNA transcription by mechanisms such as DNA methylation, histone modifications and non-coding RNAs without alteration of the original DNA sequence. For example, antidepressant-induced changes in the expression of both BDNF and P11 have been attributed to epigenetic mechanisms, specifically increased histone acetylation and reduced DNA methylation, respectively (Tsankova et al. 2006; Lopez et al. 2012; Melas et al. 2012). Evidence such as this is encouraging research into the antidepressant potential of drugs targeted at epigenetic mechanisms. Indeed, both inhibitors of histone deacetylation and DNA methylation show promise in preclinical paradigms (Menke et al. 2012), although the safety and tolerability of this approach in patients is presently uncertain.

5 A Current Molecular and Cellular Hypothesis of Antidepressant Action

Many of the same genes noted above that are increased by antidepressant treatment are found to be decreased in animal models of depression and/or postmortem tissue from subjects diagnosed with depression. For example, this is the case for each of Arc, BDNF, P11 and VGluT1 (Svenningsson et al. 2006; Covington et al. 2010; Elizalde et al. 2010; Banasr et al. 2011; Farley et al. 2012). Moreover, knockout of such genes in mouse models is associated with depressive-like or stress-vulnerable behavioural phenotypes (Lominac et al. 2005; Ren-Patterson et al. 2006; Szumlinski et al. 2008; Garcia–Garcia et al. 2009). These findings, together with the data obtained from genetic mouse studies (Table 1), add much strength to the idea that specific neural plasticity genes activated by antidepressant drugs are not only critical for the therapeutic effect of antidepressant treatment but also play a key role in depression pathophysiology.

Collectively, the above findings underpin current thinking that increased synaptic monoamine levels (presumably following desensitization of monoamine feedback) activates one or more gene programmes that leads to enhanced neural plasticity, which has failed because of the adverse effects of stress and other environmental and genetic factors (for example, see reviews by Castren 2005; Krishnan and Nestler 2008; Duman and Voleti 2012). The net effect of these changes may be an activity-dependent structural realignment and functional repair of neural circuitry that is then better able to adapt to environmental challenges. A recent advance on this thinking is that antidepressants may be even switching synapses and neural networks into a more immature developmental state, and thereby reactivating developmental-like plasticity in key brain regions. The latter idea is based on observations that repeated antidepressant administration to adult rodents induced changes in molecular and cellular markers, such as specific cell adhesion molecules and ion transporters, that are phenotypic of early developmental stages (Kobayashi et al. 2010; Karpova et al. 2011). It is also proposed that increased production of newly formed neurones is one of the endpoints of the gene expression changes that are critical to the therapeutic effect of antidepressant treatment (Santarelli et al. 2003). This theory is, however, hotly debated and is assessed in detail an accompanying chapter in this volume (Lee et al.).

These ideas on antidepressant drug action are largely based on preclinical models and are quite difficult to test experimentally in depressed patients with any degree of precision, especially at the structural and functional level of synapses. However, models of neural plasticity in humans based on psychophysiological and brain imaging methods are currently under investigation (Nathan et al. 2011), and offer much promise for the clinical testing of antidepressant mechanisms outlined above.

6 Monoamines and Antidepressant Action: Neuropsychological Approaches

The hypothesis that monoamines induce adaptive changes in neural plasticity can be considered in the context of neuropsychological theories of antidepressant drug action. A current popular theory emphasises that altered emotional processing is critical to the actions of drug treatments rather than a direct elevation in mood *per*

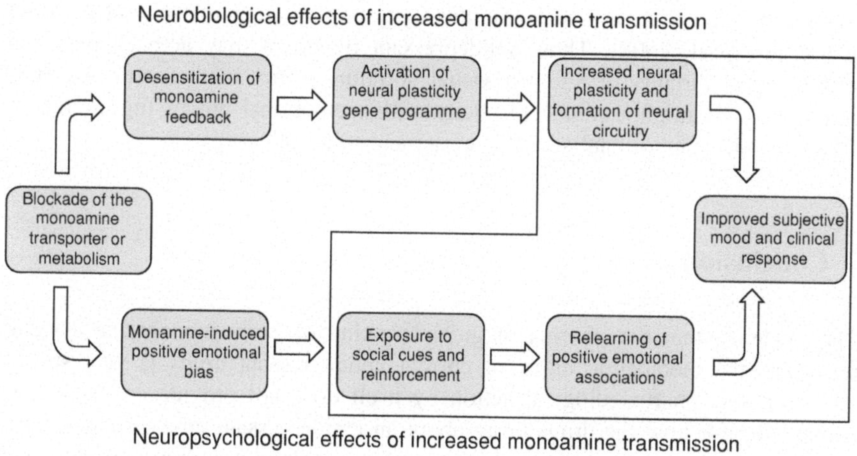

Fig. 1 Illustration of the proposed sequence of adaptive molecular, cellular and neuropsychological changes that occur during a course of several weeks treatment with antidepressant drugs. As indicated by the box, it is envisaged that the increase in neural plasticity occurs within emotional processing networks to allow positive emotional stimuli (e.g. social cues) to act and to be relearned and bring about improved mood. Figure redrafted from Sharp and Cowen (2011) to emphasise the role of both 5-HT and noradrenaline

se. In particular, it is proposed that at the neuropsychological level, antidepressants work by inducing a shift away from a cognitive state of negative affective bias, comprising negatively biased attention, memory and appraisal of internal and external emotional stimuli (Harmer et al. 2009).

Emerging evidence reviewed elsewhere in this volume (Harmer et al.) suggests that monoamines are important to this switch in emotional processing to a positive bias. For example, tryptophan depletion caused negative biases in the recognition of facial expressions in recovered depressed patients (Hayward et al. 2005), whereas SSRI treatment had the opposite effect (Harmer 2008), with both effects occurring independent of changes in mood. Interestingly, SSRIs and other antidepressants evoke a positive bias in emotional processing after a single dose, suggesting that at the neuropsychological level, antidepressants cause an immediate adjustment in negative affective bias. It is argued that this provides the basis for the relearning of new positive emotional associations eventually resulting in an improvement in subjective mood as new, positive, emotional contingencies are learned and experienced (Harmer et al. 2009).

The neural plasticity and neuropsychological hypotheses of antidepressant action are not mutually exclusive (Fig. 1). Indeed, increased neural plasticity within key neural circuits of the limbic system might provide the neural substrate through which increases in positive emotional processing lead to the relearning of positive emotional associations, which then improve mood. This idea is in accordance with recent findings that repeated antidepressant treatment brought about the functional (BDNF-dependent) recovery of circuitry of the visual cortex when combined with a visual stimulus operating through this circuitry (Maya Vetencourt et al. 2008). Thus, antidepressant treatment may activate plasticity within visual circuits to allow a visual stimulus to act in the same way that antidepressants might activate plasticity within emotional processing circuits to allow a positive emotional stimulus to act and to be relearned.

7 Conclusion

The evidence that the pharmacological targeting of brain monoamine systems underpins the therapeutic effect of current antidepressant drugs is very strong. However, this pharmacological action by itself does not provide an inadequate explanation for how the drugs bring about an antidepressant effect. Rather, what seems critical is the consequent triggering of a complex programme of molecular and cellular changes downstream of the monoamine synapse leading to increased neural plasticity. The ability of antidepressants to both facilitate the mechanisms of neural plasticity and remediate negative emotional biases, might allow convergence of neurobiological and psychological explanations of the role of monoamines in antidepressant action.

References

Artigas F, Romero L, de Montigny C, Blier P (1996) Acceleration of the effect of selected antidepressant drugs in major depression by 5-HT_{1A} antagonists. Trends Neurosci 19(9):378–383

Banasr M, Dwyer JM, Duman RS (2011) Cell atrophy and loss in depression: reversal by antidepressant treatment. Curr Opin Cell Biol 23(6):730–737

Blendy JA (2006) The role of CREB in depression and antidepressant treatment. Biol Psychiatry 59(12):1144–1150

Blier P, de Montigny C (1994) Current advances and trends in the treatment of depression. Trends Pharmacol Sci 15(7):220–226

Calabrese F, Molteni R, Gabriel C, Mocaer E, Racagni G, Riva MA (2011) Modulation of neuroplastic molecules in selected brain regions after chronic administration of the novel antidepressant agomelatine. Psychopharmacology (Berl) 215(2):267–275

Cappiello A, McDougle CJ, Malison RT, Heninger GR, Price LH (1995) Yohimbine augmentation of fluvoxamine in refractory depression: a single-blind study. Biol Psychiatry 38(11):765–767

Castren E (2005) Is mood chemistry? Nat Rev Neurosci 6(3):241–246

Castro ME, Harrison PJ, Pazos A, Sharp T (2000) Affinity of (±)-pindolol, (-)-penbutolol, and (-)-tertatolol for pre- and postsynaptic serotonin 5-HT(1A) receptors in human and rat brain. J Neurochem 75(2):755–762

Castro ME, Tordera RM, Hughes ZA, Pei Q, Sharp T (2003) Use of Arc expression as a molecular marker of increased postsynaptic 5-HT function after SSRI/5-HT_{1A} receptor antagonist co-administration. J Neurochem 85(6):1480–1487

Conti AC, Cryan JF, Dalvi A, Lucki I, Blendy JA (2002) cAMP response element-binding protein is essential for the upregulation of brain-derived neurotrophic factor transcription, but not the behavioral or endocrine responses to antidepressant drugs. J Neurosci 22(8):3262–3268

Conti B, Maier R, Barr AM, Morale MC, Lu X, Sanna PP, Bilbe G, Hoyer D, Bartfai T (2007) Region-specific transcriptional changes following the three antidepressant treatments electro convulsive therapy, sleep deprivation and fluoxetine. Mol Psychiatry 12(2):167–189
Coppen A (1967) The biochemistry of affective disorders. Br J Psychiatry 113(504):1237–1264
Covington HE III, Lobo MK, Maze I, Vialou V, Hyman JM, Zaman S, LaPlant Q, Mouzon E, Ghose S, Tamminga CA, Neve RL, Deisseroth K, Nestler EJ (2010) Antidepressant effect of optogenetic stimulation of the medial prefrontal cortex. J Neurosci 30(48):16082–16090
Cowen P, Harmer C (2009) Is it all monoamines? In: Pariante CM, Nesse RM, Nutt D, Wolpert L (eds) Understanding depression: a translational approach. OUP, Oxford, pp 171–180
Cryan JF, O'Leary OF, Jin SH, Friedland JC, Ouyang M, Hirsch BR, Page ME, Dalvi A, Thomas SA, Lucki I (2004) Norepinephrine-deficient mice lack responses to antidepressant drugs, including selective serotonin reuptake inhibitors. Proc Natl Acad Sci U S A 101(21):8186–8191
Delgado PL, Charney DS, Price LH, Aghajanian GK, Landis H, Heninger GR (1990) Serotonin function and the mechanism of antidepressant action. Reversal of antidepressant-induced remission by rapid depletion of plasma tryptophan. Arch Gen Psychiatry 47(5):411–418
Duman RS, Heninger GR, Nestler EJ (1997) A molecular and cellular theory of depression. Arch Gen Psychiatry 54(7):597–606
Duman RS, Voleti B (2012) Signaling pathways underlying the pathophysiology and treatment of depression: novel mechanisms for rapid-acting agents. Trends Neurosci 35(1):47–56
Duric V, Banasr M, Licznerski P, Schmidt HD, Stockmeier CA, Simen AA, Newton SS, Duman RS (2010) A negative regulator of MAP kinase causes depressive behavior. Nat Med 16(11): 1328–1332
Egeland M, Warner-Schmidt J, Greengard P, Svenningsson P (2010) Neurogenic effects of fluoxetine are attenuated in p11 (S100A10) knockout mice. Biol Psychiatry 67(11):1048–1056
Elizalde N, Pastor PM, Garcia–Garcia AL, Serres F, Venzala E, Huarte J, Ramirez MJ, Del Rio J, Sharp T, Tordera RM (2010) Regulation of markers of synaptic function in mouse models of depression: chronic mild stress and decreased expression of VGLUT1. J Neurochem 114(5): 1302–1314
Elsayed M, Banasr M, Duric V, Fournier NM, Licznerski P, Duman RS (2012) Antidepressant effects of fibroblast growth factor-2 in behavioral and cellular models of depression. Biol Psychiatry (in press)
Farley S, Dumas S, El Mestikawy S, Giros B (2012) Increased expression of the Vesicular Glutamate Transporter-1 (VGLUT1) in the prefrontal cortex correlates with differential vulnerability to chronic stress in various mouse strains: effects of fluoxetine and MK-801. Neuropharmacology 62(1):503–517
Garcia–Garcia AL, Elizalde N, Matrov D, Harro J, Wojcik SM, Venzala E, Ramirez MJ, Del Rio J, Tordera RM (2009) Increased vulnerability to depressive-like behavior of mice with decreased expression of VGLUT1. Biol Psychiatry 66(3):275–282
Gartside SE, Umbers V, Hajos M, Sharp T (1995) Interaction between a selective 5-HT1A receptor antagonist and an SSRI in vivo: effects on 5-HT cell firing and extracellular 5-HT. Br J Pharmacol 115(6):1064–1070
Gaska M, Kusmider M, Solich J, Faron-Gorecka A, Krawczyk MJ, Kulakowski K, Dziedzicka-Wasylewska M (2012) Analysis of region-specific changes in gene expression upon treatment with citalopram and desipramine reveals temporal dynamics in response to antidepressant drugs at the transcriptome level. Psychopharmacology (Berl) (in press)
Gelenberg AJ, Chesen CL (2000) How fast are antidepressants? J Clin Psychiatry 61(10):712–721
Greene J, Banasr M, Lee B, Warner-Schmidt J, Duman RS (2009) Vascular endothelial growth factor signaling is required for the behavioral actions of antidepressant treatment: pharmacological and cellular characterization. Neuropsychopharmacology 34(11):2459–2468
Harmer CJ (2008) Serotonin and emotional processing: does it help explain antidepressant drug action? Neuropharmacology 55(6):1023–1028
Harmer CJ, Goodwin GM, Cowen PJ (2009) Why do antidepressants take so long to work? A cognitive neuropsychological model of antidepressant drug action. Br J Psychiatry 195(2): 102–108

Hayward G, Goodwin GM, Cowen PJ, Harmer CJ (2005) Low-dose tryptophan depletion in recovered depressed patients induces changes in cognitive processing without depressive symptoms. Biol Psychiatry 57(5):517–524

Holmes A, Yang RJ, Murphy DL, Crawley JN (2002) Evaluation of antidepressant-related behavioral responses in mice lacking the serotonin transporter. Neuropsychopharmacology 27(6):914–923

Hughes ZA, Starr KR, Scott CM, Newson MJ, Sharp T, Watson JM, Hagan JJ, Dawson LA (2007) Simultaneous blockade of 5-HT$_{1A/B}$ receptors and 5-HT transporters results in acute increases in extracellular 5-HT in both rats and guinea pigs: in vivo characterization of the novel 5-HT$_{1A/B}$ receptor antagonist/5-HT transport inhibitor SB-649915-B. Psychopharmacology (Berl) 192(1):121–133

Karpova NN, Pickenhagen A, Lindholm J, Tiraboschi E, Kulesskaya N, Agustsdottir A, Antila H, Popova D, Akamine Y, Bahi A, Sullivan R, Hen R, Drew LJ, Castren E (2011) Fear erasure in mice requires synergy between antidepressant drugs and extinction training. Science 334(6063): 1731–1734

Kobayashi K, Ikeda Y, Sakai A, Yamasaki N, Haneda E, Miyakawa T, Suzuki H (2010) Reversal of hippocampal neuronal maturation by serotonergic antidepressants. Proc Natl Acad Sci U S A 107(18):8434–8439

Krishnan V, Nestler EJ (2008) The molecular neurobiology of depression. Nature 455(7215): 894–902

Larsen MH, Rosenbrock H, Sams-Dodd F, Mikkelsen JD (2007) Expression of brain derived neurotrophic factor, activity-regulated cytoskeleton protein mRNA, and enhancement of adult hippocampal neurogenesis in rats after sub-chronic and chronic treatment with the triple monoamine re-uptake inhibitor tesofensine. Eur J Pharmacol 555(2–3):115–121

Lindefors N, Brodin E, Metsis M (1995) Spatiotemporal selective effects on brain-derived neurotrophic factor and trkB messenger RNA in rat hippocampus by electroconvulsive shock. Neuroscience 65(3):661–670

Lindholm JS, Autio H, Vesa L, Antila H, Lindemann L, Hoener MC, Skolnick P, Rantamaki T, Castren E (2012) The antidepressant-like effects of glutamatergic drugs ketamine and AMPA receptor potentiator LY 451646 are preserved in bdnf(+)/(-) heterozygous null mice. Neuropharmacology 62(1):391–397

Lominac KD, Oleson EB, Pava M, Klugmann M, Schwarz MK, Seeburg PH, During MJ, Worley PF, Kalivas PW, Szumlinski KK (2005) Distinct roles for different Homer1 isoforms in behaviors and associated prefrontal cortex function. J Neurosci 25(50):11586–11594

Lopez JP, Mamdani F, Labonte B, Beaulieu MM, Yang JP, Berlim MT, Ernst C, Turecki G (2012) Epigenetic regulation of BDNF expression according to antidepressant response. Mol Psychiatry (in press)

Lucas G, Rymar VV, Du J, Mnie-Filali O, Bisgaard C, Manta S, Lambas-Senas L, Wiborg O, Haddjeri N, Pineyro G, Sadikot AF, Debonnel G (2007) Serotonin(4) (5-HT(4)) receptor agonists are putative antidepressants with a rapid onset of action. Neuron 55(5):712–725

Mannoury la Cour C, Boni C, Hanoun N, Lesch KP, Hamon M, Lanfumey L (2001) Functional consequences of 5-HT transporter gene disruption on 5-HT(1a) receptor-mediated regulation of dorsal raphe and hippocampal cell activity. J Neurosci 21(6):2178–2185

Marek GJ, Carpenter LL, McDougle CJ, Price LH (2003) Synergistic action of 5-HT$_{2A}$ antagonists and selective serotonin reuptake inhibitors in neuropsychiatric disorders. Neuropsychopharmacology 28(2):402–412

Maya Vetencourt JF, Sale A, Viegi A, Baroncelli L, De Pasquale R, O'Leary OF, Castren E, Maffei L (2008) The antidepressant fluoxetine restores plasticity in the adult visual cortex. Science 320(5874):385–388

Mayorga AJ, Dalvi A, Page ME, Zimov-Levinson S, Hen R, Lucki I (2001) Antidepressant-like behavioral effects in 5-hydroxytryptamine(1A) and 5-hydroxytryptamine(1B) receptor mutant mice. J Pharmacol Exp Ther 298(3):1101–1107

Melas PA, Rogdaki M, Lennartsson A, Bjork K, Qi H, Witasp A, Werme M, Wegener G, Mathe AA, Svenningsson P, Lavebratt C (2012) Antidepressant treatment is associated with epigenetic alterations in the promoter of P11 in a genetic model of depression. Int J Neuropsychopharmacol 15(5):669–679

Menke A, Klengel T, Binder, EB (2012) Epigenetics, depression and antidepressant treatment. Curr Pharm Des (in press)

Meyer JH, Wilson AA, Sagrati S, Hussey D, Carella A, Potter WZ, Ginovart N, Spencer EP, Cheok A, Houle S (2004) Serotonin transporter occupancy of five selective serotonin reuptake inhibitors at different doses: an [^{11}C]DASB positron emission tomography study. Am J Psychiatry 161(5):826–835

Millan MJ (2009) Dual- and triple-acting agents for treating core and co-morbid symptoms of major depression: novel concepts, new drugs. Neurotherapeutics 6(1):53–77

Mongeau R, Martin CB, Chevarin C, Maldonado R, Hamon M, Robledo P, Lanfumey L (2010) 5-HT_{2C} receptor activation prevents stress-induced enhancement of brain 5-HT turnover and extracellular levels in the mouse brain: modulation by chronic paroxetine treatment. J Neurochem 115(2):438–449

Murrough JW (2012) Ketamine as a novel antidepressant: from synapse to behavior. Clin Pharmacol Ther 91(2):303–309

Nathan PJ, Cobb SR, Lu B, Bullmore ET, Davies CH (2011) Studying synaptic plasticity in the human brain and opportunities for drug discovery. Curr Opin Pharmacol 11(5):540–548

Nibuya M, Morinobu S, Duman RS (1995) Regulation of BDNF and trkB mRNA in rat brain by chronic electroconvulsive seizure and antidepressant drug treatments. J Neurosci 15(11): 7539–7547

Ogren SO, Fuxe K (1985) Effects of antidepressant drugs on serotonin receptor mechanisms. Neuropharmacology of serotonin. Oxford University Press, Oxford 131–180

Okamoto H, Voleti B, Banasr M, Sarhan M, Duric V, Girgenti MJ, Dileone RJ, Newton SS, Duman RS (2010) Wnt2 expression and signaling is increased by different classes of antidepressant treatments. Biol Psychiatry 68(6):521–527

Pei Q, Zetterstrom TS, Sprakes M, Tordera R, Sharp T (2003) Antidepressant drug treatment induces Arc gene expression in the rat brain. Neuroscience 121(4):975–982

Portella MJ, de Diego-Adelino J, Ballesteros J, Puigdemont D, Oller S, Santos B, Alvarez E, Artigas F, Perez V (2011) Can we really accelerate and enhance the selective serotonin reuptake inhibitor antidepressant effect? A randomized clinical trial and a meta-analysis of pindolol in nonresistant depression. J Clin Psychiatry 72(7):962–969

Queree P, Peters S, Sharp T (2009) Further pharmacological characterization of 5-HT(2C) receptor agonist-induced inhibition of 5-HT neuronal activity in the dorsal raphe nucleus in vivo. Br J Pharmacol 158(6):1477–1485

Rabiner EA, Gunn RN, Castro ME, Sargent PA, Cowen PJ, Koepp MJ, Meyer JH, Bench CJ, Harrison PJ, Pazos A, Sharp T, Grasby PM (2000) beta-blocker binding to human 5-HT(1A) receptors in vivo and in vitro: implications for antidepressant therapy. Neuropsychopharmacology 23(3):285–293

Ren-Patterson RF, Cochran LW, Holmes A, Lesch KP, Lu B, Murphy DL (2006) Gender-dependent modulation of brain monoamines and anxiety-like behaviors in mice with genetic serotonin transporter and BDNF deficiencies. Cell Mol Neurobiol 26(4–6):755–780

Riad M, Zimmer L, Rbah L, Watkins KC, Hamon M, Descarries L (2004) Acute treatment with the antidepressant fluoxetine internalizes 5-HT_{1A} autoreceptors and reduces the in vivo binding of the PET radioligand [^{18}F]MPPF in the nucleus raphe dorsalis of rat. J Neurosci 24(23):5420–5426

Richardson-Jones JW, Craige CP, Guiard BP, Stephen A, Metzger KL, Kung HF, Gardier AM, Dranovsky A, David DJ, Beck SG, Hen R, Leonardo ED (2010) 5-HT_{1A} autoreceptor levels determine vulnerability to stress and response to antidepressants. Neuron 65(1):40–52

Ruhe HG, Mason NS, Schene AH (2007) Mood is indirectly related to serotonin, norepinephrine and dopamine levels in humans: a meta-analysis of monoamine depletion studies. Mol Psychiatry 12(4):331–359

Saarelainen T, Hendolin P, Lucas G, Koponen E, Sairanen M, MacDonald E, Agerman K, Haapasalo A, Nawa H, Aloyz R, Ernfors P, Castren E (2003) Activation of the TrkB neurotrophin receptor is induced by antidepressant drugs and is required for antidepressant-induced behavioral effects. J Neurosci 23(1):349–357

Sanacora G, Berman RM, Cappiello A, Oren DA, Kugaya A, Liu N, Gueorguieva R, Fasula D, Charney DS (2004) Addition of the alpha2-antagonist yohimbine to fluoxetine: effects on rate of antidepressant response. Neuropsychopharmacology 29(6):1166–1171

Santarelli L, Saxe M, Gross C, Surget A, Battaglia F, Dulawa S, Weisstaub N, Lee J, Duman R, Arancio O, Belzung C, Hen R (2003) Requirement of hippocampal neurogenesis for the behavioral effects of antidepressants. Science 301(5634):805–809

Schildkraut JJ (1965) The catecholamine hypothesis of affective disorders: a review of supporting evidence. Am J Psychiatry 122(5):509–522

Schmidt EF, Warner-Schmidt JL, Otopalik BG, Pickett SB, Greengard P, Heintz N (2012) Identification of the cortical neurons that mediate antidepressant responses. Cell 149(5):1152–1163

Scorza MC, Llado-Pelfort L, Oller S, Cortes R, Puigdemont D, Portella MJ, Perez-Egea R, Alvarez E, Celada P, Perez V, Artigas F (2012) Preclinical and clinical characterization of the selective serotonin-1A receptor antagonist DU-125530 for antidepressant treatment. Br J Pharmacol (in press)

Serres F, Millan MJ, Sharp T (2012) Molecular adaptation to chronic antidepressant treatment: evidence for a more rapid response to the novel alpha(2)-adrenoceptor antagonist/5-HT-noradrenaline reuptake inhibitor (SNRI), S35966, compared to the SNRI, venlafaxine. Int J Neuropsychopharmacol 15(5):617–629

Sharp T, Boothman L, Raley J, Queree P (2007) Important messages in the 'post': recent discoveries in 5-HT neurone feedback control. Trends Pharmacol Sci 28(12):629–636

Sharp T, Cowen PJ (2011) 5-HT and depression: is the glass half-full? Curr Opin Pharmacol 11(1):45–51

Sillaber I, Panhuysen M, Henniger MS, Ohl F, Kuhne C, Putz B, Pohl T, Deussing JM, Paez-Pereda M, Holsboer F (2008) Profiling of behavioral changes and hippocampal gene expression in mice chronically treated with the SSRI paroxetine. Psychopharmacology (Berl) 200(4):557–572

Stahl SM, Palazidou L (1986) The pharmacology of depression: studies of neurotransmitter receptors lead the search for biochemical lesions and new drug therapies. Trends Pharmacol Sci 9:349–354

Svenningsson P, Chergui K, Rachleff I, Flajolet M, Zhang X, El Yacoubi M, Vaugeois JM, Nomikos GG, Greengard P (2006) Alterations in 5-HT_{1B} receptor function by p11 in depression-like states. Science 311(5757):77–80

Svenningsson P, Tzavara ET, Qi H, Carruthers R, Witkin JM, Nomikos GG, Greengard P (2007) Biochemical and behavioral evidence for antidepressant-like effects of 5-HT_6 receptor stimulation. J Neurosci 27(15):4201–4209

Szumlinski KK, Ary AW, Lominac KD (2008) Homers regulate drug-induced neuroplasticity: implications for addiction. Biochem Pharmacol 75(1):112–133

Tardito D, Musazzi L, Tiraboschi E, Mallei A, Racagni G, Popoli M (2009) Early induction of CREB activation and CREB-regulating signalling by antidepressants. Int J Neuropsychopharmacol 12(10):1367–1381

Thome J, Sakai N, Shin K, Steffen C, Zhang YJ, Impey S, Storm D, Duman RS (2000) cAMP response element-mediated gene transcription is upregulated by chronic antidepressant treatment. J Neurosci 20(11):4030–4036

Tordera RM, Pei Q, Sharp T (2005) Evidence for increased expression of the vesicular glutamate transporter, VGLUT1, by a course of antidepressant treatment. J Neurochem 94(4):875–883

Tsankova NM, Berton O, Renthal W, Kumar A, Neve RL, Nestler EJ (2006) Sustained hippocampal chromatin regulation in a mouse model of depression and antidepressant action. Nat Neurosci 9(4):519–525

Tyrer P, Marsden CA (1985) New antidepressant drugs: is there anything new they tell us about depression? Trends Neurosci 8:427–431

Vollenweider FX, Kometer M (2010) The neurobiology of psychedelic drugs: implications for the treatment of mood disorders. Nat Rev Neurosci 11(9):642–651

Warner-Schmidt JL, Duman RS (2008) VEGF as a potential target for therapeutic intervention in depression. Curr Opin Pharmacol 8(1):14–19

Whale R, Terao T, Cowen P, Freemantle N, Geddes J (2010) Pindolol augmentation of serotonin reuptake inhibitors for the treatment of depressive disorder: a systematic review. J Psychopharmacol 24(4):513–520

Yanpallewar SU, Fernandes K, Marathe SV, Vadodaria KC, Jhaveri D, Rommelfanger K, Ladiwala U, Jha S, Muthig V, Hein L, Bartlett P, Weinshenker D, Vaidya VA (2010) Alpha2-adrenoceptor blockade accelerates the neurogenic, neurotrophic, and behavioral effects of chronic antidepressant treatment. J Neurosci 30(3):1096–1109

Neurosurgical Treatments of Depression

Yasin Temel and Lee Wei Lim

Abstract The neurosurgical treatment of treatment-resistant depression (TRD) has entered a new era with more and more patients being treated with deep brain stimulation (DBS) via surgically implanted intracerebral electrodes. Although the mechanisms of action of DBS are still not fully understood, preclinical studies are being conducted to elucidate how the treatment might work. DBS in its present form can be considered as a relatively new neurosurgical treatment for TRD. However, the use of neurosurgery in the management of depression has a much longer history particularly with ablative procedures but also vagal nerve stimulation. Here, we provide a review of the clinical neurosurgical treatments for TRD, with a main emphasis on DBS. In addition, we discuss relevant preclinical data that are revealing new information about DBS mechanisms.

Keywords Treatment resistant depression · Neurosurgery · Ablative surgery · Vagal nerve stimulation · Deep brain stimulation

---

Y. Temel · L. W. Lim
Department of Neurosurgery and Neuroscience, Maastricht University Medical Center, Maastricht, The Netherlands

L. W. Lim
School of Biological Sciences, Nanyang Technological University, Singapore, Singapore

Y. Temel (✉)
Department of Neurosurgery, University Hospital Maastricht, PO Box 5800 6202 AZ, Maastricht, The Netherlands
e-mail: y.temel@maastrichtuniversity.nl

Curr Topics Behav Neurosci (2013) 14: 327–339
DOI: 10.1007/7854_2012_222

Published Online: 4 August 2012

Contents

1 Background

Neurosurgery in patients with treatment-resistant depression (TRD) has regained interest due to recent advances in surgical approaches. The introduction of deep brain stimulation (DBS) of the subgenual cingulate gyrus (Mayberg et al. 2005) and latterly the nucleus accumbens (NAc) (Schlaepfer et al. 2007) as a potential therapy for patients with TRD, has invigorated research in this field. Since these initial reports, we have further witnessed further publications in the field. The application of DBS in patients with TRD was actually not surprising when one considers the preceding developments.

DBS in its modern form was introduced as a treatment for patients with advanced Parkinson's disease (PD) in the late 1980s (Benabid et al. 1987), and is thus far probably the most successful indication for this approach (Temel and Visser-Vandewalle 2006). This success was driven by a very solid scientific base. After extensive electrophysiological recordings in the parkinsonian non-human primate, Miller and DeLong (1987) demonstrated that the subthalamic nucleus (STN) of the basal ganglia exhibited excessive burst activity. These observations were the basis for the pioneering work of Bergman et al. (1990), who showed that lesions of the STN reduced the major motor disturbances in a primate model of PD. These results supported the hypothesis of excessive activity in the STN in PD, and that ablation of this nucleus was able to improve PD symptoms. In the meantime, Benabid and co-workers had started to use DBS as an alternative for ablative surgery in movement disorders. Motivated by the results of the lesion studies, and then by a crucial stimulation study in monkeys in the same year (Benazzouz et al. 1993), this group was the first to successfully perform STN DBS in a patient suffering from advanced PD (Pollak et al. 1993).

This successful use of electrodes to modulate abnormal neuronal activity in discrete parts of the basal ganglia of PD patients initiated the idea of using DBS to treat TRD (Ressler and Mayberg 2007). DBS in its present form is new, but electrical brain stimulation, however, has been applied earlierin the past in experiments involving psychiatric patients. Brain imaging and other evidence implicated overactivity in the subgenual cingulate gyrus, part of the anterior cingulate gyrus (ACg) (Haznedar et al. 2004) and anatomically known as the Brodmann area 25, in acute sadness (Mayberg et al. 1999; Ressler and Mayberg 2007). In addition, the antidepressant effects of drugs, electroconvulsive therapy, and ablative surgery were found to be correlated with a selective decrease in ACg activity (Mayberg et al. 2000; Nobler et al. 2001; Mottaghy et al. 2002; Dougherty et al. 2003; Mayberg et al. 2005). In animal models of depression, similar results were reported. Thus, the anterior cingulate was found to be involved both in the pathophysiology of depressive-like behaviour and in the mediation of antidepressant effects of drugs in rodents (Slattery et al. 2005; Kuipers et al. 2006).

Soon after the report of the first patients with TRD implanted with DBS electrodes in the subgenual cingulate gyrus (Mayberg et al. 2005), follow on publications appeared with one of the major targets being the nucleus accumbens (NAc). The rationale of applying DBS to the NAc in depressed patients was largely derived from evidence linking anhedonia—the inability to experience pleasure from previously pleasurable activities and a core symptom of depression—to dysfunctions of the NAc (Epstein et al. 2006). Initial reports suggest that DBS of the NAc is effective in reducing anhedonia in clinical cases (Schlaepfer et al. 2007).

With these above-mentioned developments, the neurosurgical treatment of TRD entered a new era. Although the mechanisms of action of DBS are still not fully understood, more and more patients are being implanted with DBS electrodes. In addition, preclinical studies are being conducted to elucidate how DBS might be working. A review of the clinical and preclinical data on the effects of DBS in depression follows later in this chapter. For more detailed information about the actions of DBS, the reader is referred to other reviews (Andrade et al. 2010; Hamani and Temel 2012).

DBS in its present form can be considered as a relatively new neurosurgical treatment for TRD. However, neurosurgery for depression has a much longer history with ablative procedures and vagal nerve stimulation (VNS). Occassionally experimental brain stimulation was performed.

2 Ablative Surgery

Throughout recent history many different brain regions have been targeted for ablative surgery in patients with mood disorders. The outcomes and side effects of the procedures were often not well documented. It is probable that some of the operated patients were misdiagnosed or were suffering from complex psychiatric conditions. We describe below the most frequently performed procedures with their reported outcomes and side effects. There are case reports reporting lesions of different brain areas other than the ones described below, and these are beyond the scope of this review.

2.1 Lobotomy/Leucotomy

Moniz was one of the first to use the term ‘psychosurgery’, and he was a pioneer in popularising the procedure of prefrontal leucotomy in psychiatric patients, already in the late 1940s (Moniz 1937). Freeman and Watts, with their *Freeman-Watts standard lobotomy*, performed lesions using a 5 cm long surgical instrument (‘leucotome’) which was inserted supraorbitally and directed towards the coronal suture (Freeman and Watts 1937). Lobotomy/leucotomy was at that time a widely accepted treatment in the community for patients with psychiatric disorders, including severe mood disorders, and was unfortunately been performed on a large scale. It has been estimated that at that time approximately 5,000 psychosurgery operations were conducted annually in the United States (Swayze 1995). Of note, Moniz was awarded the nobel prize for medicine for his contributions to this field in 1949.

Publications between 1951 and 1972 showed varying results from this neurosurgical approach, with improvements being reported to range from 30 to 100 % of patients treated (Elithorn 1959; Mc 1959; Birley 1964; Kelly et al. 1966; 1972). Two studies however showed no difference in the number of hospital discharges between patients who underwent leucotomy and non-operated patients (Pool 1949; Little 2009). In the early 1960s, a publication from the United Kingdom reported that in a cohort of more than 10,000 psychiatric patients with affective disorders, schizophrenia and other psychiatric illnesses, surgery induced an improvement of the core symptoms, but many patients suffered side effects such as epilepsy, personality changes, urine incontinence and there was a worrying incidence of mortality of approximately 4 % (Tooth and Newton 1961).

2.2 Orbital and Cortical Undercutting

In 1949, Scoville described the supraorbital cortical undercutting of Brodmann areas 9 and 10, cingulate gyrus and orbital surface in psychiatric patients. The idea was to make fine lesions without damaging the main vasculature. It was reported that patients undergoing this approach showed a substantial improvement (Scoville 1949, 1960). In 1955, Knight introduced a variation of Scoville’s approach, the *restricted orbital undercutting*, which was a more limited and medially located lesion. His arguments were that the lateral fibres were not involved in psychiatric disorders, and that laterally located lesions could result in undesirable personality changes. This procedure was reported to produce a reasonable improvement of symptoms in the vast majority of the patients with depression and other psychiatric conditions (Knight and Tredgold 1955). In the procedures of both Scoville and Knight, the most important side effects included personality changes, epilepsy and urine incontinence.

2.3 Cingulectomy/Cingulotomy

Le Beau was probably one of the first to describe cingulectomy as a treatment for psychiatric disorders in 1949. Cingulectomies were performed in monkeys and resulted in unresponsiveness and placidity. Le Beau (1954) found no therapeutic effects in depressive patients with anxiety and pain, but obtained a favourable outcome in patients with depression and obsessive neurosis . Later, this procedure was performed using stereotaxy when it became available for human use. The reported incidence of therapeutic effects varied from 44 to 92 % improvement in the operated patients, and side effects included personality changes, epilepsy, weight gain and urine incontinence (Brown and Lighthill 1968; Bailey et al. 1973; Spangler et al. 1996; Ridout et al. 2007).

2.4 Subcaudate Tractotomy and Limbic Leucotomy

Subcaudate tractotomy refers to lesions made in the area located anteroventrally to the head of the caudate nucleus in the substantia innominata, and was described by Knight (1964). The goal was to interrupt the fibres of the substantia innominata that run from the frontal cortex and amygdala to the hypothalamus. Therapeutic effects among patients varied from a moderate effect to major improvement. The most frequently seen side effects were fatigue, weight gain and epilepsy (Knight 1965; Bridges et al. 1973; Goktepe et al. 1975; Hodgkiss et al. 1995).

In 1973, Kelly (psychiatrist) and Richardson (neurosurgeon) described stereotactic limbic leucotomy which comprised two procedures, an anterior cingulotomy and subcaudate tractotomy. Previous studies involving both areas had shown good results when the procedures were performed separately. In the first series, most of the patients with depression showed mild to marked improvements in their core symptoms (Kelly et al. 1973a, b). Other studies reported a similar level of improvement. Side effects included drowsiness, weight gain and memory problems; but also epilepsy, urine incontinence and personality changes (Kelly et al. 1972; Montoya et al. 2002; Sachdev and Sachdev 2005).

3 Vagal Nerve Stimulation

Vagal nerve stimulation (VNS) was first introduced in 1985 by Zabara as an experimental anticonvulsant procedure in dogs (Zabara 1985, 1992), and a few years later in patients with refractory epilepsy (Penry and Dean 1990). Once it became an established therapy in refractory epilepsy at the end of the 1990s, Rush and co-workers proposed VNS as a treatment for TRD (George et al. 2000; Rush et al. 2000). The rationale was based on the observation that epileptic patients with VNS showed an improvement of mood (Elger et al. 2000; Harden et al. 2000; Hoppe et al. 2001). Studies with long-term follow-up have shown that VNS can be beneficial in patients with major depression, while the therapeutic effect at short term can be less pronounced (Rush et al. 2000, 2005; Schlaepfer et al. 2008).

At present the exact place of VNS in the management of TRD is unclear. Further clinical trials are needed to confirm the efficacy of this approach, and to identify the mechanism underlying the putative antidepressant effect. As an alternative to VNS as a stimulation approach, DBS is gaining more and more interest.

4 Deep Brain Stimulation

Below, relevant reports on the clinical application of DBS in patients suffering from TRD are outlined. These reports are categorised according to brain target for DBS. In addition, preclinical investigations are also discussed in this section. The latter have been performed to understand the mechanisms of action of DBS, to optimise the existing therapy and/or to explore new indications or brain targets for DBS.

4.1 Subgenual Cingulate Gyrus

In 2005, Mayberg and coworkers presented their first results of DBS of the subgenual cingulate gyrus in TRD (Mayberg et al. 2005). Their aim was to use high frequency stimulation to inhibit the subgenual cingulate cortex, which brain imaging studies showed to be hyperactive in chronically depressed patients. In this first report, six patients suffering from TRD received stimulating electrodes at the level of the subgenual cingulate gyrus. Striking and sustained remission of depressive symptoms was observed in four out of six patients, with a reduction of more than 50 % on the hamilton depression rating scale (HDRS). Clinically, improvement in mood was accompanied by an increase in energy, interest, psychomotor speed and a decrease of apathy and anhedonia. In addition, positron emission tomography (PET) images demonstrated normalisation of the metabolic activity of the stimulated area.

Recently, the same group published their long term findings in 20 patients with TRD (Kennedy et al. 2011). The percentage of responders (defined as more than 50 % improvement on the depression score), for 1, 2, 3 years and the last follow-up (range 3–6 years) was 62.5, 46.2, 75 and 64.3 %, respectively. Functional impairment in the areas of physical health and social functioning progressively improved up to the last follow-up visit. No significant side effects were seen. However, two patients died by suicide during depressive relapses.

4.2 Nucleus Accumbens

Schlaepfer and associates performed DBS of the NAc, initially in three patients with TRD, and observed immediate effects on depressive symptoms. Evaluations over double-blinded periods with stimulation on or off showed a beneficial effect on depressive symptoms in all patients when DBS was on (Schlaepfer et al. 2007). PET imaging showed enhanced metabolic activity in the NAc, amygdala, and dorsolateral and dorsomedial prefrontal cortices, and reduced metabolism in the ventral and ventrolateral medial prefrontal cortices. In subsequent publications, the same group described more patients with a longer follow-up period (Bewernick et al. 2010, 2012). In their latest publication, 11 patients were described, with a follow-up period of 4 years for some patients. Using a response criterion of at least 50 % improvement in the HDRS, 5 out of 11 patients met the criterion of being a responder. The responders showed a strong and sustained therapeutic effect of DBS. From the non-responder group, however, one patient committed suicide and one patient attempted suicide (Bewernick et al. 2012).

In 2009, Malone and colleagues performed a study of DBS of the ventral capsule/ventral striatum in 15 patients with TRD (Malone et al. 2009). Interestingly, the percent of patients responding was 40 % at 6 months and 53.3 % at the last follow-up (up to 4 years) examination. No serious adverse effects were reported.

4.3 Other Targets

Jimenez and co-workers described the effects of DBS of the inferior thalamic peduncle region in a 49-year-old woman with TRD. The patient had a history of recurrent episodes of major depression for over 20 years, and DBS resulted in substantial reduction of the depressive symptoms (Jimenez et al. 2005).

Kosel and associates described the case of a 62-year-old woman with TRD and comorbid neuroleptic-induced tardive dyskinesias. DBS of the globus pallidus internus not only reduced the dyskinesias, but also the HDRS-score, the latter being reduced by around 50 % (Kosel et al. 2007).

In another case report, the therapeutic effects of DBS of the lateral habenula was investigated in a patient with TRD. Interestingly, depressive symptoms reappeared when the DBS was switched off as a result of an incidental bicycle accident. After turning on the stimulator, depressive symptoms were substantially reduced (Sartorius et al. 2010).

4.4 Preclinical Studies of DBS in Depression

In rats, high frequency stimulation of the ventromedial prefrontal cortex (vmPFC), a region considered to be the equivalent of the human subgenual cingulate gyrus (Uylings et al. 2003; Vertes 2004; Hamani et al. 2011b), is reported to induce an antidepressant-like effect in depression models, specifically the forced swim test and chronic unpredictable mild stress (Hamani et al. 2010a, b). Evidence suggested that this effect was (at least in part) mediated by the serotonin (5-hydroxytryptamine; 5-HT) system because the antidepressant effect of vmPFC stimulation was attenuated in rats with a 5-HT depletion (Hamani et al. 2010b, 2011a). In contrast in rats with a noradrenaline depletion, vmPFC stimulation still induced an antidepressant-like effect (Hamani et al. 2010b). Stimulation of the vmPFC, accumbens or ventral tegmental area increased levels of brain-derived neurotrophic factor (BDNF) in rodents undergoing chronic unpredictable mild stress (Gersner et al. 2010; Hamani et al. 2011a) as well as in depression-prone flinders rats (Friedman et al. 2009). This finding is of great interest because commonly used antidepressant drugs and electroconvulsive therapy increase BDNF levels in several brain regions (Nibuya et al. 1995; Altar et al. 2003). Moreover, BDNF levels are reduced in both stressed rodents and patients with depression (Nibuya et al. 1995; Smith et al. 1995; Shimizu et al. 2003).

Animal models can also be used to find new targets and better stimulation parameters for DBS treatment of depression. For example, at settings mimicking the physiological firing properties of neurons in the ventral tegmental area, electrical stimulation of this region induced antidepressant-like effects in flinders rats (Friedman et al. 2009). A clinical trial to investigate the effects of DBS of a fibre pathway connecting the ventral tegmental area to the nucleus accumbens and prefrontal cortex (i.e., the medial forebrain bundle) has been proposed (Coenen et al. 2011).

The cellular mechanism of action of DBS can also be investigated in preclinical models. To study whether the antidepressant-like effect of vmPFC stimulation in the forced swim test was due to inhibition of the stimulated brain region, the outcome of vmPFC stimulation was compared to that observed after direct injections of muscimol into the vmPFC (which mimics actions of the inhibitory transmitter GABA) as well as radiofrequency lesion (which eliminate the neural target) or ibotenic acid lesion (which injure neuronal cell bodies while preserving local fibres) of the vmPFC (Banasr and Duman 2008; Hamani et al. 2010b; Slattery et al. 2010). Ibotenic acid lesion did not have an antidepressant-like effect in the forced swim test (Banasr and Duman 2008; Hamani et al. 2010b; Slattery et al. 2010). In comparison, muscimol injections and radiofrequency lesions (Banasr and Duman 2008; Hamani et al. 2010b; Slattery et al. 2010) showed some effect, but not to the same extent as electrical stimulation (Hamani et al. 2010b). These results suggest that the antidepressant-like effects of vmPFC stimulation in the forced swim test are not just a result of target inactivation but also rely on additional mechanisms.

5 Conclusion

The neurosurgical treatments of TRD tested thus far comprise three types, ablative surgery, VNS and DBS. Ablative surgery has been mainly performed in the beginning and the middle of the twentieth century, VNS was popularised at the end of the twentieth century, and at the beginning of the twenty-first century DBS in its current form has been introduced. The results of ablative surgery have been very diverse, and side effects have been substantial in a considerable number of patients. Ablative surgery is nowadays, according to the knowledge of the authors, rarely performed. The results of both VNS and DBS are promising, but larger scale controlled studies with long-term follow-up evaluations are lacking. On the basis of the available data, VNS does not appear as effective as DBS. However, this still needs to be confirmed in a comparative study to draw final conclusions. An advantage of VNS is that it is minimally invasive compared to ablative surgery or DBS. VNS for TRD is currently being performed in several centres worldwide.

In our opinion, on the basis of available clinical and preclinical evidence, at present the most promising procedure for the treatment of severe TRD is DBS. Thus encouraging data are accumulating on the safety and efficacy of DBS in patients with TRD. Nevertheless, a major concern is that the clinical application of DBS in TRD, and also in other psychiatric illnesses, is moving faster than the scientific evidence supporting or discouraging the application of this approach. The field is no longer waiting for case reports of DBS in a new and exciting brain region, but for well-designed clinical trials backed up with robust scientific evidence. This increases the likelihood of a therapy to become successful, as illustrated by the history behind the successful application of DBS in the treatment of severe PD.

Finally, we are witnessing the revival of neurosurgery not only in the field of TRD, but also in other psychiatric and illnesses. Perhaps with a collaborative effort of neurosurgeons, psychiatrists, scientists and ethicists, the 'negative' societal reputation of psychosurgery which has originated from experiences in the twentieth century can be changed to the benefit of severely ill patients and their carers.

Acknowledgments The scientific work of YT and LWL in the field of DBS in depression received support from the Netherlands Organization for Scientific Research (NWO Veni), Dutch Brain Foundation, and the Singapore Lee Kuan Yew Research Fellowship.

References

Altar CA, Whitehead RE, Chen R, Wortwein G, Madsen TM (2003) Effects of electroconvulsive seizures and antidepressant drugs on brain-derived neurotrophic factor protein in rat brain. Biol Psychiatry 54:703–709

Andrade P, Noblesse LH, Temel Y et al (2010) Neurostimulatory and ablative treatment options in major depressive disorder: a systematic review. Acta Neurochir (Wien) 152:565–577

Bailey HR, Dowling JL, Davies E (1973) Studies in depression. 3. The control of affective illness by cingulotractotomy: a review of 150 cases. Med J Aust 2:366–371

Banasr M, Duman RS (2008) Glial loss in the prefrontal cortex is sufficient to induce depressive-like behaviors. Biol Psychiatry 64:863–870

Benabid AL, Pollak P, Louveau A, Henry S, de Rougemont J (1987) Combined (thalamotomy and stimulation) stereotactic surgery of the VIM thalamic nucleus for bilateral parkinson disease. Appl Neurophysiol 50:344–346

Benazzouz A, Gross C, Feger J, Boraud T, Bioulac B (1993) Reversal of rigidity and improvement in motor performance by subthalamic high-frequency stimulation in MPTP-treated monkeys. Eur J Neurosci 5:382–389

Bergman H, Wichmann T, DeLong MR (1990) Reversal of experimental parkinsonism by lesions of the subthalamic nucleus. Science 249:1436–1438

Bewernick BH, Hurlemann R, Matusch A, Kayser S, Grubert C, Hadrysiewicz B, Axmacher N, Lemke M, Cooper-Mahkorn D, Cohen MX, Brockmann H, Lenartz D, Sturm V, Schlaepfer TE (2010) Nucleus accumbens deep brain stimulation decreases ratings of depression and anxiety in treatment-resistant depression. Biol Psychiatry 67:110–116

Bewernick BH, Kayser S, Sturm V et al (2012) Long-term effects of nucleus accumbens deep brain stimulation in treatment-resistant depression: evidence for sustained efficacy. Neuropsychopharmacology

Birley JL (1964) Modified frontal leucotomy: a review of 106 cases. Br J Psychiatry 110:211–221

Bridges PK, Goktepe EO, Maratos J (1973) A comparative review of patients with obsessional neurosis and with depression treated by psychosurgery. Br J Psychiatry 123:663–674

Brown MH, Lighthill JA (1968) Selective anterior cingulotomy: a psychosurgical evaluation. J Neurosurg 29:513–519

Coenen VA, Schlaepfer TE, Maedler B, Panksepp J (2011) Cross-species affective functions of the medial forebrain bundle-implications for the treatment of affective pain and depression in humans. Neurosci Biobehav Rev 35:1971–1981

Dougherty DD, Weiss AP, Cosgrove GR, Alpert NM, Cassem EH, Nierenberg AA, Price BH, Mayberg HS, Fischman AJ, Rauch SL (2003) Cerebral metabolic correlates as potential predictors of response to anterior cingulotomy for treatment of major depression. J Neurosurg 99:1010–1017

Elger G, Hoppe C, Falkai P, Rush AJ, Elger CE (2000) Vagus nerve stimulation is associated with mood improvements in epilepsy patients. Epilepsy Res 42:203–210

Elithorn A (1959) Discussion on psychosurgery; prefrontal leucotomy and depression. Proc R Soc Med 52:203–206

Epstein J, Pan H, Kocsis JH, Yang Y, Butler T, Chusid J, Hochberg H, Murrough J, Strohmayer E, Stern E, Silbersweig DA (2006) Lack of ventral striatal response to positive stimuli in depressed versus normal subjects. Am J Psychiatry 163:1784–1790

Freeman W, Watts JW (1937) Prefrontal lobotomy in the treatment of mental disorders. South Med J 30:23–31

Friedman A, Frankel M, Flaumenhaft Y, Merenlender A, Pinhasov A, Feder Y, Taler M, Gil-Ad I, Abeles M, Yadid G (2009) Programmed acute electrical stimulation of ventral tegmental area alleviates depressive-like behavior. Neuropsychopharmacology 34:1057–1066

George MS, Sackeim HA, Rush AJ, Marangell LB, Nahas Z, Husain MM, Lisanby S, Burt T, Goldman J, Ballenger JC (2000) Vagus nerve stimulation: a new tool for brain research and therapy. Biol Psychiatry 47:287–295

Gersner R, Toth E, Isserles M, Zangen A (2010) Site-specific antidepressant effects of repeated subconvulsive electrical stimulation: potential role of brain-derived neurotrophic factor. Biol Psychiatry 67:125–132

Goktepe EO, Young LB, Bridges PK (1975) A further review of the results of sterotactic subcaudate tractotomy. Br J Psychiatry 126:270–280

Hamani C, Temel Y (2012) Deep brain stimulation for psychiatric disease: contributions and validity of animal models. Sci Transl Med 4(142):142rv8

Hamani C, Diwan M, Isabella S, Lozano AM, Nobrega JN (2010a) Effects of different stimulation parameters on the antidepressant-like response of medial prefrontal cortex deep brain stimulation in rats. J Psychiatr Res 44:683–687

Hamani C, Diwan M, Macedo CE, Brandao ML, Shumake J, Gonzalez-Lima F, Raymond R, Lozano AM, Fletcher PJ, Nobrega JN (2010b) Antidepressant-like effects of medial prefrontal cortex deep brain stimulation in rats. Biol Psychiatry 67:117–124

Hamani C, Machado DC, Hipolide DC et al (2011a) Deep brain stimulation reverses anhedonic-like behavior in a chronic model of depression: role of serotonin and BDNF. Biol Psychiatry 71(1):30–35

Hamani C, Mayberg H, Stone S, Laxton A, Haber S, Lozano AM (2011b) The subcallosal cingulate gyrus in the context of major depression. Biol Psychiatry 69:301–308

Harden CL, Pulver MC, Ravdin LD, Nikolov B, Halper JP, Labar DR (2000) A pilot study of mood in epilepsy patients treated with vagus nerve stimulation. Epilepsy Behav 1:93–99

Haznedar MM, Buchsbaum MS, Hazlett EA, Shihabuddin L, New A, Siever LJ (2004) Cingulate gyrus volume and metabolism in the schizophrenia spectrum. Schizophr Res 71:249–262

Hodgkiss AD, Malizia AL, Bartlett JR, Bridges PK (1995) Outcome after the psychosurgical operation of stereotactic subcaudate tractotomy, 1979–1991. J Neuropsychiatry Clin Neurosci 7:230–234

Hoppe C, Helmstaedter C, Scherrmann J, Elger CE (2001) Self-reported mood changes following six months of vagus nerve stimulation in epilepsy patients. Epilepsy Behav 2:335–342

Jimenez F, Velasco F, Salin-Pascual R et al (2005) A patient with a resistant major depression disorder treated with deep brain stimulation in the inferior thalamic peduncle. Neurosurgery 57: 585–593

Kelly DH, Walter CJ, Sargant W (1966) Modified leucotomy assessed by forearm blood flow and other measurements. Br J Psychiatry 112:871–881

Kelly D, Walter CJ, Mitchell-Heggs N, Sargant W (1972) Modified leucotomy assessed clinically, physiologically and psychologically at six weeks and eighteen months. Br J Psychiatry 120:19–29

Kelly D, Richardson A, Mitchell-Heggs N (1973a) Stereotactic limbic leucotomy: neurophysiological aspects and operative technique. Br J Psychiatry 123:133–140

Kelly D, Richardson A, Mitchell-Heggs N, Greenup J, Chen C, Hafner RJ (1973b) Stereotactic limbic leucotomy: a preliminary report on forty patients. Br J Psychiatry 123:141–148

Kennedy SH, Giacobbe P, Rizvi SJ, Placenza FM, Nishikawa Y, Mayberg HS, Lozano AM (2011) Deep brain stimulation for treatment-resistant depression: follow-up after 3 to 6 years. Am J Psychiatry 168:502–510

Knight G (1965) Stereotactic tractotomy in the surgical treatment of mental illness. J Neurol Neurosurg Psychiatry 28:304–310

Knight GC, Tredgold RF (1955) Orbital leucotomy; a review of 52 cases. Lancet 268:981–986

Kosel M, Sturm V, Frick C, Lenartz D, Zeidler G, Brodesser D, Schlaepfer TE (2007) Mood improvement after deep brain stimulation of the internal globus pallidus for tardive dyskinesia in a patient suffering from major depression. J Psychiatr Res 41:801–803

Kuipers SD, Trentani A, Westenbroek C, Bramham CR, Korf J, Kema IP, Ter Horst GJ, Den Boer JA (2006) Unique patterns of FOS, phospho-CREB and BrdU immunoreactivity in the female rat brain following chronic stress and citalopram treatment. Neuropharmacology 50:428–440

Le Beau J (1954) Anterior cingulectomy in man. J Neurosurg 11:268–276

Little A (2009) Treatment-resistant depression. Am Fam Physician 80:167–172

Malone DA Jr, Dougherty DD, Rezai AR, Carpenter LL, Friehs GM, Eskandar EN, Rauch SL, Rasmussen SA, Machado AG, Kubu CS, Tyrka AR, Price LH, Stypulkowski PH, Giftakis JE, Rise MT, Malloy PF, Salloway SP, Greenberg BD (2009) Deep brain stimulation of the ventral capsule/ventral striatum for treatment-resistant depression. Biol Psychiatry 65:267–275

Mayberg HS, Liotti M, Brannan SK, McGinnis S, Mahurin RK, Jerabek PA, Silva JA, Tekell JL, Martin CC, Lancaster JL, Fox PT (1999) Reciprocal limbic-cortical function and negative mood: converging PET findings in depression and normal sadness. Am J Psychiatry 156:675–682

Mayberg HS, Brannan SK, Tekell JL, Silva JA, Mahurin RK, McGinnis S, Jerabek PA (2000) Regional metabolic effects of fluoxetine in major depression: serial changes and relationship to clinical response. Biol Psychiatry 48:830–843

Mayberg HS, Lozano AM, Voon V, McNeely HE, Seminowicz D, Hamani C, Schwalb JM, Kennedy SH (2005) Deep brain stimulation for treatment-resistant depression. Neuron 45:651–660

Mc KW (1959) Discussion on psychosurgery. Proc R Soc Med 52:206–209

Miller WC, DeLong MR (1987) Altered tonic activity of neurons in the globus pallidus and subthalamic nucleus in the primate model of parkinsonism. In: Carpenter MB, Jayaraman A (eds) The Basal Ganglia II: structure and function. Plenum, New York, pp 415–427

Moniz E (1937) Prefrontal leucotomy in the treatment of mental disorders. Am J Psychiatry 93:1379–1385

Montoya A, Weiss AP, Price BH et al (2002) Magnetic resonance imaging-guided stereotactic limbic leukotomy for treatment of intractable psychiatric disease. Neurosurgery 50: 1043–1049; discussion 1049–1052

Mottaghy FM, Keller CE, Gangitano M, Ly J, Thall M, Parker JA, Pascual-Leone A (2002) Correlation of cerebral blood flow and treatment effects of repetitive transcranial magnetic stimulation in depressed patients. Psychiatry Res 115:1–14

Nibuya M, Morinobu S, Duman RS (1995) Regulation of BDNF and trkB mRNA in rat brain by chronic electroconvulsive seizure and antidepressant drug treatments. J Neurosci 15:7539–7547

Nobler MS, Oquendo MA, Kegeles LS, Malone KM, Campbell CC, Sackeim HA, Mann JJ (2001) Decreased regional brain metabolism after ect. Am J Psychiatry 158:305–308

Penry JK, Dean JC (1990) Prevention of intractable partial seizures by intermittent vagal stimulatioin in humans: preliminary results. Epilepsia 31(Suppl 2):S40–S43

Pollak P, Benabid AL, Gross C, Gao DM, Laurent A, Benazzouz A, Hoffmann D, Gentil M, Perret J (1993) Effects of the stimulation of the subthalamic nucleus in parkinson disease. Rev Neurol 149:175–176

Pool JL (1949) Topectomy; the treatment of mental illness by frontal gyrectomy or bilateral subtotal ablation of frontal cortex. Lancet 2:776–781

Ressler KJ, Mayberg HS (2007) Targeting abnormal neural circuits in mood and anxiety disorders: from the laboratory to the clinic. Nat Neurosci 10:1116–1124

Ridout N, O'Carroll RE, Dritschel B, Christmas D, Eljamel M, Matthews K (2007) Emotion recognition from dynamic emotional displays following anterior cingulotomy and anterior capsulotomy for chronic depression. Neuropsychologia 45:1735–1743

Rush AJ, George MS, Sackeim HA, Marangell LB, Husain MM, Giller C, Nahas Z, Haines S, Simpson RK Jr, Goodman R (2000) Vagus nerve stimulation (VNS) for treatment-resistant depressions: a multicenter study. Biol Psychiatry 47:276–286

Rush AJ, Sackeim HA, Marangell LB, George MS, Brannan SK, Davis SM, Lavori P, Howland R, Kling MA, Rittberg B, Carpenter L, Ninan P, Moreno F, Schwartz T, Conway C, Burke M, Barry JJ (2005) Effects of 12 months of vagus nerve stimulation in treatment-resistant depression: a naturalistic study. Biol Psychiatry 58:355–363

Sachdev PS, Sachdev J (2005) Long-term outcome of neurosurgery for the treatment of resistant depression. J Neuropsychiatry Clin Neurosci 17:478–485

Sartorius A, Kiening KL, Kirsch P, von Gall CC, Haberkorn U, Unterberg AW, Henn FA, Meyer-Lindenberg A (2010) Remission of major depression under deep brain stimulation of the lateral habenula in a therapy-refractory patient. Biol Psychiatry 67:e9–e11

Schlaepfer TE, Cohen MX, Frick C et al (2007) Deep brain stimulation to reward circuitry alleviates anhedonia in refractory major depression. Neuropsychopharmacology

Schlaepfer TE, Frick C, Zobel A, Maier W, Heuser I, Bajbouj M, O'Keane V, Corcoran C, Adolfsson R, Trimble M, Rau H, Hoff HJ, Padberg F, Muller-Siecheneder F, Audenaert K, van den Abbeele D, Stanga Z, Hasdemir M (2008) Vagus nerve stimulation for depression: efficacy and safety in a European study. Psychol Med 38:651–661

Scoville WB (1949) Selective cortical undercutting as a means of modifying and studying frontal lobe function in man; preliminary report of 43 operative cases. J Neurosurg 6:65–73

Scoville WB (1960) Late results of orbital undercutting. Report of 76 patients undergoing quantitative selective lobotomies. Am J Psychiatry 117:525–532

Shimizu E, Hashimoto K, Okamura N, Koike K, Komatsu N, Kumakiri C, Nakazato M, Watanabe H, Shinoda N, Okada S, Iyo M (2003) Alterations of serum levels of brain-derived neurotrophic factor (BDNF) in depressed patients with or without antidepressants. Biol Psychiatry 54:70–75

Slattery DA, Morrow JA, Hudson AL, Hill DR, Nutt DJ, Henry B (2005) Comparison of alterations in c-fos and Egr-1 (zif268) expression throughout the rat brain following acute administration of different classes of antidepressant compounds. Neuropsychopharmacology 30:1278–1287

Slattery DA, Neumann I, Cryan JF (2010) Transient inactivation of the infralimbic cortex induces antidepressant-like effects in the rat. J Psychopharmacol 25(10):1295–1303

Smith MA, Makino S, Kvetnansky R, Post RM (1995) Stress and glucocorticoids affect the expression of brain-derived neurotrophic factor and neurotrophin-3 mRNAs in the hippocampus. J Neurosci 15:1768–1777

Spangler WJ, Cosgrove GR, Ballantine HT et al (1996) Magnetic resonance image-guided stereotactic cingulotomy for intractable psychiatric disease. Neurosurgery 38:1071–1076; discussion 1076–1078

Swayze VW 2nd (1995) Frontal leukotomy and related psychosurgical procedures in the era before antipsychotics (1935–1954): a historical overview. Am J Psychiatry 152:505–515

Temel Y, Visser-Vandewalle V (2006) Targets for deep brain stimulation in parkinson's disease. Expert Opin Ther Targets 10:355–362

Tooth JC, Newton MP (1961) Leucotomy in England and Wales 1942–1954. Reports on public health and medical subjects. Report no.: 104. Her Majesty's Stationery Office, London

Uylings HB, Groenewegen HJ, Kolb B (2003) Do rats have a prefrontal cortex? Behav Brain Res 146:3–17

Vertes RP (2004) Differential projections of the infralimbic and prelimbic cortex in the rat. Synapse 51:32–58

Zabara J (1985) Peripheral control of hypersynchronous discharge in epilepsy. Electroencephalogr Clin Neurophysiol 61:S162

Zabara J (1992) Inhibition of experimental seizures in canines by repetitive vagal stimulation. Epilepsia 33:1005–1012

Index

Curr Topics Behav Neurosci (2013) 14: 341–344
DOI: 10.1007/978-3-642-35425-0

The manufacturer's authorised representative in the EU is Springer Nature Customer Service Centre GmbH, Europaplatz 3, 69115 Heidelberg, Germany. If you have any concerns regarding our products, please contact ProductSafety@springernature.com

Printed and bound by CPI Group (UK) Ltd, Croydon, CR0 4YY
22/09/2026
02221055-0015